Language Functions and Brain Organization

This is a volume in

PERSPECTIVES IN
NEUROLINGUISTICS, NEUROPSYCHOLOGY, AND PSYCHO-
LINGUISTICS

A Series of Monographs and Treatises

A complete list of titles in this series appears at the end of this volume.

Language Functions and Brain Organization

EDITED BY

Sidney J. Segalowitz

Department of Psychology
Brock University
St. Catharines, Ontario
Canada

1983

ACADEMIC PRESS
A Subsidiary of Harcourt Brace Jovanovich, Publishers
New York   London
Paris   San Diego   San Francisco   São Paulo   Sydney   Tokyo   Toronto

ACADEMIC PRESS, INC.
111 Fifth Avenue, New York, New York 10003

United Kingdom Edition published by
ACADEMIC PRESS, INC. (LONDON) LTD.
24/28 Oval Road, London NW1 7DX

Library of Congress Cataloging in Publication Data

Main entry under title:

Language functions and brain organization.

    Includes index.
    1. Neurolinguistics. 2. Brain--Localization of
functions. I. Segalowitz, S. J. [DNLM: 1. Brain--
Physiology. 2. Language. 3. Speech--Physiology   WV
501 L287]
QP399.L37 1982   153.6   82-11589
ISBN 0-12-635640-8

PRINTED IN THE UNITED STATES OF AMERICA

83 84 85 86   9 8 7 6 5 4 3 2 1

Contents

Contributors

Numbers in parentheses indicate the pages on which the authors' contributions begin.

Robert M. Anderson, Jr. (193), Department of Psychology, University of Hawaii at Manoa, Honolulu, Hawaii 96822

Rita Sloan Berndt (5), Department of Psychology, The Johns Hopkins University, Baltimore, Maryland 21218

M. P. Bryden (341), Department of Psychology, University of Waterloo, Waterloo, Ontario N2L 3G1, Canada

Alfonso Caramazza (5), Department of Psychology, The Johns Hopkins University, Baltimore, Maryland 21218

Michael Cicone (51), Psychology Service, Veterans Administration Medical Center, Boston, Massachusetts 02130

Nancy S. Foldi[1] (51), Department of Psychology, Clark University, Worcester, Massachusetts 01610

Howard Gardner (51), Psychology Service, Veterans Administration Medical Center, Boston, Massachusetts 02130, and Harvard Project Zero, Harvard University, Cambridge Massachusetts 02138

Demetria C. Q. Leong (193), John A. Burns School of Medicine, University of Hawaii at Manoa, Honolulu, Hawaii 96822

Catherine A. Mateer (145, 171), Departments of Neurological Surgery and Speech and Hearing Sciences, University of Washington, Seattle, Washington 98195

Janice M. Millar (87), Department of Hearing and Speech Sciences, University of Maryland, College Park, Maryland 20742

Dennis L. Molfese (29), Department of Psychology and School of Medicine, Southern Illinois University, Carbondale, Illinois 62901

Victoria J. Molfese (29), Department of Psychology, Southern Illinois University, Carbondale, Illinois 62901

C. Netley (245), Department of Psychology, The Hospital for Sick Children, Toronto, Ontario M5G 1X8, Canada

Loraine K. Obler (267), Aphasia Research Center, Department of Neurology, Boston Veterans Administration Medical Center, Boston, Massachusetts 02130

George A. Ojemann (171), Department of Neurological Surgery, University of Washington, Seattle, Washington 98195

Carl Parsons (29), School of Communication Disorders, Lincoln Institute of Health Sciences, Victoria, Australia

Phyllis Ross (287), Department of Psychology, William Patterson College, Wayne, New Jersey 07470

Joanne Rovet (245), Department of Psychology, The Hospital for Sick Children, Toronto, Ontario M5G 1X8, Canada

Sidney J. Segalowitz (221, 341), Department of Psychology, Brock University, St. Catharines, Ontario L2S 3A1, Canada

Jyotsna Vaid[2] (315), Department of Psychology, McGill University, Montreal, Quebec H3A 1B1, Canada

Jaan Valsiner (231), Department of Psychology, University of North Carolina, Chapel Hill, North Carolina 27514

Harry A. Whitaker (87), Department of Hearing and Speech Sciences, University of Maryland, College Park, Maryland 20742

[1] Present address: Department of Psychiatry, Beth Israel Hospital–Harvard Medical School, 330 Brookline Avenue, Boston, Massachusetts 02130.

[2] Present address: Department of Psychology, Michigan State University, East Lansing, Michigan 48824.

Sandra F. Witelson (117), Department of Psychiatry, McMaster University, Hamilton, Ontario L8N 325, Canada

Edgar Zurif[3] (5), Aphasia Research Center, Boston Veterans Administration Medical Center, Boston, Massachusetts 02130, and Graduate Center, City University of New York, New York, New York 10036

---

[3] Present address: Ph.D. Program in Speech and Hearing Sciences, Graduate Center of C.U.N.Y., New York, New York 10036.

Preface

Just over 120 years ago, Paul Broca formalized the notion of cerebral dominance for language, and yet it has been only in the last few decades that we have seen the acceptance of the neuropsychological bases of language as a separate discipline. Indeed, there has been an enormous growth in the number of researchers contributing to the field in the past 10 years, much of it stemming from the publication in 1967 of Eric Lenneberg's popular exposition *Biological Foundations of Language*. Interest in the topic has spread from neurology to other disciplines, which has generated complicating and diversifying themes. For example, linguists and psycholinguists examine aphasia data with respect to specific aspects of a chosen theory of grammar; cognitive psychologists use neurolinguistic data to support some divisions of intellectual function and not others; educationalists seek insights into pedagogical problems from developmental neurolinguistics; philosophers debate the adequacy of brain models as theories of mind. This potential eclecticism in the field emphasizes the multifaceted nature of the question of how language is represented in the human brain. Rather than viewing language as a single intellectual process, it has become useful to consider it as a complex of functions, and each language function can be discussed with respect to its own organization in the brain. Clearly, these functions are interrelated both in behavioral and neurological correlates, yet we must not expect a single model of neurological representation to be adequate for all aspects.

This book presents a current synthesis of neurolinguistic issues. In Part I, linguistic approaches are examined. In the first chapter, Rita Sloan Berndt, Alfonso Caramazza, and Edgar Zurif discuss the problems of understanding neuropsychological organization of semantic and syntactic knowledge. They address the state of current knowledge in this complex aspect of language functioning, as well as the caveats. In the next chapter, Victoria Molfese, Dennis Molfese, and Carl Parsons review the evidence for hemispheric asymmetries in processing phonological information. Pragmatic aspects of communication are reviewed by Nancy Foldi, Michael Cicone, and Howard Gardner in Chapter 3. These aspects of language use are often treated less formally in neurolinguistics and yet are critical for a complete overview. We close Part I with a review by Janice Millar and Harry Whitaker of the right hemisphere's contribution to language.

Part II reviews neuropsychological correlates of language. Sandra Witelson opens with a discussion of right–left asymmetries in the cerebral cortex and their implications for functional asymmetries. In Chapter 6, Catherine Mateer discusses left-hemisphere language specialization from the perspective of motor and perceptual functions. The implication here is that language representation in the left hemisphere is not independent of other skills requiring fine discriminations. In the next chapter, Mateer and George Ojemann review the evidence for hemisphere asymmetry for language functioning in the thalamus. Thalamic mechanisms give a different perspective on the neurolinguistic question. Chapter 8 addresses the philosophical problem of finding a brain basis for experience. In this exercise, Robert Anderson, Jr., and Demetria Leong examine some difficulties in building a brain theory for visual experience, the brain bases for which are better understood than those for language. The nature of the theoretical problems should be instructive for neurolinguistics as well.

In Part III, we turn to developmental issues. In Chapter 9, I review the evidence for speech lateralization in infancy. This is followed by a chapter by Jaan Valsiner on attempts to integrate the notion of hemispheric specialization into the child's overall development. In Chapter 11, Charles Netley and Joanne Rovet offer a model outlining the relationship between cerebral functional asymmetries, maturation rate, and cognitive skills, through the mediation of sex chromosomes. In Chapter 12, Loraine Obler discusses language dysfunction in dementia and correlates that with brain functioning.

In Part IV, we turn to the issue of individual differences, a factor that has consistently plagued researchers in neurolinguistics. Phyllis Ross first discusses the possibility that cerebral specialization for language is different in deaf individuals. In Chapter 14, Jyotsna Vaid similarly considers the variations produced in cases of bilingualism and the factors that may be critical for this issue. In the final chapter of this volume, Philip Bryden and I address the evidence for differing lateralization in various groups, including in left- versus right-handers and in males versus females. We then propose a method for researchers to accommodate this (usually) unwanted variance.

The resulting collection, then, is addressed to researchers and students of the neuropsychology of language, whether they call themselves psychologists, neuropsychologists, neurologists, or linguists. Workers in all these areas have contributed to the development of neurolinguistics as a field.

There were many people, too numerous to mention, who helped in this project in small and large ways, sometimes unknown to them. I would especially like to thank the contributing authors for their patience and understanding during the collection and editing of this volume. They represent the variety of specialties that make up this interdisciplinary field. I have both enjoyed and learned from their collaboration and hope this experience is shared by the readership.

PART I

Language as a Mental Organ or a Mental Complex

INTRODUCTION

Fifteen years ago Eric Lenneberg's now famous book, *Biological Foundations of Language,* appeared. In it he outlined a biological approach to the study of language that would expand the psychologist's and psycholinguist's view of the enterprise. No one denied the brain's involvement and foundation for language processing; however, Lenneberg tried to put language into perspectives involving developmental psychology, cognition, genetics, and evolution, as well as neurology and neuropsychology. The unequivocal conclusion was that human language has a biological basis, which some have misinterpreted as implying an innate component (Lenneberg, 1971a). Some physiological structures are clearly adapted for speech, such as those involved in breathing, vocal chord, and tongue coordination. But more intriguing is the notion that there is a cognitive specialization and that language can be considered an autonomous "mental organ," just as the stomach is a digestive organ. The issue of whether or not language is autonomous is critical, as otherwise language would simply be one skill among many. The arguments that favored such a conclusion included the following: (1) a rigid, swift timetable of language acquisition, seemingly without earlier, nonlinguistic precursors; (2) the uniqueness of true linguistic skills to humans; (3) the characterization of the cognitive operations involved in language as different from those in other cognitive realms; and (4) a set of neurophysiological correlates of language processes.

Over the past 15 years, these characterizations of the place of language in the human species have all eroded (or no longer seem necessary), except the fourth. Yet the notion of language as an autonomous mental organ has remained (e.g., Chomsky, 1980).

Acceptance of language as an autonomous mental organ has rather strong implications for psycholinguistics and neuropsychology: it implies that linguistic knowledge is not integrated into other cognitive systems, such as a developmental Piagetian framework; it implies that verbal skills follow rules different from other (presumably equally autonomous) mental organs, such as visual–spatial thinking and emotional thought; it implies that the attempts to find bases for localization of language behaviors in the brain at a more molecular level (e.g., acoustic factors or motor integration factors) are misguided. There are arguments for and against these conclusions. Broader developmental theories are occasionally useful in accounting for specific aspects of language acquisition (e.g., Bebout, Segalowitz, and White, 1980; or more globally, Moerck, 1977), yet nonsupportive conclusions are also rampant (e.g., Moore and Harris, 1978). Linguistic and nonlinguistic thought are distinguishable in formal and certainly in substantive aspects (e.g., Paivio, 1971), yet the distinction is often considered to be either an arbitrary one or a simplistic one (Kosslyn, Pinker, Smith, and Shwartz, 1979). Although it is tempting to localize linguistic as opposed to more general neuropsychological skills (e.g., Whitaker, 1971; Schnitzer, 1974), some consider this the wrong level for such an exercise (e.g., Luria, 1966; Mateer, this volume; Segalowitz and Gruber, 1977).

The issue, then, of whether or not language is a "mental organ" is not an empirical one at all, but rather is decided by the level of this discussion. There may be times where it is useful to use the mental organ metaphor, other times when it is not. One need not postulate the characteristic "autonomous mental organ" in order to study language in its own (linguistic) terms. The relationship between linguistic and other knowledge systems is crucial for a complete understanding of language knowledge and the linguistic system. The interfaces between the use of language and its ecological niche in human overall behavior should be a prime concern of psycholinguists and neurolinguists. To relegate language to a status of an independent mental organ would be to systematically miss these important interfaces. After all, what is it that language is to be independent from? Knowledge of the world? Information processing? Imagistic thought? Emotions? Clearly there are cases when it is convenient to consider such independence, for example, when making specific points about psycholinguistic structures, but in other cases, it is equally convenient to not attribute independence (see in this volume Foldi, Cicone, and Gardner; Molfese and Parsons; and Mateer).

LANGUAGE AS A SET OF NEUROPSYCHOLOGICAL SKILLS

One can accept a biological approach to language without being committed to the "mental organ" metaphor. For example, Lenneberg (1971b) suggested that the kinds of mental operations required in human language are similar to those in arithmetic—certain relational concepts that are essential for calculation and symbolization. Similarly, there are nonsyntactic, nonsemantic aspects of language functioning that are not generally disrupted by aphasia (see Chapter 3 in this volume). Without these aspects, language loses much of its human characteristics. Only in a restricted sense are interpersonal aspects of language less linguistic, since their integrity is essential for natural human communication.

The planning and execution of sound sequences in speech requires considerable skill and is highly complex (see Mateer, this volume). Whether or not it is more complex than piano playing (with all the nuances in good performance) is a difficult issue. That it may rely on similar neuropsychological structures is a possibility. Similarly, fine timing judgments may underlie speech (Mills and Rollman, 1979; Schwartz and Tallal, 1980) without an independent mechanism being postulated.

In a sense, the view of language in behavior has come full circle. Skinner (1957) summarized the view that verbal behavior (and this was interpreted generally as meaning language behavior) should be considered just one among many, that the formal properties are no different in essence from those of other behaviors. Chomsky's (1959) subsequent denunciation of this view led to an isolation of language from other behaviors, arguing that language is inherently too complex to conform to the same pattern as other behaviors. Biological approaches, especially the study of aphasia, reinforced this view. Yet the growth of neuropsychology in the past decade has allowed the view that many human behaviors are more complex than at first thought, that language is not alone in this, and that there are brain bases to more skills than just language.

With this perspective in mind, when we examine the brain bases for linguistic behavior, we must be sensitive to the specific aspect of language under discussion. To this end, we have tried in this volume to include sections that focus on the nature of brain representation for language, each from a different, specific perspective. In this section, the evidence for various linguistic approaches is reviewed.

REFERENCES

Bebout, L. J., Segalowitz, S. J., and White, G. J. (1980). Children's comprehension of causal constructions with "because" and "so." *Child Development, 51,* 565–568.

Chomsky, N. (1959). Review of Skinner's *Verbal behavior. Language, 35,* 26–58.

Chomsky, N. (1980). Rules and representations. *The Behavioral and Brain Sciences, 3,* 1–15.

Kosslyn, S. M., Pinker, S., Smith, G., and Shwartz, S. P. (1979). On the demystification of mental imagery. *The Behavioral and Brain Sciences, 2* (4), 535–548.

Lenneberg, E. H. (1967). *Biological foundations of language.* New York: Wiley.

Lenneberg, E. H. (1971a). Developments in biological linguistics. In R. O'Brien (Ed.), *22nd Annual Round Table.* Washington, D.C.: Georgetown Univ. Press.

Lenneberg, E. H. (1971b). Of language knowledge, apes, and brains. *Journal of Psycholinguistic Research, 1,* 1–29.

Luria, A. R. (1966). *Human brain and psychological processes.* New York: Harper.

Mills, L., and Rollman, G. B. (1979). Left hemisphere selectivity for processing duration in normal subjects. *Brain and Language, 7,* 320–335.

Moerk, E. L. (1977). *Pragmatic and semantic aspects of early language development.* Baltimore: University Park Press.

Moore, T. E., and Harris, A. E. (1978). Language and thought in Piagetian theory. In L. S. Siegel and C. J. Brainerd (Eds.), *Alternatives to Piaget.* New York: Academic Press.

Paivio, A. (1971). *Imagery and verbal processes.* New York: Holt.

Schnitzer, M. L. (1974). Aphasiological evidence for five linguistical hypotheses. *Language, 50,* 300–316.

Schwartz, J., and Tallal, P. (1980). Rate of acoustic change may underlie hemispheric specialization for speech perception. *Science, 207,* 1380–1381.

Segalowitz, S. J., and Gruber, F. A. (1977). What is it that is lateralized? In S. J. Segalowitz and F. A. Gruber (Eds.), *Language development and neurological theory.* New York: Academic Press.

Skinner, B. F. (1957). *Verbal behavior.* New York: Appleton.

Whitaker, H. A. (1971). *On the representation of language in the human brain.* Edmonton: Linguistic Research, Inc.

CHAPTER 1

Language Functions: Syntax and Semantics*

Rita Sloan Berndt, Alfonso Caramazza, and Edgar Zurif

INTRODUCTION

The study of aphasia has provided one of the primary means for investigating the organization of language functions in the brain. Working from studies of aphasic patients whose brains eventually were autopsied, nineteenth-century neurologists discovered that disruptions of speech most often resulted from damage to the left cerebral hemisphere (Broca, 1861/1950), and even more importantly, that damage to different regions of the left hemisphere resulted in different patterns of aphasic symptoms (Wernicke, 1874/1980). This latter discovery has provided the basis for 100 years of research in aphasia, as neurologists and, lately, practitioners of other disciplines have attempted to determine the manner in which the

* The preparation of this chapter was supported by National Institutes of Health (NINCDS) Research Grant Nos. NS16155 and NS14099 to The Johns Hopkins University; NS11408 to Edgar Zurif and Howard Gardner; NS15972 to Edgar Zurif, Susan Carey, and Murray Grossman; and NS06209 to Aphasia Research Center, Department of Neurology, Boston University School of Medicine.

5

particular cortical tissue affected by a cerebral insult was involved in the specific language functions that had been disrupted.

The earliest characterization of the different patterns of symptoms that can occur in aphasia focused on patients' inability to perform certain grossly defined language tasks. In particular, it was observed that some patients could no longer speak without effort, but could understand speech, whereas others could speak fluently but had difficulty understanding what was said to them. No attempts were made to analyze overt language functions such as "speaking" and "understanding" into smaller components; rather, these larger language "faculties" were taken to be the units of the system that might be localizable in the brain (see Arbib and Caplan, 1979, for discussion).

The basic distinction between aphasia types that was observed by the early practitioners has survived into modern terminology as the "expressive" versus the "receptive" aphasias (Weisenberg and McBride, 1935). In fact, the strongest generalization available today concerning the correlation between symptomatology and lesion site is that "expressive" (nonfluent) aphasics have suffered cortical left-hemisphere damage anterior to the fissure of Rolando, whereas "receptive" (fluent) aphasics have incurred damage posterior to the Rolandic fissure (Benson, 1967).

Needless to say, this dichotomy is far too simple to capture the range of phenomena found in aphasia. Further, since "receptive" aphasics clearly show productive problems in that their utterances may be largely uninterpretable, and since "expressive" aphasics have been shown to have comprehension problems, the expressive–receptive distinction may give rise to the argument that language "functions" cannot be localized in the brain.

A more recent approach, which retains the basic distinction between the two patient types and their related lesion sites, builds on developments in linguistics and psycholinguistics. It is an attempt to analyze the objective deficits observable in the aphasic syndromes into theoretically distinct components of the language system. The units of function that are potentially localizable, in this view, are taken to be the abstract structural elements and processing mechanisms that comprise the normal system. Central to this more recent approach is the assumption that similar deficits in language faculties, such as "comprehension," may occur for different reasons in different patients. While this realization by itself is certainly not a new discovery (e.g., Luria, 1947/1970), it has been only recently that the mechanisms that are postulated to underlie the faculty deficits have been articulated in terms of an independently developed theoretical vocabulary.

Progress in aphasia research thus depends in no small measure on the availability of well-developed models of normal functioning. When a reasonably precise normal model is available—as, for example, in the area of phonology (e.g., Chomsky and Halle, 1968)—attempts can be made to characterize rather precisely the structure of aphasic errors both in individual patients (e.g., Schnitzer, 1972) and within specific syndromes (e.g., Kean, 1977). When a unified theoretical model is not available for a particular component (as we shall see in our discussion of semantic deficits), the assessment of aphasic impairments must focus on a limited and carefully circumscribed range of performance and develop a testable model based on the pathological behaviors that are observed.

The purpose of this chapter is to describe recent research on the semantic and syntactic aspects of aphasic deficits that attempts to demonstrate that these components can be selectively disrupted by focal brain damage. Primary attention will be devoted to the two major syndromes of Broca's and of Wernicke's aphasia, which correspond roughly to the "expressive" and "receptive" types discussed above.

More precisely,[1] the Broca's aphasic produces short, effortful, dysprosodic utterances that are made up largely of substantive ("open-class") words, mainly nouns, verbs, and adjectives. The discussion of Broca's aphasia to be presented here will stress the "agrammatic" nature of their speech: utterances are grammatically simple and often lacking the "closed class" of grammatical morphemes that are required for generating well-structured sentences. Comprehension appears to be impaired only on structured testing with syntactically weighted sentences. Wernicke's aphasics speak fluently and prosodically, but often utter incorrect word-like segments that may be phonemically or semantically related to the target ("paraphasias") or completely unrecognizable ("neologisms"). Comprehension is moderately to severely impaired.

Another syndrome to be discussed here is anomic aphasia, which is characterized by an inordinate difficulty in naming objects to confrontation and finding the correct substantive words in spontaneous speech. Comprehension is relatively good, and speech is fluent though often marked by word-finding pauses and some paraphasias. In studies that group patients on the basis of fluency or by "posterior" lesion site, anomic aphasics and Wernicke's aphasics are sometimes discussed together. Although this confusion of patient types undoubtedly masks interesting group differences that may ultimately prove neurologically important, a discussion of the literature relevant to semantic deficits in aphasia requires some grouping of the two types.

---

[1] See Goodglass and Kaplan (1972) for a complete description.

SEMANTIC DEFICITS IN APHASIA

There are several factors that contribute to the difficulties that arise in attempting to characterize semantic deficits. Most importantly, there are many diverse processes that might be considered as part of the semantic component of the language; as mentioned above, there is no unified theory of semantics available (either from linguistics or cognitive psychology) that relates these processes. For example, a semantic deficit (1) might manifest itself at the level of sentence comprehension, in which case a patient may be unable to integrate the various elements of a sentence to produce a unified semantic representation (Berndt and Caramazza, 1980b); (2) may undermine patients' lexical comprehension abilities; (3) may interfere with the ability to produce semantically well-formed utterances; or (4) may be limited to a difficulty naming objects. Little attention has been given to the problem of explicating how impaired performance in these different areas might relate to each other. That is, research has focused on each of these areas separately, with little regard for the possibility that a single underlying deficit may contribute to a decline in performance in each of these four areas. Consequently, it is no easy matter to provide a comprehensive account of semantic disorders through a unified theory of impairments of the semantic component of language processing.

Our discussion of semantic factors in aphasia will focus concurrently on deficits of lexical comprehension and of naming. More specifically, we will elaborate a theoretical position that attempts to explain naming and lexical comprehension disorders (sentence comprehension disorders will not be included) by reference to a common set of principles and assumptions. The reader is warned, however, that the data now available are only indirectly related to these critical issues; our discussion of these problems is commensurately preliminary in nature. In addition, it should be noted that the focus on naming and lexical comprehension (name recognition) necessarily limits the discussion of semantic deficits to a subset of the semantic system—namely, to disorders affecting the recovery and production of concrete, referential nouns.

Lexical Comprehension and Naming

The term *semantic deficit* will refer here to a disturbance of a theoretically definable component of the language system that underlies the various disorders that are subsumed within this category. Thus, an attempt to provide an account of lexical comprehension and naming disorders forces us to establish the common mechanisms that underlie performance of these two types of tasks. There are several elements that are transparently shared by the two tasks in question. These include a lexicon that specifies lexical forms and meanings, some type of mechanism that relates the lexical form through its meaning to the referents of the lexical form, and mechanisms of access to the lexical form (both graphemic and phonological). An analysis of these processes, even if not too detailed at this point, is a necessary step that must be taken in order to address the question of lexical–semantic deficits. This analysis will be developed through a description of the naming process first proposed by Caramazza and Berndt (1978) and subsequently developed by Caramazza, Berndt, and Brownell (1982).

Let us first consider the conditions to be satisfied for the assignment of a name to an object. We assume that a name is assigned to an object if the object satisfies a set of conditions or criteria—the meaning of the name. These conditions are of two types: perceptual and functional.[2] Simply stated, the semantic description associated with a name specifies both the set of perceptual attributes that characterizes the members of the set labeled by the name, as well as the functional attributes of members of the set. For example, the concept *cup* may be defined by the following list of properties: "artifact, upwardly concave, may have a handle, height about equal to top diameter, used as container from which to drink hot liquids." An object that satisfies these conditions can be called *cup*.

As indicated, an important function of the meaning of a word is to relate the word to its referents (including pictures) through a semantic description. The semantic description also specifies the relationships among various lexical items. Thus, for example, the observation that the words *cup* and *bowl* are semantically related merely summarizes the fact that the semantic descriptions of these words share certain components of meaning. The nature of the relationship between particular words is defined by the specific semantic elements that are shared. For example, the words *couch* and *chair* share components of meaning different from those shared by *couch* and *bed*.

These observations already suggest the possibility of a close link between disorders of naming and disorders of lexical comprehension if the semantic component of the lexicon is disrupted. In order to account for performance in various types of tasks, however, it is necessary to elabo-

---

[2] Since our discussion is limited to object names, a simple featural approach seems the most appropriate framework for analyzing the available semantic descriptions. See Miller and Johnson-Laird (1976) and Fodor, Garrett, Walker and Parkes (1980) for alternative approaches.

rate the processes that interact with the representation of semantic information. Specifically, a more detailed description is required of the component process involved in naming and in lexical comprehension.

Caramazza, Berndt, and Brownell (1982) have proposed a description of the naming process that attempts to describe the access mechanisms to the name of an object. The focus in this work is on the interaction between the system responsible for the perceptual analysis of an object, and the object's assignment to a semantic category. The major "stages" of processing in this description include the following: (1) a set of perceptual analyses of the object, (2) modality-specific analyses that produce a semantically constrained parsing of the perceptual input, (3) the application of a (modality-independent) classification algorithm to determine the category membership of the object and to allow mapping onto a specific lexical form, and (4) the execution of the phonological information specified for the selected lexical form. In a word–object matching task (lexical comprehension), a reverse sequence of processes is assumed to take place. In this case, the lexical access mechanisms presumably require modality-specific perceptual analyses for graphic and auditory inputs, and the semantic description stage is assumed to be modality independent. We will use this general characterization of naming and word–object matching tasks to interpret selected reports on disorders of lexical comprehension and naming. We will then describe briefly some of our research on aphasic patients' ability to classify objects.

The Case for Selective Disruption

The assumption that the lexical–semantic component of the language system can be disrupted selectively has received the clearest support from studies of patients with diffuse brain damage (Schwartz, Marin, and Saffran, 1979; Warrington, 1975). Warrington's three patients had severe difficulties in semantic tasks (identifying attributes and common associates of objects), but otherwise showed relatively intact intellectual abilities. Schwartz, Marin, and Saffran (1979) reported an analysis of a patient with a marked dissociation between lexical–semantic abilities (which were impaired) and phonological and syntactic abilities (which were intact). This patient (with senile dementia) manifested a selective and severe progressive deterioration in the ability to carry out semantic analyses of lexical information. Because these reports involve patients with diffuse brain damage, it is not possible to make any strong claims about localization of the selectively impaired function. Nonetheless, these cases demonstrate clearly the neurologically based separability of the components of the language system.

Studies of patients with focal lesions in the left hemisphere also support the possibility of dissociation of semantic processes from other language processes, although the evidence here is less clear. Relevant studies can be classified loosely into three groups based on the major issues they address. A first group consists of those reports that have focused on lexical comprehension; a second group comprises reports of lexical production errors in spontaneous speech; a third group represents studies of deficits in confrontation naming. In addition, several studies have attempted to relate confrontation naming abilities to performance on lexical comprehension and other production tasks. This work is the most critical for our purposes, since it is the pattern of co-occurrence of particular deficits that helps establish the common mechanisms for the performance of a set of tasks.

One relatively well-established result concerns the pattern of lexical comprehension found in patients with left posterior lesions—patients who are clinically classified as either Wernicke's or anomic aphasics. These patients have severe difficulties in carrying out lexical–semantic analyses.

Lhermitte, Derouesne, and Lecours (1971) assessed the performance of two groups of patients in a word sorting task. Patients with posterior lesions violated semantic restrictions by grouping together words that were clearly unrelated semantically, or by failing to group together those words that were semantically related (see also Grober, Perecman, Kellar, and Brown, 1980; Grossman, 1978). Zurif, Caramazza, Myerson, and Galvin (1974) asked left anterior-damaged and left posterior-damaged patients to judge the semantic relatedness among a set of words. The posterior-damaged patients' performance in this task was clearly worse than the performance of the anterior-damaged patients. In fact, just as in the study reported by Lhermitte, Derouesne, and Lecours, the patients with left posterior damage failed to use the same semantic criteria in making their classification decisions as normal controls or patients with left anterior damage.

The analysis of paraphasic productions (including those produced in confrontation naming situations), and the analysis of the factors that influence naming performance, also contribute to our understanding of the breakdown of the lexical–semantic processing component. The analysis of error patterns in paraphasic productions (Buckingham and Rekart, 1979; Rinnert and Whitaker, 1973; Schuell and Jenkins, 1961) indicates that the paraphasic production is often semantically related to the target word. Investigations of the factors that affect naming performance have identified semantic variables as determinants of performance. Thus, it has been shown that "picturability" (Goodglass, Hyde, and Blumstein, 1969) and "operativity" (Gardner, 1973) are dimensions predictive of naming performance.

Finally, it should be noted that naming defects are independent of the modality of stimulus input—visual, tactile, auditory, and olfactory (Goodglass, Barton, and Kaplan, 1968; but see Geschwind, 1965; Spreen, Benton, and Van Allen, 1966). These results, taken together, suggest that a modality-independent lexical–semantic system can be impaired. It is unclear from the reports reviewed, however, whether a selective impairment of this system has occurred in the patients tested. In addition, these studies provide no information as to whether a common deficit underlies naming and other lexical–semantic defects. To address this last issue, in particular, investigations can be considered that concern patterns of impairments, rather than deficits on single dimensions of performance.

Several studies have addressed the relationship between semantic deficits and naming disorders. Goodglass and Baker (1976) related patients' ability to name a set of objects to the patients' performance in evaluating the associational structure for each of these objects. Two groups of patients (with "good" and "poor" comprehension) were asked to name a set of objects, and then to decide which words of a set of words were related to the target object. The type of associational relationships between the object and the words included (1) the name of the object, (2) the superordinate category to which the object belonged, (3) an attribute that typically characterized the object, (4) a coordinate member of the same category, (5) an action associated with the object (functional associate), and (6) the situation or context in which the object could be found (functional context). Two results are of interest. First, the authors report that the group of patients with poor comprehension (presumed to have posterior left-hemisphere damage) exhibited a marked difficulty recognizing functional contexts and functional associates. Second, there appeared to be a relationship in this group between naming difficulty and a failure to recognize associates of the target item. Goodglass and Baker interpret these results as evidence that a disruption of the semantic network may be in part responsible for the naming deficits in the poor comprehension group.

The possibility that naming and lexical–semantic disorders may have a common basis has been supported in a series of reports by Gainotti and his associates (Gainotti, 1976; Gainotti, Caltagirone, and Ibba, 1975; Gainotti, Caltagirone, and Miceli, 1979). The major result that emerges from these reports is that patients with a lexical–semantic deficit (indexed by poor performance on lexical–semantic discrimination tasks in both auditory and visual modalities) present anomic and other production deficits involving semantic information (verbal and semantic paraphasias). In contrast, there appears to be no relationship between naming deficits characterized by a prevalence of phonetic and phonemic errors, and performance on lexical–semantic discrimination tasks (see also Coughlan and Warrington, 1978).

This highly selective review of research on lexical–semantic disorders allows two conclusions: (1) that the lexical–semantic component can be disrupted independently of phonological and syntactic processing mechanism and (2) that a disruption of the lexical–semantic system affects performance in both expressive (naming) and receptive (comprehension) modalities. These conclusions, when related to the naming model we have proposed, allow us to confirm the assumption of a modality-independent classification algorithm. For a more direct assessment of the model, we must turn to research especially designed for this purpose.

Perceptual Analysis and Object Classification

The model developed by Caramazza, Berndt, and Brownell (1982), though sketchy, is explicit enough to allow the construction of experimental tasks to assess the locus of functional impairment in naming. Some of the nonobvious predictions generated by the model can be elucidated by considering the line drawings of cup-like objects presented in Figure 1. These items vary along two dimensions: width (diameter-to-height ratio) and the presence or absence of a handle. Some of these items could clearly be labeled "cups," others "bowls," but for others the application of one of these two labels is less compelling. Furthermore, if we were to

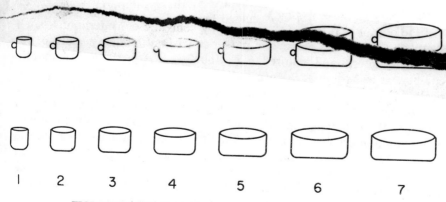

FIGURE 1. Stimulus items with two components of variation.

provide a functional context (e.g., a depiction of coffee or cereal pouring into one of the objects), this too would be a factor in determining the choice of a label (see Labov, 1973).

A minimum sequence of events in the assignment of a label to these objects requires (1) a perceptual parsing to recover the dimensions of relevance, (2) the integration of perceptual and functional information, and (3) the categorization of the object (on the basis of the recovered information) into one of the two categories "cup" and "bowl." In other words, proper classification of the objects requires a successful match between the features contained in the semantic description of the categories "cup" and "bowl," and the output of the perceptual analysis of the objects. Normal performance thus requires normal perceptual analysis and an intact semantic description. In cases of impairment to the lexical–semantic system, performance in the object classification task described here should be impaired. Whitehouse, Caramazza, and Zurif (1978) tested this prediction by comparing the classification performance of five anomic and five Broca's aphasics. If anomic aphasia reflects a disturbance of the lexical–semantic system, patients of this type should perform markedly worse than Broca's aphasics. The results reported by Whitehouse, Caramazza, and Zurif strongly confirmed this prediction. Anomic aphasics could not reliably use the perceptual dimensions that defined the categories, nor could they successfully integrate the functional context information in classifying the cup-like objects. In contrast, the Broca's aphasics performed in a qualitatively similar manner to nonaphasic controls. Since there were no indications of strictly perceptual defects in these anomic aphasics, it seems appropriate to conclude that their deficit must be at the level of the semantic representation of the perceptual information that defines the categories tested.

The model we have described allows an even stronger, less obvious prediction. This prediction is motivated by the assumption that the perceptual parsing of an object is guided by semantic considerations, such that the output of the perceptual parser consists of modality-specific, semantically interpreted features (e.g., "handle"). If the lexical–semantic component is disrupted, performance of the parser in producing a semantically interpretable output should be impaired, since the parser interacts with the semantic component in carrying out the analysis of an object. To return to the example at hand, a deficit of the lexical–semantic component should interfere with patients' ability to extract and to integrate the "handle" and "height–width" features of the stimulus set. Thus, a deficit in the lexical–semantic component should impair performance in tasks that involve judgments of perceptual similarity between objects, even when an explicit categorization of the objects in question is not required.

This prediction is motivated jointly by the description of the naming process we have advanced and the empirical fact that normal subjects' performance on judgments-of-similarity tasks reflects the categorical structure of the stimulus set and the underlying dimensions shared across categories (Caramazza, Hersh, and Torgerson, 1976; Degerman, 1970). We tested this prediction by comparing the performance of groups of aphasic, right-hemisphere damaged and neurologically intact patients on two tasks. In one task, patients were asked to judge the perceptual similarity of pairs of cup-like objects similar to those depicted in Figure 1. In a second task, subjects were asked to label the same set of stimuli. The specific prediction tested was that the aphasic patients (who had difficulty labeling the stimuli) were also impaired in their performance of the similarity task. The performance of the right-hemisphere damaged patients, as well as the performance of the aphasic patients who had no difficulty in the labeling task, was expected to be within the normal range on the similarity judgments task.

The results were clearly consistent with the predictions outlined. The right-hemisphere damaged patients performed normally on both the labeling and the similarity judgments tasks. The aphasic patients were subdivided into two groups on the basis of an inverse principal components analysis of the similarity judgments. One of these two groups of patients performed normally on both the labeling and the similarity judgments tasks; the other group of aphasic patients performed poorly on both tasks. That is, they failed to integrate the two dimensions of perceptual variation (width and handle–no handle) in judging the perceptual similarity of the items, and they failed to integrate these perceptual features with the functional context information in labeling the stimuli. Thus, it appears that patients who exhibit a deficit in object categorization (as indexed by our labeling task) also perform poorly in a perceptual similarity task when the dimensions that characterize the stimulus set are category-dependent (that is, when the dimensional variation defines the category membership of the objects represented). In other words, we are arguing that a unitary deficit of the lexical–semantic component of language processing underlies poor performance in all tasks that involve the manipulation of object category-level information.

In this section we have reviewed a selected subset of the experimental literature on naming and lexical comprehension deficits. The selected material was chosen for its relevance to the issue of whether a lexical–semantic component of processing could be disrupted selectively. The conclusion we would like to draw from the material reviewed is that a lexical–semantic deficit underlies an interesting range of performances in aphasic patients. We are painfully aware of the considerable vagueness that characterizes our description of the naming process and of the lexi-cal–semantic system. Efforts should be directed toward explicating fur-ther both the representation of lexical–semantic information and the pro-cessing mechanisms that interact with this information base. Research on the breakdown of naming and of lexical–semantic processing in general can contribute significantly to the elaboration of progressively more ex-plicit models of lexical–semantic processing.

SYNTACTIC DEFICITS IN APHASIA

The material in this section complements the preceding analyses of the fate of lexical meaning representations in the aphasias by focusing on aphasic patients' ability to process features of sentence form. In this re-spect, we will first briefly review the bases for the claim that damage to left anterior cortex selectively limits the ability to implement syntactic features, whatever the linguistic task. We will then examine this claim ex-plicitly from a processing perspective, and finally, we will raise some issues for future research that have stemmed from this work.

The Recovery and Production of Features of Sentence Form in Broca's Aphasia

The starting point here is the clinical fact that relatively intact compre-hension appears to coexist with telegraphic and otherwise syntactically impoverished speech in Broca's aphasia. The relevant investigations are those that have sought to characterize the bases for this apparent discrep-ancy. The approach taken in these investigations has been to focus on the less clinically accessible—the less "public"—comprehension capacities of these patients, and in doing so, to decompose comprehension in a man-ner that roughly corresponds to the distinction between sentence form and meaning.

As a representative example of this effort, consider an early study that required patients to indicate the correct depiction of the semantic rela-tions expressed by center-embedded sentences delivered at normal speaking rates (Caramazza and Zurif, 1976). For those sentences in which semantic or pragmatic cues dictated only one possible interpretation, as in, *The worm that the bird eats is brown,* the Broca's aphasics unerringly pointed to the picture that correctly represented the semantic roles. How-ever, when such cues were absent—when either the relativized or matrix nouns were equally likely to perform the action specified by the verb, as in, *The girl that the boy chases is tall*—the Broca's aphasics were no more likely to point to the correct drawing than to an incorrect depiction in which the subject and object of the action were reversed.

This and other similar findings (for reviews see Berndt and Caramazza, 1980a; Blumstein, 1980; Goodglass, 1976; Marin and Gordon, 1979; Zurif and Blumstein, 1978) indicate that although Broca's aphasics seem able to infer meaning directly from the major lexical items in a sentence (sampling from among them and combining their meanings in terms of what is likely to make factual sense), they are unable to appreciate syntactic indications to meaning relations. In effect, the picture that emerges is that production and comprehension in Broca's aphasia are less separable than that suggested clinically. Broadly construed, just as they are agrammatic speakers, so too can they be viewed as agrammatic listeners.

Further, the syntactic limitation in comprehension—as in production—seems at least to implicate the closed class of grammatical morphemes. It is important to emphasize at this point that although our discussion of syntactic breakdown will largely involve an analysis of the closed-class vocabulary elements, it is explicitly not being claimed that disruptions in the processing of such elements are the sole cause of syntactic breakdown under conditions of left-sided anterior brain damage. Processing involving the closed class may be only one of the structure-building operations that are affected, or even only a reflection of a larger problem. These possibilities are taken up again in the last part of this section.

The Broca patient's apparent problem with the grammatical morphemes can be inferred from the sentence comprehension study mentioned above (Caramazza and Zurif, 1976), but this problem has been demonstrated more directly with metalinguistic paradigms that shift the patients' attention from what the sentence means (or refers to) to the form of the sentence itself. The simplest metalinguistic task is to require patients to judge the grammatical well-formedness of sentences containing syntactic or semantic errors. The use of this paradigm (with languages as different as Russian and English) has demonstrated that although Broca's aphasics readily recognize "semantic" aberrations (those involving violations of selection restrictions or real-world knowledge), they have difficulty detecting errors involving closed-class items—whether inflections or free-standing functors (Gardner, Denes and Zurif, 1975; Luria, 1975; but see Linebarger, Saffran and Schwartz, 1981).

This pattern emerges, moreover, not only with respect to the recognition of deviance, but also for judgments about the structure of nondeviant sentences (e.g., von Stockert and Bader, 1976; Zurif, Caramazza, and Myerson, 1972). One example here is our assessment of the performance of Broca's aphasic patients and of a control group of neurologically intact

patients on a within-sentence word relatedness sorting task (Zurif, Caramazza, and Myerson, 1972; Zurif and Caramazza, 1976). Having presented the Broca's aphasics with all possible three-word combinations of
words from single written sentences, and for each triad having asked them
which two words "go best together" in the sentence, we observed that
they were unable to integrate normally the closed-class items. In contrast
to the control group patients who judged articles and nouns (and auxil-
iaries and main verbs) as closely related, and who thereby respected phrasal constituents, the Broca's aphasics reliably linked only the content
words of a sentence, either ignoring or inappropriately grouping the functors. As a consequence, they violated the linguistic unity of noun and verb
phrases.

To be sure, there likely exist many unspecified "demand" variables,
which indeterminately influence performance on these metalinguistic
tasks. Yet, although no one task can be taken as offering a "pure" measure of a constituent linguistic capacity, these results converge with the
more straightforward comprehension tasks on an important generalization. Namely, it seems that whatever the activity—whether speaking, listening, or "commenting" in one fashion or another on grammatical well-
formedness—the Broca's aphasic patients are unable to use closed-class
elements to guide the detailed construction of sentence form. In addition,
there is some indication that syntactic-cues to constituent structure other
than those cues involving closed-class items may be unavailable to
Broca's aphasics. Using a similar metalinguistic paradigm, Kolk (1978)
found that some Broca patients were unable to generate noun phrase constituents by combining adjectives and nouns, just as they were unable to
combine determiners and nouns. This finding may indicate that Broca patients' understanding of the structure-marking function of open-class
items may be compromised, even though these items are not at risk to the
same extent in the patients' speech (but see Kean, 1980).

Despite the evidence that Broca's aphasics are impaired in their utilization of closed-class items, this does not mean that they are necessarily
also insensitive to the semantic or "functional" force of such elements. In
fact, the contrary seems to hold. In the above-mentioned sorting study
(Zurif and Caramazza, 1976), for example, there was a notable exception
to the Broca's aphasics' general inability to integrate normally closed-
class items. This exception involved prepositions, and then only when
they were clearly and nonredundantly functional—containing important
locative information, for example. In such situations, Broca's aphasics
consistently clustered the prepositions with the relevant nouns.

Friederici (1980), working with German-speaking patients, has reported a similar pattern. Using a fill-in-the-blank test, she found that Broca's aphasics were significantly more likely to produce prepositions that had obvious semantic content than those that were only syntactically relevant. This finding—gained by taking advantage of some interesting features of the German language—is the more striking for the fact that the same phonological shapes always figured in the contrast. Thus, the patients were more apt to produce the preposition *auf* when it bore important locative information, as in, *Peter steht **auf** dem Stuhl* (*Peter stands **on** the chair*), than when it had only syntactic status, as in, *Peter **hofft auf** den Sommer* (*Peter **hopes for** the summer*). In this latter case, *auf* appears as the result of the fact that the verb is obligatorily subcategorized for that preposition.

Further, there is the possibility that Broca's aphasics seem to appreciate even the semantic–pragmatic force of articles. The evidence for this has been gained from an experimental paradigm in which articles in sentences—written sentences, always in view, and with the article underlined for emphasis—are either appropriately or inappropriately specific. Thus, *the* is appropriate in those instances where the object asked for is singularly distinguishable (where, e.g., the instruction is *touch **the** white one*, and there is only one white object). The article is anomalous in those instances where the sentence (e.g., *touch the white one*) can refer to either of two objects sharing the named attribute (e.g., a white circle and a white square). We have been assessing the performance of Broca's aphasics in this situation (Zurif, Garrett, and Bradley, in progress). Testing for reactions to anomaly, however, is less straightforward than one would wish, and to date we can only be confident that three of five patients have noticed the incorrect use of articles. For these three, presumably, the appearance of the definite article forces some notion such as "ready availability."

What must be emphasized at this point, however, is that their sensitivity to the semantic–pragmatic value of articles is apparent only when real-time processing demands are minimized, as when written sentences are used and left in view. In fact, when the same experiment is carried out with normally spoken, rather than written, sentences, none of the Broca's patients process the article (see also Goodenough, Zurif, and Weintraub, 1977). Presumably, as a consequence of being unable to integrate structurally the article in real time (that is, to locate it properly in relation to the noun), they are blocked from applying its inherent information even though such information is, in principle, available to them.

To summarize, then, the grammatical limitation in Broca's aphasia seems (1) central in nature and (2) to implicate at least processes involving the syntactic role of closed-class items. This last fact is obvious for production. For comprehension it takes the form of a parsing failure: the patients seem unable to make use of these elements as structural markers.

This parsing limitation is clear under real-time conditions, but when real-time processing constraints are absent or minimized the pattern becomes more complicated. Thus, although the Broca's aphasics' performance on sorting and other written judgment tasks (where the pragmatic relevance of articles is nonexistent) suggests that they have problems integrating closed-class elements even off-line, it is also clear that the limitation does not extend equally to all closed-class categories. In particular, prepositions that fix the functional relations of nouns to verbs seem exempt. Possibly this attests to a neurologically honored distinction of processing levels, with the processing of semantically interpretable functional relations (e.g., those between a verb and its arguments) occupying one neurologically independent level, and the processing of closed-class items in the surface assignment of phrasal constituents, occupying another level (Garrett, 1975, 1976, 1980). Prepositions could be involved at both levels, depending upon the sentential context. As a result, prepositions may be impervious to the effects of left anterior brain damage, should these effects be restricted to the stage of initial parsing.

Processing Accounts of Agrammatism

Attempts to provide an analysis of the parsing failure (in terms of the operating characteristics of the language processor) have taken various forms. One such attempt locates the underlying disruption at an initial acoustic–phonetic stage of processing (Kellar, 1978; also Brown, 1979). This characterization begins with the fact that closed-class items in English neither contribute to nor accept sentential stress except for emphatic purposes. In particular, the characterization is rooted to a demonstration that manipulating stress assignment influences within-sentence word relatedness judgments.[3] Thus Kellar has reported that patients produced significantly more groupings of items when both were STRESSED, than when both were UNSTRESSED, REGARDLESS OF THEIR SYNTACTIC RELATEDNESS. It is unclear, however, just what conclusion to draw from this. Are we to infer that the Broca's patients' inability to use closed-class

---

[3] The task Kellar used was like one we had previously employed: that is, sorting words on the basis of their relatedness in a sentence. The difference between the two had to do with the way in which the stimuli were presented. Whereas we had required the patients themselves to read each sentence either aloud or silently, Kellar read them aloud to the patients so that she could manipulate stress.

items is a function of the fact that they are generally unstressed and, therefore, unattended to, and that it is for this reason that the Broca's aphasics show a degraded "syntactic comprehension?" Or is their over-reliance on stress not the CAUSE but rather the RESULT of their problem? —the result, that is, of a differently determined inability to parse on the basis of closed-class items? So far as we can tell the latter seems the more likely alternative. The evidence for this is that (1) as a general point the aphasic patients' capacity to process acoustic input at the segmental level does not reliably predict utterance comprehension skills (Blumstein, Cooper, Zurif, and Caramazza, 1977) and (2) closed-class items appear to place an extra load on the Broca's aphasics' processing capacity quite apart from acoustic factors. This last point has been made in a word-monitoring experiment in which it was observed that unlike neurologically intact subjects, Broca's aphasics responded faster to open-class than closed-class words regardless of stress (Swinney, Zurif, and Cutler, 1980).

Granting, then, that the Broca's problem with closed-class items is not likely to be accounted for in processing terms solely at the point of assigning phonetic representations to those items, how ought it to be character-ized? One possibility first suggested by Bradley (1978; see also Bradley, Garrett, and Zurif, 1980; Zurif, 1980) is that the open–closed class distinc-tion is realized in processing at the stage of lexical access.

The initial observations supporting this claim stem from a number of lexical decision tasks. One such observation—the others have been fully detailed elsewhere (e.g., Bradley, Garrett, and Zurif, 1980)—is that while normal subjects responded in systematically different ways to open-class and closed-class items, yielding a correlation between judgment time and frequency of occurrence for open-class, but not for closed-class items, the Broca's patients treated both alike, showing a frequency effect for both. By hypothesis, the special closed-class access route, which the Broca's aphasics seemed unable to exploit, serves a syntactic function. As input to a parser, the closed-class items signal, for example, the introduction of a noun phrase, the distinctions between main and subordinate clauses, the difference between active and passive sentences, and so on.

This then, for the moment, provides an explanation for the Broca's aphasics' inability to use closed-class elements as syntactic place holders, and correspondingly, for their agrammatic comprehension. These same data also provide a perspective on the semantic–syntactic dissociation observed with respect to individual closed-class vocabulary elements. Thus, since Broca's aphasics recognized the closed-class elements as items in their language, yet failed to treat them differently from open-class items, it seems reasonable to suppose that in addition to the special closed-class route, grammatical morphemes can also be accessed—along with open-class vocabulary items—via the frequency-sensitive system, and further, that this frequency-sensitive system complements the parsing function of the closed-class route by contacting semantic information.

Interest in these findings has generated attempts at replication, and a number of them have been unsuccessful (e.g., Gordon and Caramazza, 1982). Clearly, then, we cannot yet bank on these findings; a more thorough empirical examination of the two-recognition device system must first be undertaken. Still, it is worth remarking that whether this model turns out to be correct, or is in need of revision, or is even wrong, it has specified the processing components of the system in enough detail to allow the types of predictions to be made about functional decomposition that are required for a detailed characterization of brain–language relations.

Some Current Issues

There are two final issues we wish to raise in this section. First, we return to the question of whether or not there are structure-building operations in addition to those involving closed-class items that depend upon the integrity of left-anterior cortex. In this respect, a study by Schwartz, Saffran, and Marin (1980) indicates that Broca's aphasics are unable to comprehend sentences that ostensibly do not involve closed-class items as structural markers—sentences comprised of single noun-verb-noun sequences. The fact that Broca's aphasics were unable to assign consistently the grammatical roles of subject and object to the two noun phrases of these simple sentences suggests that their syntactic deficit includes an inability to interpret word order cues to sentence structure.

This finding—as well as related evidence that Broca's aphasics may (in structured testing) demonstrate difficulty in correctly ordering elements in sentence production (Saffran, Schwartz, and Marin, 1980b)—conflicts with the well-established observation that Broca's patients do not noticeably misorder words in spontaneous speech (Goodglass, 1968). Further, they seem to plan speech in supralexical units (Cooper and Zurif, in press; Danly, deVilliers, and Cooper, 1979), although it is not yet certain that these planning units are syntactic in nature. While this question remains unresolved, evidence is accumulating that the problems experienced by Broca's aphasics with sentence structure extend to their interpretation of the sequential organization of sentence elements (Caramazza, Berndt, Basili, and Koller, 1981; see Berndt and Caramazza, 1981; Saffran, Schwartz, and Marin, 1980a, for discussion). If this possibility turns out to be correct, it will suggest that the Broca's patients are incapable of representing even open-class category information as embodied in syntactic labels such as noun, verb, and adjective.

The second—and final—issue to be covered here concerns the delimitation of the locus of the brain's functional commitment to syntactic processing.[4] Ostensibly, the contrast between Broca's and Wernicke's aphasias seems to turn on the distinction between sentence form and meaning: Broca's patients show a relative problem in the production and recovery of features of sentence form; Wernicke's patients have a relative problem in the production and recovery of word meanings. The problem with this distinction, however, is that the Wernicke's patients' comprehension and production deficits cannot be viewed unequivocally as "local" lexical problems; on the contrary, syntactic structure seems to be implicated as well. Thus, in comprehension, Wernicke's aphasics are clearly less capable than neurologically intact patients of processing relative clause constructions (Caramazza and Zurif, 1976). In production, they are far less likely to depart from simple constructions than are normal speakers (Gleason, Goodglass, Obler, Green, Hyde, and Weintraub, 1980). And to the extent that they do produce relative-clause constructions, these often contain striking semantic discontinuities across clausal boundaries—as if the constructions were prefabricated routines unresponsive to the structural arrangements they are normally meant to accommodate (Delis, Foldi, Hamby, Gardner, and Zurif, 1979). Is syntax a "weak link" in the processing chain—the first to go, wherever the site of the left-sided damage?

Any reasonable answer to this question is likely to be complicated for there are clearly a number of distinct levels concerned with syntactic processing (Garrett, 1975, 1976, and 1980). The clinically contrasting syndromes of Broca's and Wernicke's aphasias, though possibly both involving syntactic disruptions, may yet interestingly serve to distinguish these levels. But, again, the tasks of determining what counts as evidence on this issue, and then gaining the relevant data, remain before us.

CONCLUSION

We have reviewed selected literature that supports two points that are central to the neuropsychology of language. First, the syntactic and semantic structures and processes that are theoretically part of the normal

---

[4] By the term *locus* we are referring simply to the site of damage which selectively undermines a particular linguistic component, however complex the relation between the neuroanatomical location and the neurological organization of the computational unit it indexes.

system have been shown to be useful in characterizing various perform- ances that define particular aphasic syndromes. This demonstration not only lends support to the hypothesis that aphasia reflects disruption of ab- stract linguistic components as they function in different tasks and in all modalities, but it also supports the psychological reality of the theoretical components as they have been described.

The second point is that this selective dissociation of linguistic func- tions is strongly related to anterior versus posterior location of focal dam- age in the left hemisphere. Although it is possible to make somewhat stronger claims concerning the precise areas that must be involved to produce a particular syndrome (e.g., Mohr, 1976), this gross dichotomy is at present the best available generalization concerning the localization of these components. One reason that more precise correlations between le- sion site and symptoms are not yet possible is the conflation of patient types in studies of particular deficits—as is apparent in our discussion of semantic deficits. But perhaps the most serious obstacle to providing more definitive brain–function correlations is the lack of unified neuro- physiological theories that can be mapped onto increasingly specific models of linguistic functioning. What is needed is an "information pro- cessing neurology" (Caplan, 1981).

As long as this is unavailable, neurolinguistic investigations of aphasia will continue to probe the symptoms that result from neurological insult, assuming that the dissociations of function produced are traceable to a re- coverable pattern of organization in the brain. This enterprise is greatly enhanced by developments in linguistics and psycholinguistics, and many of the unresolved questions confronting aphasia researchers may be most easily answered in the normal laboratory. For example, the suggestion that Broca's aphasics fail to utilize the normal specialized access route for closed-class items must proceed from strong evidence that such a special- ized route exists in normal speakers. In fact, the role of the closed class in providing normal language users with structural cues is central to psycho- linguistic models of sentence processing (see Frazier, 1979, for a review), and any data gathered on normal utilization of these items should contrib- ute importantly to our understanding of the Broca syndrome.

Research on language functions and brain organization must thus be viewed as a cooperative enterprise. Aphasiologists will continue to rely on and borrow from linguistics and psycholinguistics for methods and models, and should begin more active hypothesis-testing in the normal setting. A connection with neuroanatomy and neurophysiology (though important) may be more difficult, as there is at present no theoretical vo- cabulary that can be used to relate cognitive–linguistic functions to neu- rological processes. Nonetheless, the growing interest in bridging this gap suggests that there is reason to be optimistic. Progress has been made in the past 100 years; there is considerable evidence that it will continue.

REFERENCES

Arbib, M. A., and Caplan, D. (1979). Neurolinguistics must be computational. *Behavioral and Brain Sciences, 2,* 449–483.

Benson, D. F. (1967). Fluency in aphasia: Correlation with radioactive scan location. *Cortex, 3,* 373–394.

Berndt, R. S., and Caramazza, A. (1980a). A redefinition of the syndrome of Broca's aphasia: Implications for a neuropsychological model of language. *Applied Psycholinguistics, 1,* 225–278.

Berndt, R. S., and Caramazza, A. (1980b). Semantic operations deficits in sentence comprehension. *Psychological Research, 41,* 169–177.

Berndt, R. S., and Caramazza, A. (1981). Syntactic aspects of aphasia. In Martha Taylor Sarno (Ed.), *Acquired aphasia.* New York: Academic Press.

Blumstein, S. E. (1980). Neurolinguistics: Language–brain relationships. In S. B. Filskov and T. J. Boll (Eds.), *Handbook of clinical neuropsychology.* New York: Wiley.

Blumstein, S. E., Cooper, W. E., Zurif, E. B., and Caramazza, A. (1977). The perception and production of voice-onset time in aphasia. *Neuropsychologia, 15,* 371–383.

Bradley, D. C. (1978). *Computational distinctions of vocabulary type.* Unpublished doctoral dissertation, Massachusetts Institute of Technology.

Bradley, D. C., Garrett, M. E., and Zurif, E. B. (1980). Syntactic deficits in Broca's aphasia. In D. Caplan (Ed.), *Biological studies of mental processes.* Cambridge, Mass.: MIT Press.

Broca, P. (1950). [Remarques sur le siège de la faculté du language articulé, suivi d'une observation d'aphémie]. (J. Kann, trans.) *Journal of Speech and Hearing Disorders, 15,* 16–20. (Originally published, 1861.)

Brown, J. (1979). Comments on Arbib paper. Paper presented at University of Massachusetts Conference on Neural Models of Language. Amherst, Massachusetts, 1979.

Buckingham, H. W., and Rekart, D. M. (1979). Semantic paraphasia. *Journal of Communication Disorders, 12,* 197–209.

Caplan, D. (1981). On the cerebral localization of linguistic functions. *Brain and Language, 14,* 120–137.

Caramazza, A., and Berndt, R. S. (1978). Semantic and syntactic processes aphasia: A review of the literature. *Psychological Bulletin, 85*(4), 898–918.

Caramazza, A., Berndt, R. S., Basili, A. G., and Koller, J. J. (1981). Syntactic processing deficits in aphasia. *Cortex, 17,* 333–348.

Caramazza, A., Berndt, R. S., and Brownell, H. H. (1982). The semantic deficit hypothesis of the naming defect: Perceptual parsing and object classification by asphasic patients. *Brain and Language, 15,* 161–189.

Caramazza, A., Hersh, H., and Torgerson, W. S. (1976). Subjective structures and operations in semantic memory. *Journal of Verbal Learning and Verbal Behavior, 15,* 103–117.

Caramazza, A., and Zurif, E. B. (1976). Dissociation of algorithmic and heuristic processes in language comprehension: Evidence from aphasia. *Brain and Language, 3,* 572–582.

Chomsky, N., and Halle, M. (1968). *The sound pattern of english.* New York: Harper.

Cooper, W., and Zurif, E. Aphasia: Information-processing in language production and reception. In B. Butterworth (Ed.), *Language production*, Vol. 2. London: Academic Press, in press.

Coughlan, A. K., and Warrington, E. (1978). Word-comprehension and word-retrieval in patients with localized cerebral lesions. *Brain, 101*, 163–185.

Danly, M., deVilliers, J. G., and Cooper, W. E. (1979). The control of speech prosody in Broca's aphasia. In J. J. Wolf and D. H. Klatt (Eds.), *Speech communication papers presented at the 97th meeting of the Acoustical Society of America*. New York: Acoustical Society of America.

Degerman, R. L. (1970). Multidimensional analysis of complex structure: Mixtures of class and quantitative variation. *Psychometrika, 35*, 475–491.

Delis, D., Foldi, N. S., Hamby, S., Gardner, H., and Zurif, E. (1979). A note on temporal relations between language and gestures. *Brain and Language, 8*, 350–354.

Fodor, J. A., Garrett, M. F., Walker, E. C., and Parkes, C. H. (1980). Against definitions. *Cognition, 8*, 263–367.

Frazier, L. (1979). On comprehending sentences: Syntactic parsing strategies. Bloomington, Indiana: Indiana University Linguistics Club.

Friederici, A., (1982). Syntactic and Semantic Processess in Aphasic Deficits: The Availability of Prepositions. *Brain and Language, 15*, 249–258.

Gainotti, G. (1976). The relationship between semantic impairment in comprehension and naming in aphasic patients. *British Journal of Disorders of Communication, 11*, 57–61.

Gainotti, G., Caltagirone, C., and Ibba, A. (1975). Semantic and phonemic aspects of auditory comprehension in aphasia. *Linguistics, 154/155*, 15–29.

Gainotti, G., Caltagirone, C., and Miceli, G. (1979). Semantic disorders of auditory language comprehension in right-damaged patients. *Journal of Psycholinguistic Research, 8*, 13–20.

Gardner, H. (1973). The contribution of operativity to naming capacity in aphasic patients. *Neuropsychologia, 11*, 213–220.

Gardner, H., Denes, G., and Zurif, E. (1975). Critical reading at the sentence level in aphasia. *Cortex, 11*, 60–72.

Garrett, M. F. (1975). The analysis of sentence production. In G. Bower (Ed.), *The psychology of learning and motivation: Advances in research and theory*. New York: Academic Press.

Garrett, M. F. (1976). Syntactic processes in sentence production. In E. Walker and R. Wales (Eds.), *New approaches to language mechanisms*. Amsterdam: North-Holland Publ.

Garrett, M. F. (1980). Levels of processing in sentence production. In B. Butterworth (Ed.), *Language Production*, Vol. 1. London: Academic Press. Pp. 177–220.

Geschwind, N. (1965). Disconnection syndromes in animals and man. *Brain, 88*, 237–294, 585–644.

Gleason, J., Goodglass, H., Obler, L., Green, E., Hyde, M., and Weintraub, S. (1980). Narrative strategies of aphasic and normal-speaking subjects. *Journal of Speech and Hearing Research, 23*, 370–382.

Goodenough, C., Zurif, E., and Weintraub, S. (1977). Aphasics' attention to grammatical morphemes. *Language and Speech, 20*, 11–19.

Goodglass, H. (1968). Studies on the grammar of aphasics. In S. Rosenberg and K. Joplin (Eds.), *Developments in applied psycholinguistics research*. New York: Macmillan.

Goodglass, H. (1976). Agrammatism. In H. Whitaker and H. A. Whitaker (Eds.), *Studies in Neurolinguistics*, Vol. 1. New York: Academic Press.

Goodglass, H., and Baker, E. (1976). Semantic field, naming and auditory comprehension in aphasia. *Brain and Language, 3*, 359–374.

Goodglass, H., Barton, M. I., and Kaplan, E. F. (1968). Sensory modality and object-naming in aphasia. *Journal of Speech and Hearing Research, 11*, 488–496.

Goodglass, H., Hyde, M. R., and Blumstein, S. (1969). Frequency, picturability, and the availability of nouns in aphasia. *Cortex, 5*, 104–119.

Goodglass, H., and Kaplan, E. (1972). *The assessment of aphasia and related disorders.* Philadelphia: Lea and Febiger.

Gordon, B., and Caramazza, A. (1982). Lexical decision for open- and closed-class items: Failure to replicate differential frequency sensitivity. *Brain and Language, 15*, 143–160.

Grober, E., Perecman, E., Kellar, L., and Brown, J. (1980). Lexical knowledge in anterior and posterior aphasics. *Brain and Language, 10*, 318–330.

Grossman, M. (1978). The game of the name: An examination of linguistic reference after brain damage. *Brain and Language, 6*, 112–119.

Kean, M. L. (1977). The linguistic interpretation of aphasic syndromes: Agrammatism in Broca's aphasia, an example. *Cognition, 5*, 9–46.

Kean, M. L. (1980). A note on Kolk's "Judgments of sentence structure in Broca's aphasia." *Neuropsychologia, 18*, 357–360.

Kellar, L. (1978). Stress and syntax in aphasia. Paper presented at the meeting of the Academy of Aphasia, Chicago, 1978.

Kolk, H. H. (1978). Judgments of sentence structure in Broca's aphasia. *Neuropsychologia, 16*, 617–625.

Labov, W. (1973). The boundaries of words and their meanings. In C. J. N. Bailey and R. W. Shuy (Eds.), *New ways of analyzing variation in English.* Washington, D. C.: Georgetown Univ. Press.

Lhermitte, F., Derouesne, J., and Lecours, A. R. (1971). Contribution à l'etude des troubles semantiques dans l'aphasie. *Revue Neurologigue, 125*, 81–101.

Linebarger, M. C., Saffran, E. M., and Schwartz, M. F. (1981). Judgments of grammaticality in agrammatic patients. Presented at the BABBLE Conference, Niagara Falls, Ontario.

Luria, A. R. (1970). Traumatic Aphasia. (English translation) The Hague: Mouton. (Originally published 1947.)

Luria, A. R. (1975). Two kinds of disorders in the comprehension of grammatical constructions. *Linguistics, 154/155*, 47–56.

Marin, O. S. M., and Gordon, B. (1979). Neuropsychologic aspects of aphasia. In H. R. Tyler and D. M. Dawson (Eds.), *Current neurology*, Vol. 2. Boston: Houghton.

Miller, G. A., and Johnson-Laird, P. N. (1976). *Language and perception.* Cambridge, Mass. The Belknap Press of Harvard Univ. Press.

Mohr, J. R. (1976). Broca's area and Broca's aphasia. In H. Whitaker and H. A. Whitaker (Eds.), *Studies in Neurolinguistics*, Vol. 1. New York: Academic Press.

Rinnert, C., and Whitaker, H. A. (1973). Semantic confusions by aphasic patients. *Cortex, 9*, 56–81.

Saffran, E. M., Schwartz, M. F., and Marin, O. S. M. (1980a). Evidence from aphasia: Isolating the components of a production model. In B. Butterworth (Ed.), *Language Production*, Vol. 1. London: Academic Press.

Saffran, E. M., Schwartz, M. F., and Marin. O. S. M. (1980b). The word order problem in agrammatism, II: Production. *Brain and Language, 10*, 263–280.

Schnitzer, M. L. (1972). Generative phonology—evidence from aphasia. University Park, Penn.: Penn. State Univ. Press.

Schuell, H., and Jenkins, J. (1961). Reduction of vocabulary in aphasia. *Brain, 84*, 243–261.

Schwartz, M. F., Marin, O. S. M., and Saffran, E. M. (1979). Dissociations of language function in dementia: A case study. *Brain and Language, 7*, 277–306.

Schwartz, M. F., Saffran, E. M., and Marin, O. S. M. (1980). The word order problem in agrammatism: I, Comprehension. *Brain and Language, 10,* 249–262.

Spreen, O., Benton, A. L., and Van Allen, M. W. (1966). Dissociation of visual and tactile naming in amnesic aphasia. *Neurology, 16,* 807–814.

Swinney, D., Zurif, E., and Cutler, A. (1980). Effects of sentential stress and word class upon comprehension in Broca's aphasics. *Brain and Language, 10,* 132–144.

von Stockert, T. R., and Bader, L. (1976). Some relations of grammar and lexicon in aphasia. *Cortex, 12,* 49–60.

Warrington, E. K. (1975). The selective impairment of semantic memory. *Quarterly Journal of Experimental Psychology, 27,* 635–657.

Weisenberg, T., and McBride, K. (1935). *Aphasia.* New York: Commonwealth Fund.

Wernicke, C. (1908). [The symptom-complex of aphasia]. In A. Church (Ed.), *Disorders of the nervous system.* New York: Appleton. (Originally published 1874.)

Whitehouse, P., Caramazza, A., and Zurif, E. B. (1978). Naming in aphasia: Interacting effects of form and function. *Brain and Language, 6,* 63–74.

Zurif, E. B. (1980). Language mechanisms: A neuropsychological perspective. *American Scientist, 68,* 305–311.

Zurif, E. B., and Blumstein, S. (1978). Language and the brain. In M. Halle, J. Bresnan, and G. Miller (Eds.), *Linguistic theory and psychological reality.* Cambridge, Mass.: MIT Press.

Zurif, E. B., and Caramazza, A. (1976). Psycholinguistic structures in aphasia: Studies in syntax and semantics. In H. Whitaker and H. A. Whitaker (Eds.), *Studies in neurolinguistics,* Vol. 1. New York: Academic Press.

Zurif, E. B., Caramazza, A., and Myerson, R. (1972). Grammatical judgments of agrammatic aphasics. *Neuropsychologia, 10,* 405–417.

Zurif, E. B., Caramazza, A., Myerson, R., and Galvin, J. (1974). Semantic feature representations for normal and aphasic language. *Brain and Language, 1,* 167–187.

CHAPTER 2

Hemisphere Processing of Phonological Information

Victoria J. Molfese, Dennis L. Molfese, and Carl Parsons

INTRODUCTION

Since Kimura's (1961a) demonstration that pairs of language stimuli presented simultaneously are more accurately reported from the right ear than the left, the dichotic technique has become a popular method for those researchers investigating questions of hemispheric specialization. Subsequent investigations have shown a right-ear advantage for words, nonsense syllables, backward speech, and synthetic syllables (Kimura and Folb, 1968; Shankweiler and Studdert-Kennedy, 1966, 1967), and a left-ear advantage for nonspeech material such as vocal nonspeech sounds, music, white noise, and environmental sounds (Kimura, 1964; Curry, 1967; Spellacy, 1970; Gordon, 1970, King and Kimura, 1972). These findings, along with those from other methodologies such as split brain procedures (Gazzaniga, 1970), clinical populations (Geschwind, 1965; Luria, 1966), anatomical procedures (Geschwind and Levitsky, 1968), and electrophysiology (Morrell and Salamy, 1971; Donchin, Kutas, and McCarthy, 1977; Molfese, 1979), provide compelling evidence that

29

LANGUAGE FUNCTIONS
AND BRAIN ORGANIZATION

language abilities are left-lateralized and nonlanguage abilities are right-lateralized.

The mechanisms responsible for ear advantages on dichotic tasks have been described in different ways but there tend to be some general agreements. According to Kimura (1961b), ear advantages are due to two components: (1) the greater number and effectiveness of contralateral over ipsilateral pathways in the transmission of auditory signals and (2) the dominance of the left hemisphere for language and the right hemisphere for nonlanguage processing. In an expansion of Kimura's model, Sparks and Geschwind (1968) hypothesized that auditory signals (such as speech sounds) arriving at the ipsilateral ear are suppressed by input from the contralateral ear during dichotic presentations. Speech sounds from the left ear arrive at the left hemisphere after first following the decussating pathway to the right hemisphere and then arriving at the left hemisphere via the transcallosal pathway. Furthermore, speech sounds from the left ear are believed to arrive as "degraded" signals at the left hemisphere due to the extra synaptic step. Thus, the right-ear advantage for speech is due to the stronger signal arriving at the hemisphere specialized for language processing.

Other researchers, while agreeing with the basic model described above, have found the model too simplistic. Several variables (including those involving procedure, subject characteristics, and stimulus types) have been found to have the effect of shifting ear advantages on dichotic tasks. For example, attentional sets arising from prior or current task demands have been found to shift the ear advantage from the left to the right ear for nonverbal stimuli containing linguistic information (Tsunoda, 1969), and for vowels when they are embedded in a word versus a melody series (Spellacy and Blumstein, 1970). The ear advantage also shifts from the right to the left ear when subjects are instructed to attend to the emotional tone rather than linguistic cues of a sentence (Haggard and Parkinson, 1971).

Several researchers (e.g., Studdert-Kennedy and Shankweiler, 1970; Blumstein and Cooper, 1972; Schwartz and Tallal, 1980) have been investigating the acoustic and phonetic mechanisms which seem to trigger left-hemisphere versus right-hemisphere processing of dichotically presented materials. Studdert-Kennedy and Shankweiler (1970; Studdert-Kennedy, 1975) believe that both hemispheres have the ability to extract the acoustic properties in speech. However, only the left hemisphere (or the language dominant hemisphere) extracts phonetic features. Two studies (Shankweiler and Studdert-Kennedy, 1967; Studdert-Kennedy and Shankweiler, 1970) demonstrated such laterality in the perception of dichotically presented synthetic or natural speech syllables differing only in phonetic features. Adults were presented with syllables which paired all combinations of six stop consonants (/b, d, g, p, t, k/) with six vowels (/i, ε, æ, a, ɔ, u/). The syllable pairs were constructed such that either the initial consonant or the final consonant varied. The two syllables in the pair always had different vowels. The subject's task was to identify both vowels presented in the pair for the vowel tests and both consonants in the pair for the consonant tests. A significant right-ear advantage was found for consonants but no significant ear advantage was found for vowels. The articulatory features of voicing and place of articulation, which were present in the syllables, produced right-ear advantages. Studdert-Kennedy and Shankweiler (1970) attributed this feature laterality effect to the specialization of the left hemisphere for the "separation and sorting of a complex of auditory parameters into phonological features" (p. 590). Studdert-Kennedy, Shankweiler, and Pisoni (1972), Blumstein and Cooper (1972), and Cutting (1974) have also found right-ear advantages in the identification of consonants differing in voice and place of articulation.

Interestingly, Studdert-Kennedy and Shankweiler (1970) have also found that voiced and voiceless consonants are equal in ease of correct identification but voiced consonants elicit stronger right-ear advantages. Berlin, Lowe-Bell, Cullen, Thompson, and Loovis (1973), however, have found that voiceless consonants are much more potent stimuli than voiced consonants. In fact, Berlin (1977) sees the voiceless effect as "so strong that if one were consistently to put a voiceless CV in the left ear in competition with a voiced CV in the right ear, there would invariably be a left-ear advantage" (p. 83). The reason for the different findings from the two studies is unclear, but may in part stem from the use of synthetic speech syllables by Studdert-Kennedy and Shankweiler, while Berlin, Lowe-Bell, Cullen, Thompson, and Loovis (1973) have used natural speech stimuli.

Several other phonetic features of consonants have also produced right-ear advantages. Blumstein, Tartter, and Michel (cited in Blumstein, 1974) used the dichotic technique to present pairs of natural speech syllables varying in stop, nasal, and fricative features. Significant right ear advantages were found for stops and fricatives but not for nasals. The nonsignificant effect for nasals was attributed to the high level of correct responses (88 and 92% for left and right ears, respectively, compared with stops and fricatives which had 10—20% fewer correct responses) and to the similarity of nasals to vowels and liquids, which have also shown small ear effects. Cutting (1974) investigated the dichotic perception of stops, liquids, and vowels. He found a right-ear advantage for liquids (/r/, /l/) that was less strong that that for stops (/g/, /k/) but stronger than that for

Table 1

NORMAL POPULATIONS

Study	Stimuli	Task	Subjects	Findings
Shankweiler and Studdert-Kennedy (1967)	Synthetic CV syllables Consonants: /b, d, g, p, t, k/ Vowels: /i, ɛ, æ, a, ɔ, u/	Dichotic identification	Normals	Consonant: right-ear advantage Vowel: no ear advantage
Studdert-Kennedy and Shankweiler (1970)	Natural speech CVC syllables Consonants: /b, d, g, p, t, k/ Vowels: /i, ɛ, æ, a, ɔ, u/	Dichotic identification	Normals	Consonant: right-ear advantage Vowel: no ear advantage Place: right-ear advantage Voice: right-ear advantage
Blumstein and Cooper (1972)	Natural speech CV syllables Consonants: /b, d, g, p, t, k/ Vowel: /a/	Dichotic identification	Normals	Consonant: right-ear advantage
Cutting (1974)	4 experiments: Synthetic CCV syllables (consonants: /g, k, l, r/; vowels: /ɛ, æ/) Synthetic CV, VC syllables (consonants: /b, p, l, r/; vowels: /i, æ, ɔ/) Synthetic steady-state vowels (/i, æ/) Sinewave CV syllables	Dichotic identification	Normals	Initial liquids: right-ear advantage Initial and final stops: right-ear advantage Steady-state vowels: right-ear advantage No ear advantage for final liquids, vowels, sinewave vowels, and sinewave CVs
Studdert-Kennedy, Shankweiler, and Pisoni (1972)	Synthetic CV syllables Consonants: /b, d, p, t/ Vowels: /i, u/	Dichotic identification	Normals	Consonants: right-ear advantage Right-ear advantage greatest for alveolar pairs with same vowel, and labial pairs with different vowels
Blumstein, Tartter, and Michel (1973)	Natural speech CV syllables Consonants: /b, d, m, n, v, z/ Vowel: /a/	Dichotic identification	Normals	Stops: right-ear advantage Fricatives: right-ear advantage Nasal: no ear advantage
Schwartz and Tallal (1980)	Synthetic CV syllables Consonants: /b, d, g, p, t, k/ Vowel: /a/	Dichotic identification	Normals	Stronger right-ear advantage to syllables with 40-msec than with 80-msec formant transitions

vowels (/æ/, /ɛ/). Interestingly, Cutting also found that the position of the consonant in the CV, VC pairs (either initial or final) affected ear advantage. Initial and final stops and initial liquids yielded a right-ear advantage but final liquids showed no significant ear advantage.

Dichotic perception of vowel stimuli has also been studied but the findings have been less clear with regard to ear advantage and laterality. Most researchers have found nonsignificant right-ear advantages (Shankweiler and Studdert-Kennedy, 1967; Studdert-Kennedy and Shankweiler, 1970; Cutting, 1974; Tsunoda, 1975). However, significant right-ear advantages have been found for steady state vowels (Cutting, 1974); for vowels presented while embedded in noise (Weiss and House, 1973); in a word series (Spellacy and Blumstein, 1970); or to certain ethnic groups (e.g., native Japanese, Tsunoda, 1975).

On the basis of their dichotic research with vowels, nasals, and final liquids in CV syllables, Blumstein (1974) and Cutting (1974) have argued that these speech sounds may have acoustical cues which are similar, and which cause them to be processed differently from stop consonants and initial liquids. Blumstein writes that "it has been suggested that these differential right ear effects reflect the degree to which various speech sounds are encoded. Thus, those sounds which require less acoustic restructuring as, for example, vowels (and nasals and final liquids) may be less dependent upon specialized left hemisphere mechanisms" (1974, p. 345).

The common conclusion that the right-ear advantage reflects a phonetic rather than an acoustic processing has not gone unchallenged. Schwartz and Tallal (1980) point out that variations in temporal acoustic cues, as well as variations in phonetic cues, are present in the speech sounds used in research. They investigated the possibility that the right-ear advantage might reflect acoustic as well as phonetic processing. Thirty adults participated in a dichotic task that required them to identify the CV syllables they heard on each trial. The stimuli were two sets of synthetic stop consonant–vowel syllables (/ba, da, ga, pa, ta, ka/). One set was characterized by normal 40-msec duration formant transitions while the other set had extended 80-msec duration formant transitions. Both sets of stimuli were perceived by the subjects with the same level of accuracy but the right-ear advantage for the 40-msec stimulus set was larger than that for the 80-msec stimulus set. These findings show that the left hemisphere is also sensitive to rapidly changing acoustic cues. The authors point out that while these rapidly changing cues are frequent characteristics of linguistic stimuli, it is not the linguistic stimuli per se that trigger left hemisphere processing.

In a later study, Tallal and Newcombe (1978) used verbal and nonverbal stimuli to demonstrate the role of the left hemisphere in the processing of rapidly changing acoustic stimuli. Adults with left- and right-hemisphere lesions were presented with a series of stimuli during a repetition task developed by Tallal and Piercy (1973). The verbal stimuli were /ɛ/ and /æ/ steady-state vowels and synthetic CV syllables (/ba/, /da/) with 40-msec or 80-msec transitions. The nonverbal stimuli were two complex tones with interstimulus intervals (ISI) between the tones of 428 msec or short intervals varying from 8 to 305 msec. The results showed that only subjects with intact left hemispheres were able to respond accurately to rapidly changing acoustic information, whether contained in verbal or nonverbal stimuli. The performance of the left-hemisphere damaged group was comparable to that of the control subjects only when the tone stimuli were separated by long ISI, and the CV stimuli had extended formant transitions. Thus, the left hemisphere seems to play a specific role in acoustic processing.

DICHOTIC LISTENING
WITH BRAIN-DAMAGED POPULATIONS

One means of verifying the role played by hemisphere dominance in the processing of dichotically presented speech stimuli is to study the speech perception of brain-damaged populations. There has been considerable research on the perception of auditory stimuli by adults with various types of brain damage. Unfortunately, the results have often been contradictory. For example, both left- and right-ear advantages have been reported in studies of dichotic digit perception by aphasics (Dobie and Simmons, 1971; Schulhoff and Goodglass, 1969; Sparks, Goodglass, and Nickel, 1970). The differences found in these and other studies have been difficult to explain, but stem in part from methodological weaknesses involving (1) lack of similarity in subject characteristics (particularly confirmation of the type and extent of brain damage), (2) confounding due to differences in stimulus type and construction, (3) variability in task requirements, and (4) inappropriate or lack of statistical analyses. Though these problems have existed in past work, research with brain-damaged populations continues to offer a unique means of examining hemispheric functioning.

The work reviewed in this section will center on the extensive studies into the perception of stimuli varying in phonetic features by brain-damaged adults. Shanks and Ryan (1976) studied the perception of nonsense CV syllables (/pa, ba, ta, da, ka, ga/) by left-hemisphere brain-damaged and normal adults. The brain-damaged adults were all aphasic (mean

PICA score: 71%) as a result of vascular (N = 8) or traumatic (N = 3) left-hemisphere damage. The control subjects were matched by age, sex, handedness, education, and bilingualism to the brain-damaged subjects. Results of the dichotic task showed significant differences between the two groups, with only the normal controls producing a significant right-ear advantage. Shanks and Ryan compared aphasics with a right-ear advantage (N = 6) with those demonstrating a left-ear advantage (N = 5) and found the overall accuracy level of the two aphasic subgroups to be about equal. These two aphasic subgroups were hypothesized as representing different lesion sites. Following the model of Sparks, Goodglass, and Nickel (1970), Shanks and Ryan reasoned that the left-ear advantage subgroup had a lesion which interferred with the processing of the right-ear signal and which allowed better ipsilateral processing. The aphasic subgroup with the right-ear advantage was thought to have a lesion which blocked the input from the left ear through the corpus callosum. This latter group's performance was reported to be comparable to that reported by Milner, Taylor, and Sperry (1968) with corpus callosal patients. Shanks and Ryan did not attempt to confirm the existence of the damage which was hypothesized to differentiate their two aphasic subgroups.

Johnson, Sommers, and Weidner (1977) also studied left-damaged adults but reported results different from those of Shanks and Ryan. Johnson, Sommers and Weidner studied the dichotic performance of 20 aphasics with left-hemisphere posterior lesions to determine the effects of initial severity, time since onset, and language recovery in relation to direction and magnitude of ear advantage. The aphasics' performance was compared with that of 20 normal subjects. The stimuli were CVC words which differed only in the initial segment by contrasts in place, voicing, or both. The results showed that while the control subjects showed a right-ear advantage, 18 of the 20 aphasic subjects showed left-ear advantages. Those aphasics with the greater initial severity showed the largest left-ear advantages. Magnitude of language improvement was also related to ear preference, though time since onset of aphasia was not. Unfortunately, neither the results of this study nor those of Shanks and Ryan were discussed in relation to place or voicing features of the stimuli.

A study by Oscar-Berman, Zurif, and Blumstein (1975) does report on the perception of specific phonetic features by brain-damaged subjects. They presented CV syllables varying in place, voice, or both to 25 brain-damaged and 25 normal subjects. The brain-damaged subjects each had unilateral damage (16 left-brain, 9 right-brain). Nine left-brain and 3 right-brain subjects had anterior lesions, and the remaining 7 left-brain and 6 right-brain subjects had posterior damage. The results of the dichotic task showed that the brain-damaged subjects made more errors than the nor-

Table 2

BRAIN-DAMAGED POPULATIONS

Study	Stimuli	Task	Subjects	Findings
Oscar-Berman, Zurif, and Blumstein (1975)	Synthetic CV syllables Consonants: /b, d, g, p, t, k/ Vowel: /a/	Dichotic identification	LH damaged RH damaged Normal controls	Normals: right-ear advantage Damaged: ear advantage ipsilateral to damage LH damaged: better with voiced consonants. All other subjects better with shared features
Shanks and Ryan (1976)	CV syllables Consonants: /b, d, g, p, t, k/ Vowel: /a/	Dichotic identification	LH damaged Normal controls	Normals: right-ear advantage 6 LH damaged: right-ear advantage 5 LH damaged: left-ear advantage
Basso, Casati, and Vignolo (1977)	Synthetic CV syllables with VOT values from −150 to +150 msec (/da, ta/)	Identification	LH damaged aphasics LH damaged nonaphasics RH damaged Normal controls	RH damaged and LH damaged nonaphasic: normal identification LH damaged aphasic: 66% show poor identification; 26% show normal identification
Blumstein, Baker, and Goodglass (1977)	Real words and nonsense words constructed from initial consonants: /b, d, g, p, t, k/	Discrimination	LH damaged: 12 Anterior 12 Posterior Normal controls	LH posterior damage: fewest errors on voice contrasts

Study	Stimuli	Task	Groups	Results
Blumstein, Cooper, Zurif, and Caramazza (1977)	Synthetic CV syllables with VOT values from −20 to +80 msec (/da, ta/)	Discrimination Identification	LH damaged RH damaged Normal controls	RH damaged: normal perception LH damaged: 56% show normal perception, 25% show only normal discrimination, 19% no discrimination or identification
Johnson, Sommers, and Weidner (1977)	CVC words differing in place, voice, or both	Dichotic identification	LH damaged Normal controls	Normals: right-ear advantage LH damaged: 90% show left-ear advantage
Miceli, Caltagirone, Gainotti, and Payer-Rigo (1978)	Synthetic CVC syllables Initial consonants: /b, d, g, p, t, k/ CVC: /rin/	Discrimination	RH damaged LH damaged aphasics LH damaged nonaphasics	LH damaged aphasics: fewest errors with voiced stimuli All others: fewest errors with place stimuli
Tallal and Newcombe (1978)	Complex tones Synthetic steady-state vowels (/ɛ, æ/) Synthetic CV syllables (/ba, da/) varying in formant transition length	Repetition	LH damaged RH damaged Normal controls	No group differences on vowels LH damaged: poor on rapidly changing acoustic stimuli RH damaged equal to controls
Perecman and Kellar (1981)	Natural speech consonants /b, d, g, p, t, k/	Matching	LH damaged aphasics RH damaged Normal controls	LH damaged match on voice RH damaged and controls match on voice and place equally

mal controls, though the two brain-damaged groups were similar in error rate. The normal control subjects displayed a significant right-ear advantage. The brain-damaged groups showed ear advantages which were ipsilateral to the lesion site. Though contrary to Shank's and Ryan's finding of both left- and right-ear advantages in left-damaged subjects, the findings of Oscar-Berman, Zurif, and Blumstein are consistent with those reported by Johnson, Sommers, and Weidner (1977), and those reported by Schulhoff and Goodglass (1969), who found that the lesion site affects contralateral more than ipsilateral processing.

In examining the effects of the specific phonetic features, Oscar-Berman, Zurif, and Blumstein found that when the dichotically presented stimuli shared common phonetic features, the performance of normal subjects and right-hemisphere damaged individuals was better than when no features were shared. The left-hemisphere damaged subjects, however, were only benefitted when the stimulus pair shared the feature "voice." Similar results were found in a study by Miceli, Caltagirone, Gainotti, and Payer-Rigo (1978), who also used stimuli differing in place and voice. Miceli and colleagues used a nondichotic technique to present pairs of synthetic CCVC syllables to 42 normal control and 127 brain-damaged adults (43 right-brain, 48 left-brain without aphasia, 36 left-brain with aphasia). The consonants /p, t, k, b, d, g/ were paired with /rin/ to form the CCVC syllables (e.g., *prin, trin*). None of the syllables was a recognizable word in Italian, the native language of the subjects. The results of the test showed more voice than place errors for all subjects except the left-brain damaged aphasic group. That group showed fewest errors with voiced stimuli. Miceli, Caltagirone, Gainotti, and Payer-Rigo, and Oscar-Berman, Zurif, and Blumstein attributed their findings to the differences in the acoustic cues present in stimuli with place and voice features as described by Liberman, Cooper, Shankweiler, and Studdert-Kennedy (1967). The voice feature contains more invariant and fewer complex acoustical cues compared to the place feature. Because processing the voice feature requires "less restructuring of the acoustic waveform" (Oscar-Berman, Zurif, and Blumstein, 1975, p. 354), left-damaged subjects are thought to process it easier than the more complex place feature.

Blumstein, Baker, and Goodglass (1977) also studied the processing of place and voicing features by brain-damaged and normal adults. There were 25 left-brain damaged aphasic subjects, of which 6 were classified as Broca's aphasics, 6 as mixed anterior, 6 as Wernicke aphasics, and 7 as residual posteriors. Pairs of real words and nonsense words which varied in place, voice, or both were used as stimuli in a nondichotic discrimination task. Fewest errors were made by Broca's aphasics. Mixed anterior patients made the most errors. The other two groups fell in between. Only the two posterior groups (the Wernicke's aphasics and residual posteriors) showed a significant difference between the place and voicing features, with fewest errors on the voice contrasts.

The voicing feature seems to play a clearly different role than the place feature does in the speech perception of left-brain damaged adults. Several researchers have examined voicing in more detail by investigating perception of voice-onset-time (VOT) perception. Blumstein, Cooper, Zurif, and Caramazza (1977) studied VOT perception in 16 left-brain damaged aphasic subjects (8 with posterior damage and 8 with anterior damage), 4 right-damaged nonaphasics, and 4 control subjects. The stimuli were /da/ and /ta/ syllables synthesized with VOT values ranging from −20 to +80 msec in 10-msec steps. The normal subjects identified stimuli with VOT values of −20 to +20 msec as /da/ and stimuli with VOT values of +40 to +80 msec as /ta/. The usual categorical perception of the VOT stimuli was found, in which subjects could discriminate between stimuli with VOT values of +25 and +40 msec and identified them as different phonemes (/da/ and /ta/, respectively). The subjects were unable to discriminate VOT values between −20 and +25 msec or between +40 and +80 msec. The four right-brain damaged subjects showed VOT perception similar to that of the normal adults. Only 9 of the left-brain damage subjects, however, showed normal discrimination and identification of the stimuli. Of the remaining subjects, 4 showed normal discrimination between the VOT stimuli but could not label them and 3 could not normally discriminate or identify the stimuli. The only significant relationship between type of brain damage and performance was for the 4 Wernicke's aphasics, of which 3 could discriminate but not label normally.

Basso, Casati, and Vignolo (1977) studied VOT perception in a large group of brain-damaged Italian subjects. The subjects were 50 left-brain damaged with aphasia (dichotomized according to fluency and comprehension), 12 left-brain damaged nonaphasics, 22 right-brain damaged nonaphasics, and 53 controls. There were no significant differences between the groups on age, education, etiology of damage, or time since damage. The stimuli were synthetic /da/ and /ta/ syllables with VOT values ranging in 10-msec steps between −150 and +150 msec. The normal subjects identified VOT values of −150 to 0 msec as /da/ and values of +50 to +150 msec as /ta/. For purposes of analysis, the upper boundary of /da/ was set at VOT values of −30 to +30 msec. The lower boundary of /ta/ was set between VOT values of +10 to +60 msec. These boundaries were based on the performance of the normal control subjects.[1] All but one right-brain damaged and all left-brain damaged nonaphasics showed nor-

---

[1] The control subjects' responses were quite variable and led the authors to speculate that the voicing cue in Italian may play a different speech perception role than it does in English speech perception.

---

mal identification. Only 13 of the 50 left-brain damaged aphasia subjects showed normal identification. The identification performance of 33 of the remaining left-brain damaged aphasics was so poor that no voice/voiceless boundary could be determined, though about half of these subjects showed a "trend" toward roughly correct identification. No significant relationship between fluency, comprehension abilities, and phoneme identification was found, though a significant positive correlation between Token Test scores and phoneme identification deficits was found. Those subjects with lower Token Test scores tended to have poorer phoneme identification abilities.

The seemingly contradictory results from the voicing studies summarized thus far may be due to methodological differences across studies. Of the five studies discussed above, three employed same–different tasks rather than or in addition to the phoneme identification task used in the other two studies. Such task differences produce quite different results. For example, Blumstein, Cooper, Zurif, and Caramazza (1977) report that 13 of their 16 (81%) left-brain damaged aphasic subjects could DISCRIMINATE VOT stimuli in a manner comparable to their normal control subjects. Basso, Casati, and Vignolo (1977), however, reporting on a VOT phoneme IDENTIFICATION task, found that only 13 of 50 (26%) left-brain damaged aphasic subjects showed normal identification. Unfortunately, Basso, Casati, and Vignolo did not report the discrimination data. This is unfortunate since those data would make the issue of discrimination versus identification clearer. It appears as if left-brain damaged aphasic subjects show better voicing discrimination than identification abilities. Both discrimination and identification for voice contrasts in these individuals are superior to those of place contrasts.

In yet another study of voice and place features, Perecman and Kellar (1981) employed a matching task with aphasic, nonaphasic, and normal subjects. The purpose of the study was to determine if the differences between aphasic and nonaphasic subjects in the use of voice and place features were due to the time constraints involved in auditory presentation methods which interfere with linguistic features' extraction by aphasics, or whether the differences were due to linguistic representation of the features. There were 12 left-damaged aphasic (7 fluent and 5 nonfluent), 6 right-damaged nonaphasic, and 16 control subjects. Subjects were presented with two series of natural speech stop-consonant triads representing voice or place features (/b, p, d, t, g, k/). One series consisted of 18 triads in which two consonants in the triad shared the feature voice while the third consonant shared neither voice nor place with the other two consonants (e.g., /b, d, k/). The second triad series consisted of 36 triads in which one consonant pair shared voice (e.g., /b, d/), one pair shared place (e.g., /d, t/), and the third pair shared neither voice nor place (e.g., /b, d/). The consonant triads were repeated by an experimenter as often as necessary for the subject to make a judgment as to which two of the three sounds heard were most similar. Visual displays of the speech sounds heard on each trial were constantly available. The results showed that all three subject groups were able to use linguistic features to match sounds. However, when matches could be made on the basis of voice or place, aphasics matched more often using the feature voice. Normal subjects and right-damaged nonaphasics used both features equally.

Perecman and Keller conclude from their investigation that voice and place require differential hemispheric processing. They suggest that voice can be processed by either the left or right hemisphere. Damage to either hemisphere produces voice processing by the remaining undamaged hemisphere. This accounts for the similar performance on the voice feature by left- and right-damaged subjects. Place, however, seems to be processed by the left hemisphere. If the left hemisphere is damaged, place processing is affected. Support for the hypothesis of bilateral as well as differential hemispheric processing of voice and place features is provided by electrophysiological studies described in the next section.

ELECTROPHYSIOLOGICAL CORRELATES OF SPEECH PERCEPTION

Electrophysiological recording procedures have also been used to assess phonological processing by various cortical regions. These techniques involve the presentation of an auditory stimulus and the recording of the brain's auditory evoked response (AER) which is triggered by this event. Various portions of the AERs have been found to reflect different stimulus properties (Regan, 1972). In the earliest electrophysiology study on adult VOT perception, Molfese (1978) recorded AERs from the left and right temporal regions of 16 adults during a phoneme identification task. Subjects were presented with randomly ordered series of synthesized bilabial stop consonants which varied in VOT with values of 0, +20, +40, and +60 msec. AERs to each stimulus were recorded during the identification task. Subsequent analyses involving principal components analysis and analysis of variance indicated that two early AER components recorded from electrodes placed over the temporal region of the right hemisphere varied systematically as a function of the phoneme category of the evoking stimulus. Stimuli with VOT values of 0 and +20 msec elicited AER waveforms from the right hemisphere sites which were different from those elicited by the +40 and +60 msec stimuli. No differences in the AER waveforms were found between the VOT values within a phoneme category (i.e., no differences were found between the +40 and +60 msec responses or between the +0 and +20 msec responses). These AER patterns of responding were identical to the identification responses given verbally by the subjects during the testing session. Components of the left-hemisphere response reflected an ability of that hemisphere to differentiate 0 and +60 msec stimuli, as well as to differentiate 0 and +60 msec stimuli from +20 and +40 msec stimuli. In this respect, the left hemisphere did not reflect the categorical—only discriminations shown by the right hemisphere.

A subsequent study (Molfese, 1980a) was designed to determine whether the laterality effects noted with the VOT stimuli are elicited by only speech stimuli or whether similar electrophysiological effects could be found for both speech and nonspeech materials. If the latter case is found to be true, some conclusions could be reached regarding similarities in the mechanisms which underlie the perception of those different materials with similar temporal lags. Since the 0 and +20 msec VOT stimuli have been found to be processed differently in the right hemisphere than the +40 and +60 msec VOT stimuli, similar patterns of responding should be elicited by nonspeech temporal lag stimuli. No differential left-hemisphere responses should be noted.

Four tone-onset-time (TOT) stimuli from Pisoni (1977) were used in this series of studies. These stimuli contained temporal relationships comparable to the voicing contrasts of the synthesized speech syllables previously employed by Molfese (1978). Use of these four TOT stimuli facilitated comparisons with the earlier electrophysiological research. The four two-tone stimuli differed from each other in the onset time of the lower tone in relation to the higher tone. For the 0 msec stimulus, the lower tone began at the same time as the higher tone. The lower tone lagged 20 msec behind the higher tone for the +20 msec stimulus. For the +40 msec stimulus, the lower tone was delayed 40 msec after the onset of the higher tone, while for the +60 msec stimulus this delay was increased to 60 msec. Both tones ended simultaneously. In this manner overall stimulus length can be controlled. Thus, AER differences would not reflect overall differences in stimulus duration since all stimuli were 230 msec in duration.

AERs from four scalp electrode sites over each hemisphere were recorded from 16 adults in response to each TOT stimulus. These responses were later analyzed using the principal components and analysis of variances procedures noted earlier. Nine factors accounting for approximately 81% of the total variance were isolated and identified. One portion of the AER, common to all four electrode sites over the right hemisphere,

Table 3

ELECTROPHYSIOLOGY STUDIES

Study	Stimuli	Task	Subjects	Electrode sites	Findings
Molfese (1978a)	Synthetic CV syllables with 0, +20, +40, and +60 msec VOT values (/ba, pa/)	Identification	Normals	T_3, T_4	RH waveform components: categorical discrimination LH waveform components: differentiate 0 and +60 from +20 and +40 msec stimuli
Molfese (1978b)	Synthetic normal and sinewave formant CV syllables with phonetic and nonphonetic formant transitions Consonants: /b, g/ Vowel: /æ/	None	Normals	T_3, T_4	LH waveform component: changes in F_2 transition are reflected
Molfese (1980a)	Synthetic CV syllables with 0, +20, +40, and +60 msec TOT values	None	Normals	T_3, T_4 C_3, C_4 P_3, P_4 T_5, T_6	RH waveform component: categorical discrimination LH waveform component: differentiated 0 from +20 and +40 from +60 msec stimuli Bilateral waveform components: reflect categorical discrimination
Molfese (1980b)	Synthetic normal and sinewave formant CV syllables Consonants: /b, g/ Vowels: /i, æ, ɔ/	None	Normals	T_3, T_4 T_5, T_6 P_3, P_4	LH waveform component: discriminated /b/ from /g/ independent of vowel Bilateral component: /b/ from /g/ discrimination
Molfese and Erwin (1981)	Synthetic vowels /i, æ, ɔ/ and nonspeech stimuli matched to vowels' duration, rise time, peak, intensity, and mean formant frequency, but differing in bandwidth	None	Normals	T_3, T_4 T_5, T_6 P_3, P_4	No hemisphere X vowel interaction but AERs from: T_3, T_4 discriminated /æ/ from /ɔ/, T_5, T_6 discriminated /i/ from /ɔ/, P_3, P_4 discriminated /i/ from /æ/ and /i/ from /ɔ/

categorically discriminated the 0 msec and +20 msec TOT stimuli from the +40 msec and +60 msec TOT stimuli. No such changes were noted to occur over the left hemisphere electrode sites at the same latency. The left-hemisphere electrical activity did distinguish between TOT stimuli from within a category (e.g., it discriminated 0 from +20 msec and +40 msec from +60 msec). A second AER component which occurred earlier in time reflected the detection of category-like boundaries over both the left and right parietal regions. An additional component that indicated category-like discrimination of these temporal cues also occurred early in time following the onset of the stimulus over both hemispheres in the temporal and central cortical regions. Processing of temporal information, then, would appear to involve both bilateral mechanisms localized in or near the temporal and parietal regions of both hemispheres, as well as an additional right hemisphere lateralized process that occurs later in time.

Results from electrophysiological studies with additional language-relevant cues show that such cues appear to be perceived and processed differently from temporal contrast cues. Several studies were undertaken in order to identify the electrocortical correlates of acoustic and phonemic cues important to the perception of place of articulation contrasts (Molfese 1978b; Molfese, 1980b; Molfese and Molfese, 1979b). In general, these studies indicate that multiple processing mechanisms, which include left-hemisphere and bilateral processes, are involved in the perception of cues such as formant transition and formant bandwidth. These findings agree with behavioral studies that utilized dichotic temporal processing procedures (Cutting, 1974).

Molfese (1978b) attempted to isolate and localize the neuroelectrical correlates of formant structure and transition characteristics by presenting a series of computer-generated consonant–vowel syllables in which the stop consonants varied in place of articulation (/b, g/), formant structure (nonspeech-like formants composed of sinewaves 1 Hz in bandwidth or by speech-like formants with bandwidths of 60, 90, and 120 Hz for formants 1, 2, and 3 respectively), and phonemic versus nonphonemic transitions (the direction of the frequency changes for formant 1 and formant 3 were either rising to produce a phonetic transition in the sense that it could characterize human speech patterns, or these transitions were falling and therefore occurred in a manner not found in an initial position in human speech patterns). Again using the principal components analysis to isolate major features of the AERs recorded from the left-hemisphere and right-hemisphere temporal regions of 10 adults, Molfese found six major AER components that accounted for 97% of the total variance. Analysis of variance on the gain scores for these factors resulted in the identification of these components as sensitive to the various stimulus and subject variables under investigation. One AER component that chacterized only the left-hemisphere electrode site was found to vary systematically to changes in F_2 transitions.

A replication–extension study (Molfese 1980b) also found that the left hemisphere discriminated consonant place of articulation information. In this study 20 adults were presented a series of consonant–vowel syllables which varied in the initial consonant, /b, g/, and the final vowel, /i, æ, ɔ/. AERs were recorded from three scalp locations over each hemisphere. One component of the brain's AER was found to reflect the ability of only the left hemisphere to differentiate between the consonants /b/ and /g/, independent of the following vowel. A separate portion of the brain response that behaved similarly was detected by electrodes placed over both hemispheres.

There appear to be some basic differences in the organization and localization of brain mechanisms as measured by electrophysiological techniques which respond to the temporal information contained in VOT and TOT stimuli and to place-of-articulation contrasts. Although both contrasts elicit simultaneous and identical discrimination responses from both hemispheres (bilateral processes), they differ in important respects. Voicing contrasts elicit an additional right-hemisphere response while place-of-articulation contrasts evoke an additional left-hemisphere response.

These findings fit well with the results of studies involving the brain-damaged populations described above. Subjects with an INTACT hemisphere which produces the additional lateralized response to a specific cue (left hemisphere—place-of-articulation contrasts; right hemisphere—voicing contrasts), performed better when asked to discriminate that cue. For example, left-hemisphere damaged subjects (who have an additional mechanism in the right hemisphere which is sensitive to temporal-voicing contrasts) performed better with stimuli differing in voicing (Oscar-Berman, Zurif, and Blumstein, 1975; Miceli, Caltagirone, Gainotti, and Payer-Rago, 1978; Blumstein, Baker, and Goodglass, 1977). If the hemisphere which produced the additional lateralized response to a speech cue is damaged, performance is worse when subjects are asked to discriminate that cue. Thus right hemisphere damaged patients made more voicing errors (Miceli, Caltogirone, Gainotti, and Payer-Rago, 1978). Likewise, left-damaged patients performed worse when asked to discriminate place-of-articulation contrasts. In addition, left-damaged patients showed increased errors for voicing contrasts when asked to identify these stimuli. Perhaps linguistic or cognitive processes within the left hemisphere are no longer available, or are impaired.

Vowel perception, which has been shown to have unique characteristics in dichotic studies, seems to be processed quite differently from the voicing and place stimuli described above. Molfese and Erwin (1981) recorded AERs from 20 adults in response to the three synthetic vowels /i, æ, ɔ/ and three nonspeech stimuli matched to the vowel stimuli in duration, rise time, peak intensity, and mean formant frequencies but differing in formant bandwidth. As in the majority of behavioral studies, no hemisphere effects were found to interact with vowel identification. However, one particularly interesting effect was noted. No single electrode site detected differences between all three vowels. Rather, the anterior temporal electrode sites (T_3, T_4) discriminated /æ/ from /ɔ/; the posterior temporal sites (T_5, T_6), discriminated /i/ from /ɔ/; and the parietal sites (P_3, P_4) discriminated /i/ from /æ/ and /i/ from /ɔ/. These findings were interpreted as suggesting that a number of discrete mechanisms located over different regions of both hemispheres are involved in the processing of different vowel sounds. The data clearly do not support a single localized region as responsible for vowel detection. The involvement of both hemispheres in the processing of vowels may be the reason why single ear advantages for vowels have not been found on dichotic tasks.

SUMMARY

This review of speech perception, based on behavioral measures from dichotic, same–different, and phoneme identification tasks, and on electrophysiological measures, has provided several insights into hemisphere functioning. The different, perceptually important cues for speech perception seem to be subserved by different regions of the brain. Analysis of the electrophysiological responses indicates that each cue is processed by a number of distinct mechanisms, some of which are bilaterally represented, and some of which are lateralized to one cortical region. These findings suggest that there is some degree of redundancy in the cortical mechanisms involved in speech perception. This redundancy may be responsible for the relatively small ear differences reported with the dichotic technique as well as the abilities of brain-damaged populations to perform speech perception tasks with some competency even after left-hemisphere damage. Clearly, then, given the findings to date, it would appear that language perception for even relatively simple discriminations must depend on multidimensional and complex processes rather than on solely left-or right-hemisphere processing.

REFERENCES

Basso, A., Casati, G., and Vignolo, L. (1977). Phonemic identification defect in aphasia. *Cortex, 13,* 85–95.

Berlin, C. (1977). Hemispheric asymmetry in auditory tasks. In S. Harvad, R. Doty, L. Goldstein, J. Jaynes, and G. Krauthames (Eds.) *Lateralization In the Nervous System.* New York: Academic Press.

Berlin, C., Lowe-Bell, S., Cullen, J., Thompson, C., and Loovis, C. (1973). Dichotic speech perception: An interpretation of right-ear advantage and temporal offset effects. *Journal of the Acoustical Society of America, 53,* 699–709.

Blumstein, S. (1974). The use and theoretical implications of the dichotic listening technique for investigating distinctive features. *Brain & Language, 1,* 337–350.

Blumstein, S. and Cooper, W. (1972). Identification vs. discrimination of distinctive features in speech perception. *Quarterly Journal of Experimental Psychology, 24,* 207–214.

Blumstein, S., Baker, E., and Goodglass, H. (1977). Phonological factors in auditory comprehension in aphasia. *Neuropsychologia, 15,* 19–30.

Blumstein, S., Cooper, W., Zurif, E., and Caramazza, A. (1977). The perception and production of voice-onset-time in aphasia. *Neuropsychologia, 15,* 371–383.

Curry, F. (1967). A comparison of left-handed and right-handed subjects on verbal and nonverbal dichotic listening tasks. *Cortex, 3,* 343–352.

Cutting, J. (1974). Two left hemisphere mechanisms in speech perception. *Perception and Psychophysics, 16,* 601–612.

Dobie, R., and Simmons, B. (1971). A dichotic threshold test: Normal and brain damaged subjects. *Journal of Speech and Hearing Research, 14,* 71–81.

Gazzaniga, M. S. (1970). *The Bisected Brain.* New York: Appleton-Century-Crofts.

Geschwind, N. and Levitsky, W. (1968). Human brain: Left-right asymmetries in temporal speech regions. *Science,* 161, 186–187.

Gordon, H. (1970). Hemispheric asymmetries in the perception of musical chords. *Cortex,* 6, 387–398.

Haggard, M., and Parkinson, A. (1971). Stimulus and task factors as determinants of ear advantages. *Quarterly Journal of Experimental Psychology, 23,* 168–177.

Johnson, J., Sommers, R., and Weidner, W. (1977). Dichotic ear preferences in aphasia. *Journal of Speech and Hearing Research, 20,* 116–129.

Kimura, D. (1961a). Some effects of temporal lobe damage on auditory perception. *Canadian Journal of Psychology, 15,* 156–165.

Kimura, D. (1961b). Cerebral dominance and the perception of verbal stimuli. *Canadian Journal of Psychology, 15,* 166–171.

Kimura, D. (1964). Left–right differences in the perception of melodies. *Quarterly Journal of Experimental Psychology, 14,* 355–358.

Kimura, D., and Folb, S. (1968). Neural processing of backward-speech sounds. *Science,* 161, 395–396.

King, F., and Kimura, D. (1972). Left-ear superiority in dichotic perception of vocal nonverbal sounds. *Canadian Journal of Psychology, 26,* 111–116.

Miceli, G., Caltagirone, C., Gainotti, G., and Payer-Rago, P. (1978). Discrimination of voice versus place contrasts in aphasia. *Brain and Language, 6,* 47–51.

Milner, B., Taylor, L., and Sperry, R. (1968). Lateralized suppression of dichotically presented digits after commissural section in man. *Science, 161,* 184–185.

Molfese, D. L. (1978a). Neuroelectrical correlates of categorical speech perception in adults. *Brain and Language, 5,* 25–35.

Molfese, D. L. (1978b). Left and right hemisphere involvement in speech perception: Electrophysiological correlates. *Perception and Psychophysics, 23*, 237–243.

Molfese, D. L. Cortical and subcortical involvement in the processing of coarticulated cues. *Brain and Language,* 1979, 7, 86–100.

Molfese, D. L. (1980a). Hemispheric specialization for temporal information: Implications for the perception of voicing cues during speech perception. *Brain and Language, 11,* 285–299.

Molfese, D. L. (1980b). The phoneme and the engram: Electrophysiological evidence for the acoustic invariant in stop consonants. *Brain and Language, 9,* 372–376.

Molfese, D. L. & Hess, T. (1978). Hemispheric specialization for VOT perception in the preschool child. *Journal of Experimental Child Psychology, 26,* 71–84.

Molfese, D. L. & Molfese, V. J. (1979a). VOT distinctions in infants: Learned or inate? In H. A. Whitaker and H. Whitaker (Eds.), *Studies in neurolinguistics,* Vol. 4. New York: Academic Press.

Molfese, D. L. & Molfese, V. J. (1979b). Hemisphere and stimulus differences as reflected in the cortical responses of newborn infants to speech stimuli. *Developmental Psychology, 15,* 505–511.

Molfese, D. L. & Molfese, V. J. (1980). Cortical responses of preterm infants to phonetic and nonphonetic speech stimuli. *Developmental Psychology, 16,* 574–581.

Molfese, D. L. & Erwin, R. J. (1981). Intrahemispheric differentiation of vowels: Principal component analysis of auditory evoked responses to computer-synthesized vowel sounds. *Brain and Language, 13,* 333–344.

Oscar-Berman, M., Zurif, E., and Blumstein, S. (1975). Effects of unilateral brain damage on the processing speech sounds, *Brain and Language, 2,* 345–355.

Perecman, E., & Kellar, L. (1981). The effect of voice and place among aphasic, nonaphasic right-damaged and normal subjects on a metalinguistic task. *Brain and Language, 12,* 213–223.

Pisoni, D. B. (1977). Identification and discrimination of the relative onset time of two component tones: Implications for voicing perception in stops. *Journal of the Acoustical Society of America, 61,* 1352–1361.

Regan, D. (1972). *Evoked potentials in psychology, sensory physiological, and clinical medicine.* New York: Wiley.

Schulhoff, C., and Goodglass, H. (1969). Dichotic listening, side of brain injury, and cerebral dominance. *Neuropsychologia, 7,* 149–160.

Schwartz, J., and Tallal, P. (1980). Temporal acoustic analysis may underlie hemispheric specialization for speech perception. *Science, 207,* 1380–1381.

Shanks, J., and Ryan, W. (1976). A comparison of aphasic and non-brain-injured adults on a dichotic CV-syllable listening task. *Cortex, 12,* 100–112.

Shankweiler, D., and Studdert-Kennedy, M. (1966). Lateral differences in perception of dichotically presented synthetic consonant-vowel syllables and steady-state vowels. *Journal of the Acoustical Society of America, 39,* 1256.

Shankweiler, D., and Studdert-Kennedy, M. (1967). Identification of consonants and vowels presented to left and right ears. *Quarterly Journal of Experimental Psychology, 19,* 59–63.

Sparks, R., Goodglass, H., and Nickel, B. (1970). Ipsilateral versus contralateral extinction in dichotic listening resulting from hemisphere lesions. *Cortex, 6,* 249–260.

Spellacy, F., and Blumstein, S. (1970). The influence of language set on ear preference in phoneme recognition. *Cortex, 6,* 430–440.

Spellacy, I. (1970). Lateral preferences in the identification of patterned stimuli. *Journal of the Acoustical Society of America, 47,* 574–578.

Studdert-Kennedy, M. (1975). Two questions. *Brain and Language, 2,* 123–130.
Studdert-Kennedy, M., and Shankweiler, D. (1970). Hemispheric specialization for speech perception. *Journal of the Acoustical Society of America, 48,* 579–594.
Studdert-Kennedy, M., Shankweiler, D., and Pisoni, D. (1972). Auditory and phonetic processes in speech perception: Evidence from a dichotic study. *Cognitive Psychology, 3,* 455–466.
Tallal, P., and Newcombe, F. (1978). Impairment of auditory perception and language comprehension in dysphasia. *Brain and Language, 5,* 13–24.
Tallal, P., and Piercy, M. (1973). Developmental aphasia: Impaired rate of nonverbal processing as a function of sensory modality. *Neuropsychologia, 11,* 389–398.
Tsunoda, T. (1969). Contralateral shift of cerebral dominance for nonverbal sounds during speech perception. *Journal of Auditory Research, 9,* 221–229.
Tsunoda, T. (1975). Functional differences between right- and left-cerebral hemispheres detected by the key tapping method. *Brain and Language, 2,* 152–170.
Weiss, M., and House, A. (1973). Perception of dichotically presented vowels. *Journal of the Acoustical Society of America, 53,* 51–58.

CHAPTER 3

Pragmatic Aspects of Communication in Brain-Damaged Patients*

Nancy S. Foldi, Michael Cicone, and Howard Gardner

INTRODUCTION

Mr. Harvey is seated in the speech therapist's room. Several familiar objects are placed in front of him, and, as part of a standard aphasic language evaluation, the therapist ascertains his comprehension of object names: "Pick up the pencil." The patient stares at the objects, then turns to the therapist and laboriously utters in a rising tone, "Ah?" The therapist repeats the command, whereupon the patient reaches for the comb, pushes it across the table, returns his arm to his lap and looks at the therapist. The therapist scores the response and proceeds to the next item.

Such an exchange is familiar to most investigators of aphasia. While designed to determine a level of comprehension in a patient, it also reveals how the patient responds to an interactive situation. Given a command

* We wish to acknowledge the support of the National Institute of Neurological and Communication Disorders and Stroke (M 11408), the Veterans Administration and Harvard Project Zero.

LANGUAGE FUNCTIONS
AND BRAIN ORGANIZATION

("Pick up the pencil"), Mr. Harvey seems to recognize that it is his turn to do or say something. Unable to perform, he requests a repetition by uttering a simple sound with a rising intonation. Following a repeat of the command, the patient responds with an action on one of the objects, but his response is incorrect. He indicates he is finished by returning his arm to a resting position.

While this brief exchange suggests a severe comprehension deficit, we can gain another perspective of the patient's behavior by observing how he participates in an entire interaction. The following vignette demonstrates this point. Here, outside of the context of a formal test, Mr. Harvey interacts with a nurse on the ward of his hospital.

> Mr. Harvey approaches a nurse in the ward office and points to the rack of patient charts. The nurse looks in the direction that Mr. Harvey is indicating and states, "No, Mr. Harvey, you have already been to physical therapy." Mr. Harvey abruptly responds, "No!" and repeatedly points to the charts. The nurse again addresses the patient, "Do you need something?" The patient's dysarthric response is unintelligible and the nurse asks him to repeat: "What?" The patient attempts again, but to no avail. Rather, he turns to the head nurse, who has just entered the room, and points alternately first to the head nurse, then to the charts. The head nurse reminds her colleague, "Yes, remember we had talked about it this morning: Mr. Harvey needs to go to the ophthamologist this afternoon and must take his chart with him." The head nurse hands Mr. Harvey his chart, saying, "You know where to go", and he wheels himself out of the room.

In contrast to the situation described in the first vignette, the dialogue here has been initiated by the patient and involves a series of interactions. The richness of the interchange in this more natural situation can be conveyed by an indication of what will be called the "pragmatic" structures entailed in the discourse.

Initially, the patient makes reference by pointing to the charts. By this act, he presupposes that the nurse knows that the chart symbolizes "leaving the ward." Both the nurse and the patient share a point of regard, as they focus on the chart, thereby acknowledging a mutual appreciation of the reference previously indicated by Mr. Harvey. When the patient recognizes that the nurse has incorrectly interpreted the referent (namely as an appointment in physical therapy), he strategically attempts to get the nurse to pursue a different avenue and to strive for the correct interpretation. The nurse, in turn, compensates for Mr. Harvey's deficits and indeed continues to guide the patient's responses. He responds appropriately to an indirect request for information ("Do you need something?"), by attempting to provide information, rather than responding with a simple assent or denial. Yet his unintelligible response is not understood.

Next, the conversation returns to the nurse who requests repetition ("What?"), and, though the patient's attempts again fail, he still responds to the task of repetition. Mr. Harvey then avails himself of the approach of another informed person, the head nurse, and uses similar tactics as before, that is, attracting the other's point of regard to a referent present in the room. When the head nurse supplies the desired interpretation of that referent, Mr. Harvey seems content. While he himself did not supply the content of the information, his intentions and desires were made clear to both nurses and he successfully allowed others to ascertain the precise referent.

These opening vignettes raise two major issues for investigators of the communicative ability of brain-damaged patients. First, we must ask which of the two vignettes more accurately conveys the patient's communicative competence. Second, we must address ourselves to the manner in which various strategies aid the communication.

Turning to the first issue, several authors (Brookshire, 1977; Chester and Egolf, 1974; Egolf and Chester, 1977; Holland, 1978; Taylor, 1965; Ulatowska, Macaluse-Haynes, and Mendel-Richardson, 1975) have proposed that the formal aphasia test batteries overemphasized the purely linguistic aspects of communication. Their efforts have provided support for a proposed dissociation between performance on quantitative test batteries and performance on more casual and naturalistic interactions. But because some of the quasi-natural settings used by the aforementioned investigators involve pretense and other artifices (videotape and score keepers, for example), they may also fail to capture correctly the efficiency of the organic patient's communicative competence. In short, we face a kind of neuropsychological indeterminancy principle: The challenge has been to observe the phenomenon of naturalistic communication without destroying it in the process.

Nonetheless, our vignettes, as well as the observations of these other investigators, strongly suggest an answer to the first question: Communication is indeed more effective for a patient in a natural situation. This then leads to the second issue—the use of strategies by such patients. It seems evident that the brain-damaged patient has available various strategies for communicating information and that these strategies can be exploited in an appropriate setting. But a mere indication that the "context" aids in communication becomes an uninformative truism. The whole problem is to break down the communicative situation into its component parts; one must determine *which* sorts of features are of use to the patient in various circumstances and *to what extent* these features can compensate for the language deficits incurred by the brain damage.

In this chapter we adopt the hypothesis that a context is not simply an amorphous unit that is part of discourse. Rather, context is comprised of many features which must be meaningfully integrated with the verbal message. The variety of contextual features constitutes the realm of pragmatics. Broadly defined, pragmatics includes the many aspects of communication that may interface with the semantic, syntactic, or even phonological components of an exchange, but that are not purely linguistic in nature. For example, utterances are spoken with particular patterns of intonation, amplitude, and pauses; they are presented with communicative changes in posture, gesture, facial, and other nonverbal cues (Wiener, Devoe, Rubinow, and Geller, 1972). In addition to these intra-individual components, the speaker and listener must appreciate the social situation surrounding the exchange (Goffman, 1971), and adhere to previously learned rules of conversation (Grice, 1969, 1975). Communication thus becomes a subtle interplay among two or more individuals. Since all of these features are subsumed under the rubric of pragmatics, it follows that the level of analysis must proceed beyond the single word or sentence level and include elements of discourse.

In this essay, we examine each of the aforementioned features and levels of analysis, with particular reference to the ability of organic patients to use such features to communicate. Our focus falls on organic patients, but throughout we maintain a comparative perspective and concentrate especially on the comparative effects of unilateral left- and right-hemisphere damage. Following traditional analytic practice, we generally distinguish between those patients whose left-hemisphere damage is exclusively in the pre-Rolandic area (hereafter referred to as anterior aphasics) and those with damage exclusively in the post-Rolandic area (hereafter called posterior aphasics). When more specific identity of the syndrome has been established, we note this fact (e.g, Broca's aphasia within the anterior group, Wernicke's aphasia within the posterior). In the next part of the chapter, Vehicles of Communication, we focus on two primary vehicles used in communication: gesture and intonation. Then in the second major part of this chapter, Pragmatic Structures, we turn our attention to a number of pragmatic structures—conceptual constructs posited as a means of describing the communication knowledge of the interlocutors. This survey includes the areas of reference and speech acts. The last portion, Other Linguistic Forms, reviews the command of other nonlinguistic forms that may also contribute to the communication. This outline emphasizes command of humor, figurative language, and comprehension of written texts.

VEHICLES OF COMMUNICATION

Gesture

Any examination of communication in context must take into account the role of gesture in the exchange of information. Gesture is perhaps the most obvious means by which communication can take place apart from language: A traveller in a foreign country, for example, unable to communicate through talking or writing, will commonly resort to gesturing. The extent of the preservation or impairment of gestural communication in aphasics—who could be analogized with a foreign traveller—is therefore of primary interest in any consideration of paralinguistic competence after brain damage.

A number of typologies of normal gesture have been devised (e.g., Ekman and Friesen, 1969; Wiener, Devoe, Rubinow, and Geller, 1972). In general, these systems classify limb gesture into categories such as (1) emphatic and rhythmic movements, which follow and punctuate the flow of speech; (2) pointing gestures; (3) "emblems," usually simple gestures for which there exists a single, culturally accepted meaning (such as making a circle with the thumb and forefinger to signal "OK"); and (4) various types of pantomimes, which refer to actions, objects, or physical attributes, and can appear singly, or in combination, to depict complex events. The major factor which varies among these categories is the extent to which the gestures convey semantic content: for example, emblems and pantomimes always make meaningful references to some aspect of the world, while emphatic and rhythmic gestures are for the most part semantically empty. Since the semantic aspects of language can be differentially impaired in aphasia—for example, nonfluent aphasics' speech mostly consists of highly discrete content words, while the speech of fluent aphasics is largely devoid of content—one might postulate that the gesturing of aphasics may be differentially affected as well, depending on the semantic content of the gestures considered.

Gesture can also assume a number of different roles in relation to language and conversation: it can substitute for speech, convey information in addition to speech, or be totally redundant with speech. Gesture can stand by itself, or it can be understood only in terms of the ongoing situation: an upturned palm, for example, can communicate a request in one situation, a denial in another; a finger pointing in the air can make reference to a previously established object or idea. Investigations of gestural use after brain damage, therefore, must take into account not only the different types of gestures used, but also the role these gestures play with the particular language that is uttered and the particular situation where they appear.

Gestural impairments associated with aphasia have been noted by aphasiologists since the late 1800s (cf. Duffy, Duffy, and Pearson, 1975, for a brief review of earlier literature). In particular, it has been noted that aphasics do not seem to resort to increased gesturing to compensate for their difficulties with language (Liepmann, 1905; Goodglass and Kaplan, 1963). Since these observations were made, a small body of research has been conducted on the gestural capacities of aphasic and other brain-damaged patients. Because our primary concern in this chapter is the role of contextual elements in the communication of organic patients, we will consider separately those studies which investigate gesture as it appears in a conversational setting from those that do not.

THE PROBLEM OF APRAXIA   Most of the earlier literature on gesture in organic patients deals with impairments of gesture under the heading of "apraxia" (Liepmann, 1920). Apraxia is usually defined as an inability to perform accurately voluntary movements, an inability which is not, however, due to peripheral impairments. Different varieties of apraxic disturbances have been delineated (Brown, 1972; Hécaen and Albert, 1978); the type most relevant to our discussion of gesture as a vehicle for communication is ideomotor apraxia, exemplified as a difficulty in pantomiming the use of a common object when the object itself is not present. Note that, as discussed previously, such gestures convey semantic content.

In an effort to chart the communicational use of gesture by organic patients, Goodglass and Kaplan (1963) investigated the ability of aphasics and right-hemisphere damaged subjects to perform to verbal command object–action pantomimes, as well as two types of "intransitive" movements (i.e., gestures not making reference to a specific object): (1) "natural" gestures (movements which would naturally occur in certain situations, such as holding one's nose) and (2) "conventional" gestures (emblems such as saluting and waving). These authors found that aphasics performed worse than right-hemisphere patients and controls on all tasks, and that the difficulties were greater for the pantomime than for the "intransitive" gestures—an indication that gestures carrying different amounts of semantic–referential meaning may indeed be differentially impaired in aphasia. The authors, however, concluded that because they could find no correlation between gestural performance and measures of severity of aphasia, apraxic impairments should *not* be considered to be linked specifically to aphasic disturbances, but instead are due to certain left-hemisphere lesions regardless of the presence of aphasia.

In comments on Goodglass and Kaplan (1963), Bates (1980) has argued that their evidence could support the opposite interpretation as well—that apraxia *is* linked to aphasic disturbances. Indeed, she views this study as a strong demonstration of the connection between aphasia and impairments in these kinds of symbolic gestures. Moreover, there exists the possibility that, because these tasks were presented via verbal command, verbal comprehension deficits contributed to the decreased scores of the aphasics.

This problem raised by the Goodglass and Kaplan study is in fact a general issue: it is unclear whether apraxia, as usually discussed in the literature, applies to disturbances in any voluntary gestural movements produced by a patient, or only to movements which are "voluntary" in that the patient is consciously "willing" them to happen at the request of another person. Clearly Brown's category (1972) of "ideomotor" movements would be very difficult to study in any other way than by verbal command. At any rate, it seems expedient for our discussion to consider "apraxia" as dealing primarily with impairment in producing certain kinds of gestures on command, and hence not directly bearing on our concern here—the ability of organic patients to utilize gestures spontaneously as a means of conveying information in a natural setting. At the very least, however, apraxic difficulties involving symbolic gestures suggest that there may be a basic symbolic disturbance underlying the impairments of both language and certain types of semantically loaded gestural activity.

PANTOMIME RECOGNITION   Another group of studies investigating gestural ability in aphasic patients has focused on the comprehension of simple pantomimes (Duffy, Duffy, and Pearson, 1975; Gainotti and Lemmo, 1976; Seron, van der Kaa, Remitz, and van der Linden, 1979; Varney, 1978). In all four studies, patients viewed pantomimes involving the pretended use of a common object (e.g., brushing teeth, combing hair) and were required to pick from an array the picture of the associated object. Note that these gestures are the same kinds of "ideomotor" movements which are referential–symbolic in nature and are impaired in apraxia. While these studies used different methods, all four found aphasics impaired on the recognition tasks relative to nonneurological control subjects, and other organic (i.e., subcortical or right-hemisphere) groups.

As with apraxia, the question arises concerning the relationship between the gestural deficits observed, and the linguistic impairments of aphasia. All four pantomime recognition studies attempt to assess this relationship by looking for correlations between recognition scores and scores on tests of language abilities. Although Seron, van der Kaa,

Remitz, and van der Linden (1979) found no correlation between pantomime recognition scores and general ratings of severity of aphasia, they
and others have found recognition performance to be significantly correlated with more specific measures of language ability, particularly reading
comprehension (Gainotti and Lemmo, 1976; Seron, van der Kaa, Remitz,
and van der Linden, 1979; Varney, 1978). Varney reports that decreased
pantomime recognition is *always* associated with impaired reading comprehension in his subjects, but not vice versa. These correlations, as well
as the consistency with which impaired pantomime recognition appears in
the aphasic groups of these four studies, support the interpretation that
certain aphasic impairments and deficits in the comprehension of representational gestures may both reflect a central disturbance in symbolic
functioning.

REVIEW OF RESEARCH ON GESTURE "IN CONTEXT"   Cicone,
Wapner, Foldi, Zurif, and Gardner (1979) conducted a detailed investigation of the spontaneous speech and gesture of a small number of brain-
damaged patients and normal controls. Two anterior and two posterior
aphasics and four controls were videotaped during an informal conversation with an experimenter; randomly chosen segments of the tapes were
then analyzed for a number of contents, including the amount and complexity of the gestures produced; the types of gestures used (i.e., referential versus nonreferential; pantomimes, emblems, pointing, etc.); and the
clarity with which gestures conveyed information to the listener.

Analyses showed that (1) the amount and complexity of gesturing was
like that of speech: anterior patients' gestures were sparse and simple,
whereas posterior aphasics' gesturing was abundant, fluent, and complex,
similar to that of control subjects; (2) a very high proportion of the anteriors' gestures were semantically meaningful (i.e., somehow referring to an
object, event, or idea), whereas posterior aphasics and controls produced
a more even mix of semantically meaningful and nonmeaningful (rhythmic, emphatic) gestures; (3) of the semantically meaningful gestures, anterior aphasics made greater use of emblems, whereas posteriors again re-
sembled controls in the production of more general "pointing"
movements; and (4) with regard to the clarity of the information conveyed, anterior and control patients were generally successful in communicating clearly using gestures, but posterior patients produced a relatively high proportion of movements whose meanings were vague to the
observer.

The authors interpret these findings as evidence for a parallel breakdown of gesture and speech in aphasia: for anterior patients, output in
both modalities is sparse, simple, concrete, and clear; for posterior apha-

sics, both gesture and speech are abundant, fluent, and less referentially clear than for normals. Considered in another way, these patterns indicate that aphasics do not actively seem to compensate for their linguistic impairments with increased or more "content-full" gesturing. Aphasics' gestural communication in conversation is no more informative than is their speech. This is especially obvious for the anterior aphasics in the study: one might expect much increased gesturing from patients whose verbal output is so limited (like that of our foreign traveller), but instead their gestures were produced mostly as accompaniment to the few words they said, and rarely occurred alone, as a means of expanding upon what was spoken.

A number of processes could yield this striking similarity between speech and gesture in aphasia. Regarding the findings about the amount and complexity of the output, it may be that such structural aspects of gestural production are somehow dependent on speech production: posteriors are fluent and complex gesturers because their movements can "ride on" their fluent and syntactically complex verbal output; anteriors do not gesture often, and do not make elaborate rhythmic or emphatic gestures, because they cannot produce the flow of speech on which such movement depends. This view is supported by the reports of Kimura (1973a), which suggest that for normal speakers verbal agility and the motor behaviors which accompany it are initiated simultaneously by way of cortical activation in the dominant hemisphere. In other words, both speech and gesture may spring from a central cortical source and so tend to resemble each other in form and fluency.

Furthermore, the similarity in the semantic meaningfulness and informational clarity of speech and gesture in each type of aphasia suggests that the two output modalities may be centrally affected in terms of semantic–symbolic content as well as in initiation and form. This idea, taken together with the results from the research on pantomime recognition discussed above, indicates that production and comprehension of the symbolic aspects of gestural communication may be impaired in tandem with the better-known linguistic–symbolic deficits of aphasia.

Nevertheless, there are other aspects of gestural communication which are less symbolically based and which may remain relatively intact in aphasics: for example, gestures that regulate turn-taking in the conversation, that make emphasis or punctuate speech, and others that make greater use of the conversational context to convey information. In the study reported by Cicone, Wapner, Foldi, Zurif, and Gardner, it emerged that emphatic, rhythmic gestures are preserved in the posterior aphasics' performance, and "floorholding" types of gestures seem to be well-used by all of the aphasic patients studied (however, a detailed analysis of such gesturing was not included as part of the investigation). In another study, Nespoulous (1980) reports on the gestural behavior of one Wernicke's aphasic, whose gesturing he describes as primarily deictic in nature, pointing out aspects of the environment in which the patient's conversations took place. Such nonsymbolic and/or heavily contextual types of gesture may show a relative hardiness in the face of brain damage, in contrast to the more symbolic varieties studied in the research discussed previously. It is worth noting that the successful use of such nonsemantic gestures by aphasics reinforces the impression that certain aspects of communication are spared after brain damage.

Most importantly, it is likely that the gesturing behavior of aphasic patients varies with the kinds of situations in which the patients communicate. In ward or home settings, where they are voluntarily interacting with familiar people to whom they must express needs and desires, aphasics may make much more effective use of gesturing for communication than is evident in the situation presented by Cicone and colleagues. Thus, superior performance may be manifest both in the amount and the quality of the gestures used and in the creative use of gestures in combination with the ongoing events.

In summary, there is considerable redundancy of information in the linguistic and gestural channels of aphasic patients, and surprisingly little complement or enhancement by the spontaneous gestures. We have encountered isolated cases where natural gestures provide some aid in conveying new information, but the communication of these patients seems generally restricted to their ability to communicate information in linguistic terms. There is also modest anecdotal evidence that left-hemisphere patients are sensitive to, and benefit from, the gestures used by others in communicating. But the difficulty encountered by most aphasic patients in decoding the "frozen gestures" used in formal tests of gestural comprehension suggests that gesture cannot substitute for language in efforts to communicate meanings of any complexity to aphasic patients. However, there is also evidence to suggest that aphasics' limitations on gestural communication are most severe for gestures used to convey semantic–symbolic content: the ability to use gesture in the service of other kinds of communication (e.g., emphasis, punctuation, turntaking, or floorholding) may remain relatively intact.

Intonation

We turn our attention next to the use of intonation by organic patients. The review includes evidence from the production of intonation, as well as the perception and comprehension of intonational contours. In discussion of this latter category, we will, following Hughlings Jackson (1932), treat intonation both as it is utilized in propositional speech and also as it conveys emotional force.

Intonation refers to the change in pitch, or fundamental frequency, that is part of any vocal sound. In linguistic as well as nonlinguistic utterances (e.g., expletives), intonation usually does not act alone, but may be accompanied by other physical modulations, such as changes in amplitude (stress), or the lengthening of speech sounds or pauses. Unless the stimuli are elicited artificially, it is difficult to monitor precisely a change in intonation only; the impact of the other parameters is inevitable. Implications of this phenomenon will be discussed in the review of emotionally toned sentences.

Intonation patterns may carry a variety of meanings, including the communication of affect or emotion, or a presentation of the force of the proposition itself. For example, in the second vignette above, "You know where to go" is a declarative, but if the utterance had been produced with an overall rising tone, the same linguistic structure would serve as a question, demanding an assent or a denial as a response from the listener. While declaratives and such "yes–no" questions are most frequently associated with a particular contour (declining and rising patterns), other constructions, such as conditionals or Wh-questions (i.e., questions using *where, what, who, how*, etc.) can also be characterized by typical contour patterns.

Thus, our interest in intonation as a pragmatic vehicle is to specify its overall communicative value. Furthermore, when investigating the fate of intonation after brain damage, we encounter a question of laterality. Since melodic features have been associated with the right hemisphere, we can ask whether the communicative use of intonation contours is differentially affected in right and left brain-damaged patients.

PRODUCTION   Observations of intonation patterns of aphasic speech have indicated that despite their production of jargon, Wernicke's aphasics clearly maintain contour. Global and Broca's aphasics have been described as "monotonic" (Goodglass and Kaplan, 1963), but even those who have a single stereotyped utterance can superimpose different prosodic contours onto their utterances (Alajouanine and Lhermitte, 1964). Therapeutic methods (e.g., Melodic Intonation Therapy, cf. Albert, Sparks, and Helm, 1973) have indeed taken advantge of the spared intonation in aphasics and employed it as a means to foster further speech and language rehabilitation.

While the aphasic's output displays preserved intonation at least to some extent, it seems that right-hemisphere damaged patients show impaired production contours. The quality of this speech has been characterized as "flat" (Ross and Mesulam, 1979), or "lacking in emotion" (Gainotti, 1972). The deficits of the contours of right-hemisphere patients have also been associated with their impairments in musical tasks (Kimura, 1973a,b) which make demands on pitch discrimination.

These impressions themselves cannot be faulted. However, it is risky to assume that these differences in intonation proficiency herald correlative skill in communicative competence. In fact, Cooper and Danly and their coworkers (Danly, de Villiers, and Cooper, 1979; Cooper, Danly, and Hamby, 1979) have shown that Broca's aphasics although apparently monotonic, are capable of planning the intonation pattern of a sentence utterance: the contour declination across declarative sentences characteristic of normal speech is found in semantically related two-word utterances of the Broca's speech, even when the two words are separated by long pauses. In contrast, Wernicke's patients, although apparently fluent, exhibit inappropriate alteration of the melodic contour: unlike normals, they adjust the peaks of intonation in long sentences and produce higher fluctuations in contour, giving their speech a "hypermelodic" quality.

In terms of speech planning, the authors suggest that Broca's aphasics are demonstrating a preservation of fundamental mechanisms, whereas the Wernicke's aphasics exhibit appreciable deficits in this area. Thus, our perception of Broca's aphasics' speech as faulty may be a function of our own inability to perceive the contour when the speech is lengthened and the time of the intermediate pauses increased. Perhaps a better appreciation of the distorted contours would facilitate our task as listeners of the aphasics' utterances.

When listening to the Wernicke's aphasics' speech, on the other hand, the authors propose that we may tend to ignore subtle abnormalities of their intonation patterns in an effort to capture the overall pattern of the sentence. One hypothesis may be that as listeners, we adjust the patients' patterns to model our own conceptions. Further evidence to support this view would be required; however, since only declarative sentences were employed, it remains to be determined if the abnormalities of the Wernicke's aphasics' intonation patterns could significantly disrupt our perception in spontaneous output or our perception of other communicatively relevant contours, such as those found in interrogatives and conditionals.

In summary, the production of intonation patterns by aphasics provides at least a potential avenue for communication. Our strategies as listeners may also depend on attention to their intonation patterns, particularly when the linguistic deficits are severe. However, it is now apparent from the data of Danly and Cooper and their colleagues that our gross impressions of the intonation contours are not wholly accurate and that our sensitivity to the patterns requires further careful investigations. Whether similar deficits may also characterize the sentence planning of right-hemisphere patients is under investigation, (Danly, Shapiro, and Gardner, 1982).

COMPREHENSION   In what follows we will look at organic patients' comprehension of intonation as it is used in two different ways in conversation: (1) as it conveys emotion or affect and (2) as it communicates propositional meaning.

Intonation in Emotion   As mentioned previously, right-hemisphere damaged patients have typically been characterized as unmotivated and lacking emotion. Several studies have documented that their perception of emotional qualities of speech is disturbed (Heilman, Scholes, and Watson, 1975; Tucker, Watson, and Heilman, 1977). When presented with emotionally toned sentences, and asked to judge the expressed emotions, the right brain-damaged patients performed significantly worse than the aphasic patients. It was not clear, however, what deficit led to the decreased performances. At issue may have been either the patients' impoverished ability to discriminate the tonal pattern, or alternatively, a dissociated deficit in affect (affective agnosia). Moreover, in a similar experiment, probing recognition of emotional moods to sentences presented orally, Schlanger, Schlanger, and Gerstman (1976) failed to replicate the findings of differences across unilaterally injured organic patients.

In light of the fact that various studies do not yield the same results, and further, in view of the typical confounding of the factors of intonational information and emotional content in these studies, the exact nature of sensitivity to emotional information carried by intonation remains to be determined. If it can be shown that intonational contour—as it relates to information—can be discerned by patients who fail to appreciate intonational information when it carries emotional content, then it will be demonstrated that the patients' difficulty is a reflection of the content (emotional) rather than the vehicle of information transmission (intonational contours).

Intonation in Propositions   We turn now to the comprehension of the propositional use of intonation, where the melodic contour conveys information about the speaker's intention in an utterance. We can ask whether the contour can be dissociated from the verbal component and whether that contour can carry meaning of its own.

Two studies are particularly relevant to this issue. The first investigation "dissociated" intonation from language by filtering the intelligibility of the speech from the utterance, while retaining the intonation pattern (Blumstein and Cooper, 1972). Four different filtered sentence types were used (declaratives, interrogatives, conditionals, and imperatives) and were dichotically presented to normal subjects. When asked to match the stimuli to a binaurally presented foil (presented subsequently to each dichotic pair) or to a set of multiple-choice items which pictorially represented the melodic contour, the subjects showed a significant left-ear effect. However, when subjects were required to determine what sentence type the filtered stimuli represented, the left-ear effect failed to reach significance. In this condition (where the subjects had been previously trained to recognize the melodic patterns of the different sentence types), the task differed interestingly from the aforementioned ones; one might speculate that the subject was required to associate a communicative value to the particular pattern and not simply to treat the stimuli as a match in a recognition task.

To determine the influence of a "linguistic medium" on the processing of the intonation contours, these authors then superimposed nonmeaningful syllables (*pa, ta, ga*) on the same melodic patterns. While these sounds are traditionally thought to evoke a right-ear effect (Shankweiler and Studdert-Kennedy, 1967), even these stimuli (presented both in filtered and unfiltered conditions) did not reverse the overall left-ear superiority found in the above conditions. Again, the data suggest the primary role of the right hemisphere in intonation processing and that, at best, the left hemisphere is only minimally involved in these sorts of matching tasks.

In a study currently in progress (Kellar, unpublished), aphasics (predominantly Wernicke's aphasics) and non-brain-damaged, normal controls were given a dichotic presentation of both filtered and unfiltered versions of the sentence types used above, and were asked to match the stimuli with a subsequent binaural foil. In accord with the Blumstein and Cooper study, the results indicate a significant left-ear effect for the filtered condition for both aphasics and controls. In the unfiltered version (i.e., normally spoken sentences), the normal subjects demonstrated either a right-ear effect or a perfect score, thereby suggesting a strong predominance of the linguistic message. The aphasic subjects, however, again produced a left-ear effect, although there was an overall improvement from the filtered to the unfiltered condition. This amelioration in performances was a reflection of the increased left-ear performance and was not due to any change in the performance of the right ear. These results suggest that in cases of left-hemisphere damage, the right hemisphere may assume a more important role in the task. Perhaps the overall improvement might also designate participation of the right hemisphere in the integration of the contour of a sentence with the linguistic information. However, the data collection from right-hemisphere damaged patients, necessary to evaluate such hypotheses, is still in progress.

This study again seems to underscore the potential role that the right hemisphere may play in intonation. The study also addresses the previously discussed confounding of emotion and intonation; apparently the problems experienced by the right-hemisphere patient in decoding emotionally charged intonation patterns are not solely a problem of emotional content. What remains open is the possibility that these patients have difficulties with both emotional *and* intonational aspects (cf. Cicone, Wapner, Foldi, Zurif, and Gardner, 1979; Gainotti, 1972; Wechsler, 1973).

From this brief review of intonation, we can conclude that in production, aphasics have some spared ability to utilize intonation for communicative purposes, though the Wernicke's aphasics' fluency may to some extent mask intonational impairments. Pilot work by Danly, Shapiro, and Gardner (1982), suggests severe limitations in patients with right anterior damage.

The decoding of intonation is somewhat problematic. On the one hand, we are led to conclude that when the intonation pattern stands alone (i.e., filtered from the linguistic stream), a right-hemisphere superiority becomes possible, at least in a masking task. What remains unclear, however, is how the right hemisphere functions when the task is either more demanding, or when the contour is woven into a dialogue. The issues then turn on whether the right hemisphere is necessary for integrating all the melodic and linguistic information, or whether, in fact, both hemispheres are simultaneously involved in exploiting the functional value of the intonation patterns.

PRAGMATIC STRUCTURES

In the previous sections, we have focused on those vehicles of communication which can carry various kinds of information. We now turn our attention to the pragmatic structures which enter into, and make possible, communication at the level of discourse. While constructs such as conversational postulates and turn-taking (Grice, 1969 and 1975) and contingent queries (Garvey, 1977 and 1979) have been observed in patients with organic brain damage, this necessarily selective survey begins with the most specific pragmatic structure: the capacity to make reference. We then move to a consideration of more general aspects of communication —the speech act. Our survey concludes with a brief look at two somewhat more remote, but important aspects of communication: the capacity to appreciate nonliteral meanings, and the ability to deal with suprasentential information as found in narrative texts.

Reference

One of the most important aspects of communication is to make its referents clear to the other conversants. Reference to an action or an object can be communicated in any number of ways, through language, complex gestures (e.g. emblems), or more simple means like deixis. In precise communication, referents are unambiguous and clearly specified. But when two speakers share knowledge about their environment, reference can be less specific and still achieve effective communication. For instance, surrounding or presupposed information, like that used by Mr. Harvey when he addresses the head nurse, can influence the success of the communication of the referent.

In this section we will focus on two aspects of reference. First, given their language impairment, how do aphasics make their referents known to their listeners? The second aspect focuses on the role of the listener: what meaning does the listener attribute to such referents, and are such interpretations justified?

In response to the first question, the aphasic has two potential vehicles available: language and gesture. Naturally, if a patient can verbally name an item, the referent is clearly conveyed. But given the language impairment, this is neither facilely achieved, nor is it always appropriate to continue naming an item that has already occurred in the situation. In this context, Bates, Hamby, and Zurif (1982) have undertaken a study to investigate whether aphasics are sensitive to *given* or *new* distinctions when describing reference. In their experiment, Broca's and Wernicke's patients were asked to describe the events shown on three separate pictures, where one subject (or action) remained *constant* and where one object (or action) was *variable* across the three scenes. Preliminary findings suggest that the Broca's aphasics' ability to lexicalize is pragmatically appropriate, whereas the Wernicke's aphasics' ability is not. Thus, from the data so far, Broca's patients tend to lexicalize or otherwise indicate *variable* referents more than *constant* referents (i.e., "new" as opposed to "shared"), while Wernicke's patients are not sensitive to this distinction.

For example, when the Broca's patients are shown a sequence of three different animals each eating a banana, they lexicalize the agent in each instance (bear, monkey, rabbit) and once they have mentioned the common activity of eating a banana, they will not repeat this common element. The Wernicke's patients, however, are as likely to mention the common as well as the new information.

The clarity of reference occurring in aphasics' informal conversation was also analyzed by Cicone and colleagues (1979). The success both of verbally and gesturally conveyed referents was observed. The clarity of each verbal referent made by the Wernicke's patients was similar to that of normal controls, suggesting that the listener's difficulty in comprehending the speech of fluent aphasics presumably derives from the incoherent manner in which the patients incorporate these referents into their spoken language. Gestural movements made by Wernicke's patients were frequently of little communicative value; a high proportion of the gestures which seemed to be attempts at making reference (i.e., were clearly not simply emphatic or rhythmic) were judged as unclear.

Whenever Broca's aphasics specified referents, they were clear to the listener both in speech and in gesture. However, in the verbal realm, the clarity was frequently a function of the examiners' conversation. In the gestural cases most gestures (apart from the assents and denials) were deictic. Thus, both the verbal and gestural referents relied heavily on the surroundings and the dialogue of the other conversant. If indeed the patients depend heavily on their environment, it remains to be determined empirically whether the patients actively manipulate their environment in order to adopt such strategies.

We now turn to the role of the listener in aphasic communication. Frequently, as exemplified in our second vignette, the conversation is clarified, not by the patient, but by the other conversant. The listener thus assumes a critical role, inferring much of the content of the dialogue. To be sure, such influences are common in normal conversation as well. But because of the aphasics' deficit, listeners may, if unconsciously, exaggerate this constructive activity.

But even if we, as listeners, in fact must supplement and reconstruct the aphasics' messages, evidence is accumulating that this reconstruction is based largely on information actually contained in the aphasics' speech. The aforementioned studies of Danly and Cooper, for example, indicate that even telegrammatic utterances by Broca's aphasics are apparently planned to communicate meaningful propositions. A second source of corroborative evidence comes from work by Schönle (1979) on the respiration patterns of aphasic patients. This investigator found that the patterns of respiration exhibited by Broca's aphasics indicated that the individuals were attempting to utter connected discourse and not simply unrelated lists of elements.

In conclusion, since making reference is essential for effective communication, the demands for clarity are crucial. Obviously, aphasics are disadvantaged by their impaired ability to draw on the primary symbolic vehicle of language. The Wernicke's aphasics seem less resourceful in adopting alternate strategies, presumably because their own comprehension deficits mask any awareness of their poor communicability. Gesture in the form of complex pantomimes may be employed by certain patients, but such pantomimes are found to be equally perplexing to other individuals. Broca's patients, however, aware of their deficit, will draw on their surroundings to facilitate the communication—by simply pointing to a present object (highly concrete) or by pointing to another object which is representative of the desired referent (e.g., Mr. Harvey's use of the chart rack to represent his appointment). Although this may be characterized as a more abstract way of making reference, the method may not always be successful. Effectiveness is most likely achieved when the shared knowledge of the referent is already understood by the other conversant (e.g., the nurse). Given their impoverished syntactic abilities to make reference, these aphasics are obliged to draw on the more concrete strategies of exploiting the other conversant and the surrounding props to facilitate the communication of different referents.

Speech Acts

USES OF SPEECH ACTS   Moving from the designation of an object or topic, we now focus upon the meaning of a complete utterance within the discourse. As the concept of the speech act originates from neighboring fields of philosophy (Austin, 1962; Searle, 1969, 1975; Wittgenstein, 1953) and linguistics (Gordon and Lakoff, 1971; Ross, 1970; Sadock, 1974), we will briefly review some major theoretical points. The details of specific empirical investigations with organic patients will follow this brief introduction.

The speech act refers to *how* an utterance is used in a communicative exchange. Its importance in a pragmatic analysis is that the utterance is not considered solely on its own; rather the situation at hand and the stance of the speakers may also contribute to the meaning of the utterance.

Since utterances may be used in a variety of fashions, there are different types of speech acts. For example, an utterance may serve to make a command, to make a request, to promise, or to remind. The different acts are the commands, the requests, the promises, etc. Relevant here is the concept of Austin's "illocutionary force"—what the speaker hopes to accomplish by making the utterance. In turn, the role of the listener is to infer these different utterances as different acts and to act, or react (the "perlocutionary force"), accordingly. This ability has been studied in organic patients by Green and Boller (1974), and by Prinz (1980).

A question central to an analysis of communication is how the full meaning of these acts is conveyed in conjunction with the literal meaning of the utterance itself. Several options are available to the speaker. One option has been clarified through a performative analysis (Fraser, 1976; Gordon and Lakoff, 1971; Ross, 1970). In this analysis the "performative clause" (e.g., *I command, I remind, I promise*) states the action of an agent and purportedly precedes every utterance, if not in the surface structure, then at least in the deep structure. When present in the surface form, these utterances are described as "explicit performatives" (e.g., *I promise you I will arrive by noon*). If the clause is not present in the surface structure (e.g., *I will arrive by noon.*), the "implicit performative" exemplifies cases where the clause has been deleted—semantic and syntactic arguments are proposed as supportive evidence for their deletion. Thus, the explicit performatives have captured one way by which to convey the full intended meaning, or act, of the utterance. These forms have been investigated in aphasic patients by McCurdy (1979).

Another alternative for conveying full meaning is for the syntactic form of the utterance to reflect the intended speech act. That is, the declarative form can indicate a command, while an interrogative is more closely associated with a request. By comprehending the utterance alone, the listener may directly interpret the full meaning. Those instances where the utterance, by itself, conveys the act (that is, where the literal meaning concurs with the full connoted meaning), are referred to as "direct speech acts." Often, however, the literal interpretations of the surface forms do not clearly reflect the acts intended by the speaker. The form of the utterance may be a declarative (e.g., *You have just closed the car door on my finger.*), yet the intention be a request or a command (e.g., *Open the car door!*). Similarly, although the syntactic structure may demand an assent or denial (*Can you tell me who closed the car door on my finger?* Or, as the nurse says to Mr. Harvey in the second vignette, "Do you need something?"), the utterance is posed as a request to gain pertinent information (***Who** closed the car door?*, or ***What** does Mr. Harvey need?*). These are instances of "indirect speech acts"—another alternative of expressing the intention. Here, the full interpretation is derived by considering information from the literal utterance (i.e., the interrogative), and from the paralinguistic and situational cues.

Indirect acts take on several forms. Some are highly idiomatic expressions (e.g., *Do you know the time?*). Others abide by conventions (Clark, 1979) that we use in English as means of making requests: questioning one's ability (e.g., *Can you tell the time?*) or one's possession (e.g., *Do you have a watch?*). Still other indirect forms are used, for example, to indicate to someone that they should hurry: an ironic or sarcastic form (e.g., *It's real early.*) or even a metaphoric form (e.g., *Time is marching on.*) can be employed. The major point in all these cases, however, is that the nonlinguistic features are salient and in some instances even critical (Cook-Gumperz and Gumperz, 1978) for the full interpretation or appropriate use of the utterance. It is this mandatory use of the paralinguistic and situational cues that has led investigators (e.g., Heeschen and Reisches, 1979; Wilcox, Davis, and Leonard, 1978; Green and Boller, 1974) to use indirect speech acts as vehicles for exploring a subject's appreciation of the surrounding context. It is proposed (Foldi, in preparation) that if subjects do not extend their interpretation beyond the literal meaning at the sentence level, the appreciation of the indirect act will suffer.

Our review of the literature indicates that the use of speech acts has thus far addressed three questions: (1) whether brain-damaged subjects can distinguish among different types of speech acts in comprehension and production, (2) whether the aphasic appreciates the role of the speech act in the form of an explicit performative, and (3) whether the patient can implement the contextual information to derive the full meaning of an utterance.

REVIEW OF THE LITERATURE   Two studies address the ability of the aphasic patient to distinguish among types of speech acts, one in comprehension (Green and Boller, 1974), and the second in production (Prinz, 1980). While Green and Boller (1974) did not explicitly discuss different speech acts, their study clearly addresses this pragmatic domain. Green and Boller investigated the nature of the aphasics' response to different acts imposed by the examiner (commands, yes–no questions, requests for information). The study also addressed whether the form of the act (e.g., direct, indirect, or preceded by an introductory sentence) or the mode of presentation (with or without visual cues; live versus audio tape recording) influenced the patients' responses. In analyzing the patients' performance, Green and Boller made an important distinction between those responses which were totally "correct" and those which were unsuccessful but were at least "appropriate" to the type of speech act presented; for example, uttering a string of jargon in response to a Wh-question, or producing a gestural response to a performance command. These appropriate responses could be interpreted as the subjects' recognition of the broader conversational demands, and therefore reflect the appreciation of the ensuing speech act.

When responses were scored on the dimension of *correctness,* the responses to information questions were significantly inferior to the responses to command or yes–no questions. However, when the same responses were scored on the dimension of *appropriateness,* the performance improved and the difference among these three stimulus types was negligible. Moreover, the appropriateness scores also yielded differences related to the *form* of the stimuli presentation; that is, whether the stimulus was direct, indirect, or given with an introductory clause. The indirectly worded stimuli were responded to significantly worse than the other forms. This difference was not captured when the scoring was solely based on the "correct" responses. Thus, the different scoring dimensions allowed the examiners to be more attuned to the responses of the patients and permitted them to give credit to some of the patients' spared communicative abilities.

Other cues, such as the live presence of the examiner, were also identified as beneficial for patients, in view of their poor performance with the taped stimuli. As confirmed in a more recent study (Boller, Cole, Vrtunski, Patterson, and Kim, 1979), the tape-recorded message introduced a more artificial situation, lacking in natural cues; the poor, even inappropriate responses (e.g., imitation of the stimuli item) suggest the crucial part played by "live" cues in creating the environment of natural conversation.

Thus, the authors were able to demonstrate several aspects of the performance of the aphasics: their preserved ability to be "appropriate," their positive response to live situations, and their difficulties with stimuli presented in indirect forms. It is of interest that while Green and Boller recognized that intonation might play an important role in the aphasics' performance (as it provides some nonlinguistic cues for the patient), they did not articulate what that role might be.

A second study focuses on the production of specific speech acts by aphasics. By addressing a series of questions to the individual in an informal one-to-one conversational setting, Prinz (1980) sought to elicit requests from three prototypical aphasics (one Broca's, one Wernicke's and one global). The familiarity between patient and experimenter, the props in the room, and the overall environment were carefully constructed to provide a situation optimal for eliciting requests. For instance, the patients were required to respond to some environmental anomaly (e.g., being asked to sign their names, while the only pencil present was in the experimenter's pocket). They were then scored on the discernibility of their illocution (e.g., a request), and on the adequacy of the propositional content of their response. While the aphasics' propositions were frequently unclear, the experimenters could reliably identify the patients' actions as requests, regardless of the severity or type of aphasic impairment.

The above studies indicated that, despite several language deficits, the aphasic can draw on certain devices to respond to different types of speech acts. Our ability as researchers to arrive at this finding rests on the important distinction between "appropriate" versus "correct" responses. As Green and Boller explain, the utterance must be viewed in terms of both form and content; the appropriateness dimension allows the experimenter to address the form. Patients are thereby given credit for their active role in an exchange and for their spared pragmatic abilities, rather than simply faulted for their inaccuracy and their limitations in language.

Turning to another aspect of speech acts, McCurdy (1979) conducted a detailed study of aphasics' appreciation of explicit performatives. Aphasic patients of differing severity were presented with a battery of tasks in order to ascertain their syntactic, semantic, and pragmatic appreciation of the performative. The battery of tasks devised by McCurdy employed numerous devices (e.g., role-playing, sentence completion, repetition) in order to tap different levels of analysis. For instance, to explore a "situational setting," the patient was required to role-play: the examiner read a short paragraph creating a specific situation and then awaited an utterance from the patient. As an example, the examiner would first inform the patient of her (the examiner's) impending dentist appointment and would then ask the patient how he would remind the examiner of the appointment.

In general, McCurdy found surprising sensitivity to pragmatic uses of the performative. Subjects responded with appropriate utterances which conveyed the semantic and pragmatic force required by the context (e.g., the patient responded "don't forget" as a means of *reminding* the examiner). In contrast, when the patient was required to complete a sentence which might convey a similar idea, but was presented in the absence of any contextual cues, the patient's performance suffered. Specifically, the patient would be read the initial phrase of a performative structure (e.g., *I remind you . . .*) and would be asked to complete the sentence in any way possible. The patients (particularly the more severely damaged ones) responded inappropriately and failed to adhere either to the pragmatic (i.e., presuppositional) or to the semantic constraints imposed by the performative verb.

In comparing these two items of the performative test battery (situational and completion), McCurdy suggests that an intention is more aptly communicated when it is elicited within a meaningful context. When a task requiring performatives is presented in isolation, the patients suffer from not being able to formulate a similar intention. While the nature of these tasks clearly differ, these results again indicate the extent to which the presence of a plausible situation aids the patients' production. Furthermore, it demonstrates a level of pragmatic competence independent of the production of complex sentence constructions, such as those of the explicit performatives.

Another set of studies also addresses the issue of context, but focuses instead on the comparison of direct and indirect speech acts. In the experiment of Wilcox, Davis, and Leonard (1978), the patient viewed a videotaped sequence of two individuals involved in a conversation. One of the speakers then addressed the other with an indirect request ("Can you . . . " or, "Will you . . . ") and, in turn, the listener would respond. The "meta-communicative" task of the aphasic patient was to determine whether the listener's final response was "appropriate" or "inappropriate." All patients received a score of their appreciation of conversational appropriateness in these test situations as well as a score for their comprehension of narrower linguistic tasks on a standardized test battery.

The results showed that virtually all patients scored higher in the experimental task than in the standardized tests; however, the improvement was only modestly correlated with the linguistic comprehension scores as measured in the standard tests. Again, the dimension of appropriateness (as in the Green and Boller study) allowed these investigators to assess the aphasics' awareness of the contextual setting. In other words, the patients appreciated the impact of the conversation even when they were apparently unaware of the content of the transpiring conversation.

In another experiment (Heeschen and Reisches, 1979), the investigators adopted a similar technique to explore the patients' use of context. In this case the subjects (Broca's and Wernicke's aphasics, right-hemisphere damaged patients and non-brain-damaged controls) were presented with a story composed of several sentences, the last of which (the target sentence) was a statement made by one of the two characters. Half of the items were constructed such that the illocutionary force of the target sentence would be clear without knowledge of the context of the surrounding story (direct speech act), for example,

> *In an arithmetic test, Fritz received a mark of 5* [equivalent to an American
> "D"]. *He came home with a tear-stained face to his mother, who said: "Oh,
> Fritz, that's not so bad."*

In other items, the context was essential to understanding the appropriate
intention of the target sentence (indirect speech act), for example,

> *A student has an interview with his professor. The window is wide open. The
> student says to the professor: "I have a very bad cold."*

In both cases, the patient was to choose from four multiple choice items
the one which described possible uses of the target sentence. For
example,

For the direct act:

> a. *The mother comforts Fritz.*
> b. *The mother reproaches Fritz.*
> c. *The mother tells Fritz her troubles.*
> d. *The mother tells Fritz a "white lie."*

For the indirect act:

> a. *The student requests something from the professor.*
> b. *The student complains about the state of his health.*
> c. *The student contradicts the professor.*
> d. *The student gives a command.*

In general, the aphasics' performance was significantly worse than the
performance of the normal controls or the right-hemisphere damaged patients. Moreover, the indirect speech act condition proved more problematic for all groups. But, although the aphasics' performance was poorest,
the profile of their performance ("direct" more accessible than "indirect" conditions) mirrored that of the normal controls.

The performance of the right-hemisphere group, on the other hand, revealed considerable deviations from the other two groups. The difference
in their errors between the direct and indirect conditions differed significantly from any of the other three groups; not only did their performance
significantly deteriorate from the direct condition to the indirect condition, but further analysis showed that they tended (unlike the aphasics) to
choose the *literal* interpretation of the indirect request. The authors con-

clude that the right-hemisphere damaged patients failed to take into account the contextual information that would lead to appropriate interpretations of the indirect act. Certainly, as suggested by the intonational studies reviewed before, the right hemisphere may not be able to benefit from certain paralinguistic cues. Indeed, there may be other nonlinguistic cues that these patients have difficulty in encoding, such as facial expressions and emotions. In addition, however, the deficiency may lie in the fact that the right hemisphere cannot integrate the various subtle cues into a single concept and thus fails to organize the whole situation in an appropriate way.

These studies document that even severely damaged left-hemisphere patients have preserved the ability to resort to certain elements of the context; their responses tend to be appropriate to the situations, even if they are semantically incorrect. These results suggest that the form (direct versus indirect) of the exchange is important to the patients' understanding and that despite their language impairment, the aphasics can assume the role of a conversant. Thus, the findings support the anecdotal evidence of aphasics performing adequately in social situations.

It is not clear, however, what strategic devices are being adopted to achieve this adequacy. As the previous section outlined, intonation contours can provide some structural information, particularly in instances of direct speech acts; some aphasics can use intonation patterns and can comprehend meaningfully intoned patterns. A second device to aid the communication may be gesture. Here, the literature suggests that simple deictic gestures can be successfully used. More abstract uses of gestures, such as complex pantomime, appear to be less common among patients.

The studies of Wilcox, Davis, and Leonard, and of Heeschen and Reisches point out further vehicles of conversation, apart from intonation and gesture, which may also constitute integral parts of the aphasic's spared communication. We can designate these as general pragmatic knowledge about the world and rules of conversation. Specifically, a single utterance takes place in a particular situation, where speaker and listener play particular roles with respect to one another (husband–wife, physician–patient); where certain circumstances surround the conversations (often-repeated interactions at bedside rounds; an outing with family members); where the initiator of the interaction is perceived; and where the interlocutions are subject to continuous monitoring by both members. In short, pragmatic information about an utterance is critical for the speaker and listener to gain the full meanings of the utterances. Systematic investigations of these dimensions have yet to be undertaken, but in light of this review, they are likely to play a necessary role in the understanding of strategies available to organic patients.

The work with speech acts has also addressed the importance of another factor, that of redundancy. If a whole exchange is viewed as a set of separate elements, then the direct acts exemplify situations where at least some elements are redundant. For example, a request for information may be expressed with rising intonation, and in the interrogative syntactic form; the speaker's facial expression can display inquisitiveness, and both interlocutors are aware that only the addressee possesses the desired information. These elements by themselves would be sufficient to make the act of request known, but together supply the listener with redundant information about the nature of the intended act. In contrast, indirect speech acts lack the redundancy of the concurring elements. Not surprisingly, these studies have shown that performance on tasks with the direct acts are superior than with indirect acts. Thus the possibility exists that not only are the indirect forms more demanding to comprehend and produce, but also that the redundancy of the direct forms provides a facilitative effect in successful communication. Redundancy has also been cited as a reason for greater comprehension of stories by aphasics (Stachowiak, Huber, Poeck, and Kerschensteiner, 1977).

Finally, Heeschen and Reisches's study also provides provocative hypotheses about laterality. When required to interpret a sentence in a given context, the right-hemisphere damaged patients did not consider the pragmatic information; in contrast, the left brain-damaged patients were relatively sensitive to the extrasentential information crucial to the successful interpretation of indirect forms. This behavioral dissociation is suggestive of an emerging model of communication: it delineates, admittedly in a very gross fashion, the difference between the purely linguistic and nonlinguistic elements of communication (Foldi, in preparation). The nonlinguistic portion of the exchange allows aphasic patients to engage in conversation despite their langue deficits; conversely it thwarts subtle interactions with right brain-damaged patients despite their preservation of language.

OTHER LINGUISTIC FORMS

A large part of communicative competence, including many of the richest and most valued aspects of communication, entails the capacity to deal with more complex and less literal forms of language. In this final section, we consider the small amount of information which has been accumulated on the ability of brain-damaged patients to process three different linguistic forms: humor (primarily jokes), figurative language (primarily metaphor), and narrative texts (chiefly stories). The findings from these three lines of research point to a picture of the use of contextual elements consistent with the research on speech acts, and highlight the role played by the right hemisphere in such broader forms of linguistic and paralingustic communication.

Humor

In the appreciation of a joke, it is necessary not only to attend to the specific information—be it linguistic or pictorial—but also to attend to those features central to the "point of a joke" and to synthesize them properly so as to appreciate that point. An early study by Gardner, Ling, Flamm, and Silverman (1975) indicated that right-hemisphere patients were able to appreciate the point of a cartoon when a caption was provided, but fared poorly when they had to appreciate the point of a joke on the basis of pictorial information alone. Moreover, and of more interest for the present review, right-hemisphere patients exhibited peculiar humorous responses, and their response to specific items differed from those favored by left-hemisphere patients and by normal controls. These findings suggested some difficulties on the part of right-hemisphere patients in appreciating humorous material.

A set of studies being carried out by Brownell, Michel, Powelson and Gardner (in press) supplements this picture. Of special interest is the finding that, given a set of endings for a joke, right-hemisphere patients show a peculiar predilection for non sequiturs: endings which, like good joke endings, are a surprise but which, unlike appropriate joke endings, do not fit the preceding context. Thus given a set of jokes and asked to indicate which were funny, the patients exhibited an anomalous set of preferences. Then, in another condition, patients were asked to indicate which jokes they thought other persons would consider funny. Documenting that their bizzare preferences were not simply an idiosyncratic set, they indicated that other individuals would share *their* sense of humor.

Naturally, on most verbal tests of humor, aphasic patients perform poorly. Yet a number of lines of evidence suggest that they have retained an understanding of what sorts of information and elements are in fact humorous and that, when they are able to appreciate a joke at all, they do so in the same fashion as normal control patients. In contrast, right-hemisphere patients seem to be characterized by an inappropriate sense of humor; they seem oblivious to the attitudes of other individuals and they have a well-documented penchant for making inappropriate jokes, and for doing so in inappropriate contexts (Geschwind, 1976; Weinstein and Kahn 1955). Once asked to go beyond the literal, right-hemisphere patients seem unable to incorporate the surrounding context properly; they thus may be considered to be pragmatically "at risk" in the realm of humor.

Figurative Language

In a study of the comprehension of figurative language conducted by Winner and Gardner (1977), the investigators probed the capacity of unilaterally brain-damaged patients to appreciate the meaning of common figures of speech, such as *He had a heavy heart* or *It was a loud tie*. Asked to provide a paraphrase of the phrase, aphasic patients had great difficulties and generally proved unable to make the necessary shift to a higher level of abstraction. Surprisingly, right-hemisphere patients were also reluctant to offer an appropriate interpretation ("You wouldn't say something like that" was a frequent rejoinder); yet, when pressed, they usually were able to offer a proper paraphrase of these familiar idiomatic expressions.

When patients were required to match the phrases with the appropriate picture from a set of four, however, a different and far more surprising picture emerged. Despite their avowed language difficulties, aphasic patients did extremely well; in about 75% of the cases they made correct choices and they never selected the literal response (an individual lugging a heart about; a tie with a bugle blaring forth from it); indeed, like the normal controls, they found these literal depictions very humorous. In sharp contrast, the right-hemisphere patients were as likely to choose the literal depictions as the correct, metaphoric depictions (a sad person; a garish tie). The correct choice was a matter of complete indifference to them and they failed to detect anything humorous in the literal choices. Yet, it could be shown that they were not answering randomly, for, like the aphasic patients, they also spurned the other foils.

While these results have proved difficult to explain completely, they seem to point to the same conclusion proposed earlier: right-hemisphere patients may have preserved a sense of the literal meaning of language and yet experience great difficulty in determining the context in which language might be used, or in matching language to its appropriate context. It is as if their literal language capacity has been preserved but their use of surrounding, pragmatic features has been destroyed. Conversely, aphasic patients, while crippled in literal language, have a surprisingly spared knowledge of contexts of communication—knowledge which stands them in good stead when they are attempting to determine how to act, or respond, in a given context (cf. Wapner, Hamby, and Gardner, 1981).

Comprehension of Text

We turn finally to some research on the ability of organic patients to appreciate language of the sort encountered in text. The most extensive completed investigation of texts after brain damge is that of Stachowiak, Huber, Poeck, and Kerschensteiner (1977). In this study, subjects heard a six-sentence story and then were asked to pick from an array of five drawings the one which best described what had happened to the main character. A central element in this study was the inclusion in each story of one sentence which was a metaphorical idiom. The common interpretation of this idiomatic expression was appropriate to the story, but the idioms varied in the extent to which the literal interpretations were also plausible within the context of the narrative. That is, half were idioms such as *They stripped him right down to his shirt* which could be interpreted literally as well as in the metaphorical sense (e.g., he lost all his money); the remaining half were expressions such as *He filled his soup with pieces of bread* (the German equivalent of *He got himself into a mess*) which would and can only be anomalous in the context if interpreted literally. Possible incorrect choices included one depicting a literal interpretation of the metaphor in the story, as well as three pictures which did not fit the story semantically, that is showed an incorrect agent, action, or situation.

Four aphasic groups (globals, Wernicke's, Broca's, and amnesics), right-hemisphere patients, and normals were tested. The authors found that the aphasics' ability to comprehend the general context of the story was surprisingly well preserved. They base their conclusion on two results: (1) the aphasics' overall performance did not differ from that of the right-hemisphere damaged patients and the controls and (2) the pattern in which the aphasics chose the "literal" representation was the same as that of other groups. Specifically, the literal picture was chosen more often in those items which featured an idiom that was contextually plausible in its literal interpretation than in the second type of items where only the metaphorical interpretation could represent the main event of the story. This pattern can occur, argue the authors, only if the subjects can comprehend the general outlines of the story well enough to recognize that the literal interpretation could also be a plausible choice in the former type of item.

The early stages of research on narrative comprehension conducted by Delis, and by Gardner and his colleagues suggest tentative support for Stachowiak's claims of relatively preserved story comprehension in aphasics (Delis, 1980; Gardner, Brownell, Wapner, and Michelow, in press). In one line of testing, subjects are shown a 7-minute silent movie, which depicts a simple parable-like story and are also read a verbally matched version (in length and complexity) of the movie. In each instance, patients are first asked to retell the story; then they receive a series of questions probing the events and details of the story, inferences about motivations of the characters, and knowledge of the story's moral. Not surprisingly, both anterior and posterior aphasics perform better in retelling the movie version than in relating the verbal presentation; their responses indicate the awareness of major theme and important contents, and they can provide (or identify) an appropriate moral. Furthermore, there are some preliminary indications that the right-hemisphere damaged patients perform poorly on the verbal version of the story. While they are generally able to relate the major events of the story, they seem to have trouble inferring the motivations and emotions of the characters and are unexpectedly inept at extracting an appropriate moral.

Additional tests indicate that, while able to recall important facts, the right brain-damaged patients have difficulty in apprehending the overall structure of the narrative. In one study in which three short narratives had to be recalled (Wapner, Hamby, and Gardner, 1981), these patients experienced difficulty in judging the plausibility of individual events within the context of the narrative. At times, perfectly appropriate events in the story were challenged or embellished by right-hemisphere patients; at other times, incongruous elements, which had been purposely inserted (and were generally laughed at or ignored by other subjects) were accepted and their presence was rationalized. Possibly, the right-damaged patients could not appreciate the sequence of the story sufficiently well to judge when a given event fit the context and when it did not. In another study (Delis, 1980), patients were asked to sequence randomly presented sentences in order to produce a coherent, logically formed paragraph. The hypothesis, predicting that holistic processing would be impaired in right-hemisphere damaged patients, was supported: these patients, despite their presumed "intact language abilities" failed to organize this linguistic material to produce the appropriate sequence for the paragraph.

Further evidence suggesting that right-hemisphere patients may have problems appreciating the framework of a narrative text comes from their tendency to inject irrelevant personal elements into their retellings, confabulate events which did not happen in the story, and misinterpret the emotions felt by the characters (Wechsler, 1973; Cicone, Wapner, and Gardner, 1980). It is worthy of note in these studies that misascribed emotions are often justified by the patient, who produced an interpretation of the story which *could* make such emotions appropriate. Or, patients could accurately infer how a character *might* have been feeling in the situation, but often could not appreciate the emotions or motivations of the characters agreed on by normal controls. This apparent inability to comprehend the "whole situation at hand" and to make appropriate inferences on that basis may occur in real life contexts as well as narrative ones, and so possibly account for some of the social inappropriateness observed in right-hemisphere damaged patients (Gardner, 1975; Geschwind, 1976).

The various strands of research from this new area of study all point to a suggestive conclusion. Literal competence in langue clearly depends upon an intact left hemisphere and accordingly, one can expect aphasic patients to have difficulties with all extended instances of text. Nonetheless, particularly when redundancy is present, or when the narrative is designed to highlight a few central points or conclusions, aphasic patients show a surprising preservation of the ability to detect the sense of a passage, or the underlying message or effect that is intended. In sharp contrast, right brain-damaged patients can rarely be faulted on literal comprehension but often show an inability to go beyond the literal: they take figures of speech at face value, are unable to appreciate or choose the punch lines for jokes, and have difficulty in sorting out essential from adventitious information in extended texts. Accordingly, the brain damage has interfered with their ability to decode extensive and subtle forms of linguistic information, and to integrate such linguistic information with the situations for which they are suitable.

CONCLUSION

In this review we have considered several aspects of communication in brain-damaged patients. Our survey of the vehicles of communication revealed that the aphasics show some preserved use of intonation, but only a modest use of gesture. Intonation serves a stronger communicative function for these patients than does gesture. While we do not know the role of gesture in the right-hemisphere patients, we do know that certain tasks involving intonation pose problems for them. At least for normal speakers, we might infer that the right hemisphere is, to some extent, involved in the propositional facets of intonation.

Next, we find the following trend in the pragmatic structures. Broca's aphasics are generally able to make clear reference, but rely largely on cues from their conversants or from items in the surrounding environment. Wernicke's aphasics are less likely to adopt such strategies and, while they may verbally or even gesturally express objects or actions, they are less successful at coherently communicating clear ideas about these referents.

In terms of speech acts, the right-hemisphere patients seem able to derive the literal meanings and can produce a variety of different acts. Left brain-damaged patients can often not appreciate the linguistic aspects of the speech act but surprisingly they can infer the main point of the conversation. We might conjecture that any one act usually contains several redundant features and that the left-hemisphere damaged patient can usefully draw on this information; this information does not benefit the right-hemisphere damaged patients, who will tend to focus solely on the literal linguistic portion of these acts.

Turning finally to other complex aspects of language, we encounter here a surprising yet instructive picture. Despite their competence with literal language, right-hemisphere patients are severely disadvantaged. They often miss the points of jokes and metaphors and they often prove unable to deal comfortably in the narrative realm. In contrast, while often failing on the exact content of metaphors and jokes, aphasic patients clearly abstract crucial points about the operations of these forms of language and are amused by violations. When it comes to texts, they take advantage of the redundancy and often get the point or the moral; the right-hemisphere patients are likely to focus excessively on a single point and thus may fail to integrate the content with the surrounding information.

The review suggests a preliminary working model of pragmatic factors in brain-damaged individuals. Certainly, the left hemisphere is crucial for literal language—phonology, syntax, and low-level semantics. Moreover, it seems to control the substantives and the detail of other symbolic modes of communication, particularly the vehicle of gesture, thereby suggesting its role as a central organizer in communication.

Nonetheless, patients with right-brain damage are apparently not successful in processing more complex aspects of language: those involved in nonliteral meanings, or in more extensive narratives, such as in texts, or in the subtler meanings conveyed by paralinguistic cues. As we have seen, the right hemisphere seems important in aspects of intonation and emotional content; thus utterances which exploit these paralinguistic aspects will be missed by right-hemisphere patients and they themselves will exhibit difficulties in conveying such information through other than literal language channels. Thus, in the case of more complex instances of language use, we have found that right-hemisphere damaged patients have difficulty in appreciating nonliteral meanings, as in jokes and metaphors, and in getting the points or morals of extended narratives.

Just why this is so, and just how to characterize the role of the right hemisphere in communication remains obscure and we can do little more than to describe the difficulty. It may be that, as in other realms (for example, in visual-spatial tasks, Goodglass and Kaplan, 1977), the right hemisphere is important for simultaneously organizing and integrating different elements of conversation.

At the very least, however, our review should allow us to go further than simply saying that "context" helps communication. Of course it does. But it is only by testing out its various components and manifestations—ranging from vehicles like intonation and gesture to pragmatic structures like speech acts, presuppositions, turntaking, or the redundancies of narrative—that the notion of context can be given significant meaning as a whole integrated construct. It remains for future research to cleave out the precise roles of each of these contextual factors.

We return finally to our central problem—the communicative capacities of aphasic patients, as exemplified by our first subject, Mr. Harvey. We have seen that the more we rely on standardized testing, the worse aphasics like Mr. Harvey are likely to perform; and correlatively, it is on those messages which are relatively devoid of surrounding information that right-hemisphere damaged patients will seem correspondingly intact. But the more we deal with complex linguistic entities, with redundancies in the material, with meanings which extend beyond the literal, and with various paralinguistic cues in the environment, the better the aphasic performance will become and the more the deficits of the right-hemisphere patients will become patent. Communicative competence is not a static or fixed cipher; the sensitivity of one's measures, and the circumstances in which they are administered, can make a significant difference for the individuals involved in a communicative encounter no less than for investigators bent upon understanding such encounters.

ACKNOWLEDGMENTS

We are indebted to Edgar Zurif for commenting on the manuscript and informing us of some recent findings. We would also like to thank Lucia Kellar for providing the results of a recent study.

REFERENCES

Alajouanine, T., and Lhermitte, F. (1964). Nonverbal communication in aphasia. In A. de Reuck and M. O'Connor (Eds.), *Disorders of Language,* Vol. 1, Ciba-Geigy Foundation. London: Churchill.

Albert, M. L., Sparks, R., and Helm, N. (1973). Melodic intonation therapy for aphasia. *Archives of Neurology, 29,* 130–131.

Austin, J. L. (1962). *How to do things with words.* Cambridge: Harvard Univ. Press.

Bates, E. (1980). Intentions, conventions, and symbols. In E. Bates (Ed.), *The emergence of symbols: Cognition and communication in infancy.* New York: Academic Press. Pp. 33–68.

Bates, E., Hamby, S., and Zurif, E. B. (1982). *Effects of focal brain damage on pragmatic expression.* Manuscript, Aphasia Research Center, Boston, Veterans Administration Medical Center.

Blumstein, S., and Cooper, W. E. (1972). Hemispheric processing of intonational contours. *Cortex, 10,* 146–158.

Boller, F., Cole, M., Vrtunski, P., Patterson, M., and Kim, Y. (1979). Aspects of auditory comprehension in aphasia. *Brain and Language, 7,* 164–172.

Brookshire, R. (1977). Communicative behavior in the clinic: A coding of patients' behavior in treatment. Paper presented at the Academy of Aphasia, Montreal.

Brown, J. W. (1972). *Aphasia, apraxia, and agnosia.* Springfield, Ill.: Thomas.

Brownell, H. H., Michel, D., Powelson, J., and Gardner, H. (in press). Surprise and coherence: sensitivity to verbal humor in right hemisphere patients. *Brain and Cognition.*

Chester, S., and Egolf, D. (1974). Nonverbal communication and aphasic therapy. *Rehabilitation Literature, 35*(8), 231–233.

Cicone, M., Wapner, W., Foldi, N. S., Zurif, E., and Gardner, H. (1979). The relation between gesture and language in aphasic communication. *Brain and Language, 8,* 342–349.

Cicone, M., Wapner, W., and Gardner, H. (1980). Sensitivity to emotional expressions and situations in organic patients. *Cortex, 16,* 145–158.

Clark, H. (1979). Responding to indirect speech acts. *Cognitive Psychology, 11,* 430–477.

Cook-Gumperz, J., and Gumperz, J. (1978). Context in children's speech. In N. Waterson and C. Snow (Eds.), *The development of communication.* New York: Wiley.

Cooper, W., Danly, M., and Hamby, S. (1979). Fundamental frequency (F_o) attributes in the speech of Wernicke's aphasics. In J. J. Wolf and D. H. Klatt (Eds.), *Speech communication papers presented at the 97th meeting of the Acoustical Society of America.* New York: Acoustical Society of America.

Danly, M., deVilliers, J., and Cooper, W. (1979). The control of speech prosody in Broca's aphasia. In J. J. Wolf and D. H. Klatt (Eds.), *Speech communication papers presented at the 97th meeting of the Acoustical Society of America.* New York: Acoustical Society of America.

Danly, M., Shapiro, B. E., and Gardner, H. (1982). Dysprosody in right brain-damaged patients: Linguistic and emotional components. Paper presented at the Academy of Aphasia. Lake Mohonk, New York.

Delis, D. (1980). *Hemispheric processing of discourse.* Unpublished doctoral dissertation, University of Wyoming.

Duffy, R. J., Duffy, J. R., and Pearson, K. L. (1975). Pantomime recognition in aphasics. *Journal of Speech and Hearing Research, 18*(1), 115–132.

Egolf, D., and Chester, S. L. (1977). A comparison of aphasics' verbal performance in the language clinic with their verbal performance in other program areas of a comprehensive rehabilitation center. *Rehabilitation Literature, 38*(1), 9–11.

Ekman, P., and Friesen, W. W. (1969). The repertoire of nonverbal behavior: Categories, origin, usage and coding. *Semiotica, 1,* 49–98.

Foldi, N. S. (in preparation). *Comprehension of indirect and direct commands by right and left brain–damaged patients.* Doctoral dissertation, Clark University.

Fraser, B. (1976). *On requesting: An essay in pragmatics.* Unpublished manuscript.

Gainotti, G. (1972). Emotional behavior and hemispheric side of the lesion. *Cortex, 8,* 41–55.

Gainotti, G., and Lemmo, M. A. (1976). Comprehension of symbolic gestures in aphasia. *Brain and Language, 3,* 451–460.

Gardner, H. (1975). *The shattered mind.* New York: Knopf.

Gardner, H., Brownell, H. H., Wapner, W., and Michelow, D. (in press). Missing the point: the role of the right hemisphere in the processing of complex linguistic materials. In E. Perecman (Ed.), *Cognitive processing in the right hemisphere.* New York: Academic Press.

Gardner, H., Ling, P. K., Flamm, L., and Silverman, J. (1975). Comprehension and appreciation of humorous material following brain damage. *Brain, 98,* 399–412.

Garvey, C. (1977). The contingent query: A dependent act in conversation. In M. Lewis and L. Rosenblum (Eds.), *Interaction, conversation, and the development of language: The origin of behavior,* Vol. 5. New York: Wiley.

Garvey, C. (1979). Contingent queries and their relations in discourse. In E. Ochs and B. Schieffelin (Eds.), *Developmental pragmatics.* New York: Academic Press.

Geschwind, N. (1976). Approach to a theory of localization of emotion in the human brain. Paper presented at the International Neuropsychological Symposium, Roc-Amadour, France.

Goffman, E. (1971). *Relations in public.* New York: Harper.

Goodglass, H., and Kaplan, E. (1963). Disturbance of gesture and pantomime in aphasia. *Brain, 86,* 703–720.

Goodglass, H., and Kaplan, E. (1977). Assessment of cognition deficits in brain-damaged patients. In M. Gazzaniga (Ed.), *Handbook of behavioral neurobiology,* Vol. 2 *Neuropsychology.* New York: Plenum.

Gordon, D., and Lakoff, G. (1971). Conversational Postulates. In *Papers from the seventh regional meeting.* Chicago, Illinois: Chicago Linguistic Society.

Green, E., and Boller, F. (1974). Features of auditory comprehension in severely impaired aphasics. *Cortex, 10,* 133–145.

Grice, H. P. (1969). Utterer's meaning and intentions. *The Philosophical Review, 78,* 147–177.

Grice, H. P. (1975). Logic and conversation. In P. Cole and J. Morgan (Eds.), *Syntax and Semantics: Speech Acts.* New York: Academic Press.

Hécaen, H., and Albert, M. (1978). *Human neuropsychology.* New York: Wiley.

Heeschen, C., and Reisches, F. (1979). *On the ability of brain-damaged patients to understand indirect speech acts.* Unpublished manuscript.

Heilman, K., Scholes, R., and Watson, R. T. (1975). Auditory affective agnosia. *Journal of Neurology. Neurosurgery and Psychiatry, 38,* 69–72.

Holland, A. (1978). Factors affecting functional communication skills of aphasic and non-aphasic individuals. Paper presented at the American Speech and Hearing Association Convention, San Francisco.

Jackson, J. H. (1932). On affections of speech from diseases of the brain. In J. Taylor (Ed.), *Selected writings of John Hughlings Jackson,* Vol. 2: *Evolution and dissolution of the nervous system.* London: Hodder and Stoughton, Ltd.

Kellar, L. (1980). Personal communication.

Kimura, D. (1973a). Manual activity during speaking: Right handers. *Neuropsychologia, 11,* 45–50.

Kimura, D. (1973b). Manual activity during speaking: Left handers. *Neuropsychologia, 11,* 51–55.

Liepmann, H. (1905). Die linke Hemisphäre und das Handeln. *Münchener Medizinische Wochenschrift, 2,* 2375–2378.

Liepmann, H. (1920). Apraxie. *Ergebnisse der gesamten Medizin, 1,* 516–543.

McCurdy, P. (1979). *Selected performative utterances in aphasic language.* Unpublished doctoral dissertation, University of Texas.

Nespoulous, A. (1980). Geste et discours. Etude du comportement gestural spontané d'un aphasique en situation de dialogue. *Etudes de linguistique appliquée.* Paris: Didier.

Prinz, P. (1980). A note on requesting strategies in adult aphasics. *Journal of Communication Disorders, 13,* 65–73.

Ross, J. R. (1970). On declarative sentences. In R. Jacobs and P. Rosenbaum (Eds.), *Readings in English Transformational Grammar.* Waltham, Mass.: Ginn and Company.

Ross, E. D., and Mesulam, M. (1979). Dominant language functions of the right hemisphere? Prosody and emotional gesture. *Archives of Neurology, 36,* 144–148.

Sadock, J. (1974). *Toward a linguistic theory of speech acts.* New York: Academic Press.

Schlanger, B. B., Schlanger, P., and Gerstman, L. (1976). The perception of emotionally toned sentences by right-hemisphere damaged and aphasic patients. *Brain and Language, 3,* 396–403.

Schönle, P. (1979). Speech and respiration in normals and aphasic patients. Paper presented at the Academy of Aphasia, San Diego, California.

Searle, J. (1969). *Speech Acts.* Cambridge: Cambridge Univ. Press.

Searle, J. (1975). Speech acts and recent linguistics. In D. Aaronson and R. Rieber (Eds.), *Developmental Psycholinguistics and Communication Disorders.* New York: New York Academy of Science.

Seron, X., van der Kaa, M. A., Remitz, A., and van der Linden, M. (1979). Pantomime interpretation and aphasia. *Neuropsychologia, 17,* 661–668.

Shankweiler, D., and Studdert-Kennedy, M. (1967). Identification of consonants and vowels presented to left and right ears. *Quarterly Journal of Experimental Psychology, 19,* 59–63.

Stachowiak, F., Huber, W., Poeck, K., and Kerschensteiner, M. (1977). Text comprehension in aphasia. *Brain and Language, 4,* 177–195.

Taylor, M. (1965). A measurement of functional communication in aphasia. *Archives of Physical Medicine and Rehabilitation, 46,* 101–107.

Tucker, D., Watson, R. T., and Heilman, K. M. (1977). Discrimination and evocation of affectively intoned speech in patients with right parietal disease. *Neurology, 27,* 947–950.

Ulatowska, H., Macaluse-Haynes, S., and Mendel-Richardson, S. (1975). The assessment of communicative competence in aphasia. Unpublished paper, University of Texas.

Varney, N. R. (1978). Linguistic correlates of pantomime recognition in aphasic patients. *Journal of Neurology, Neurosurgery and Psychiatry, 41,* 564–568.

Wapner, W., Hamby, S., and Gardner, H. (1981). The role of the right hemisphere in the apprehension of complex linguistic materials. *Brain and Language, 14,* 15–33.

Wechsler, A. F. (1973). The effect of organic brain disease on the recall of emotionally charged versus neutral narrative texts. *Neurology, 23,* 130–135.

Weinstein, E., and Kahn, R. L. (1955). *Denial of illness.* Springfield, Ill.: Thomas.

Wiener, M., Devoe, S., Rubinow, S., and Geller, J. (1972). Nonverbal behavior and nonverbal communication. *Psychology Review, 79*(3), 185–213.

Wilcox, M. J., Davis, G. A., and Leonard, L. B. (1978). Aphasics' comprehension of contextually conveyed meaning. *Brain and Language, 6,* 362–377.

Winner, E., and Gardner, H. (1977). The comprehension of metaphor in brain-damaged patients. *Brain, 100,* 719–727.

Wittgenstein, L. (1953). *Philosophical Investigations.* Oxford: Basil, Blackwell and Mott.

CHAPTER 4

The Right Hemisphere's Contribution to Language: A Review of the Evidence from Brain-Damaged Subjects

Janice M. Millar and Harry A. Whitaker

INTRODUCTION

One of the most dramatic clinical observations encountered is the difference between brain-damaged patients who have sustained relatively extensive left-hemisphere damage compared to right-hemisphere damage. After medical stabilization, the former typically have obvious disturbances in language, whereas the latter typically do not. For the few cases that do not follow this pattern, there is usually a history of early brain damage, atypical handedness or some other factor that may be expected to influence lateralization of brain functions. Zangwill (1967) reviewed 2133 cases of brain damage previously reported in the literature. Of those with left-hemisphere damage, 59.7% of the right-handers were aphasic, while 54.9% of the left-handers were aphasic. Of those with right-hemisphere damage, only 1.8% of the right-handers were aphasic, and 29.2% of the left-handers were aphasic. This familiar clinical observation has been supported again and again in the published neurolinguistics literature; however, it has been and continues to be questioned, challenged,

87

and modified in terms of its deeper level and more sophisticated details.

Of the two experimental populations used to research the question of the right hemisphere's contribution to language processing, it seems to us that only the brain-damaged population at this time is likely to contribute significant data. The use of normal, non-brain-damaged subjects, studied by methods such as dichotic or tachistoscopic presentation, or lateralized motor or sensory assessments, presents a problem that is not yet amenable to either a theoretical or empirical solution. Specifically, the human brain processes information at a higher rate of speed than the reaction time (RT) within which data samples are taken. This implies, in our opinion, that any dependent measure from normal, non-brain-damaged subjects is a product of multiple interactions of the left and the right hemisphere, rendering it problematic at best to figure out how each hemisphere makes its specific contribution.

It is, for example, well known that analogous tachistoscopic experiments have produced quite different results (or no results at all), a situation which is likely to be due, in part, to uncontrolled features of the experimental paradigm. There is good reason to believe that the cognitive strategies best handled by each hemisphere dynamically interact in tachistoscopic and dichotic experiments; over the time span of 500–800 msec, there is surely a number of such interactions as signals traverse the corpus callosum. When these strategies are better understood, it may be possible to integrate the results of the many tachistoscopic and dichotic studies published in the literature. At this point, however, it is not clear to us that either contrary data (e.g., a left visual field [LVF] advantage in word–letter processing) or supportive data (e.g., a right visual field [RVF] advantage in word–letter processing) directly and unequivocally address the matter of the brain lateralization of linguistic processing.

Scalp electroencephalograph (EEG) techniques can avoid the temporal problem mentioned above, in that on-going brain activity is assessed in the time frame in which it occurs. But these techniques face the problems of relating a relatively global neural activity of large populations of neurons to specific components of behavior (and some EEG paradigms face still-unresolved problems of electrical artifacts). At the moment, these problems are of sufficient magnitude as to call into question those (few) EEG studies which contradict (or fail to support) the basic clinical facts noted above.

For all these reasons, this review and critique of the right hemisphere's contribution to language processing will focus on studies of brain-damaged subjects. Although neurolinguistic studies have their own set of problems, many of which we will discuss, we believe that at present there is more of a possibility that this research directly addresses brain–behavior correlates. There is certainly no doubt that this research most directly deals with the original clinical observation of left-hemisphere language representation.

LESION STUDIES

It would be helpful to begin with a few general observations about the lesion studies which have argued in favor of a right-hemisphere role in language processing. Some studies have been less than optimal in providing neurological data as to etiology, locus, or extent of the lesion. Mixing subjects who have stable disease processes with those who have on-going disease processes is likely to confound results from the presence of edema or multiple lesion sites, for example. Some studies have looked at subjects who clearly have had early left-hemisphere damage, raising the obvious question of transference of some functions from the left to the right hemisphere. Some studies have not accounted for those associated neuropsychological deficits which are likely to contribute to impaired performance. Some studies have not used appropriate control groups when the experimental design clearly called for them, and some studies have used a very small number of subjects, being essentially case history reports. If one finds a single case of a restricted, well-defined, exclusively right-hemisphere lesion that produces aphasia, it certainly proves that this can occur; such a result does not generalize to other subjects, however. On the other hand, finding a restricted, well-defined, exclusively right-hemisphere lesion that does *not* produce aphasia, proves that *this*, too, is a possible state of affairs. The latter case also proves that in at least one subject the right hemisphere makes no contribution to language, that is, lateralization of language can be a complete phenomenon. Within the above framework, and considering the caveats just offered, there are two principal kinds of brain-damaged subject populations we will review (1) the neurological cases, or those with neurological damage inferred from testing, behavioral, or radiological exam, and (2) the neurosurgical cases, or those with the neurological damage directly assessed by surgery, pharmacological intervention (Wada testing), or electrical stimulation.

Neglect

Recently, visuospatial neglect, or hemi-inattention, has received some attention in the literature (Friedland and Weinstein, 1977; Weinstein, 1964; Weinstein and Friedland, 1977; Heilman and Valenstein, 1979; Heilman and Van Den Abell, 1979).

When patients manifest hemi-inattention, it appears they do not search the physical environment for information through which they could orient themselves. Hemi-inattention is seen more often in right-hemisphere lesions than in left-hemisphere damage, although severe language deficits and right-side hemiparesis (in the favored writing hand) may mask right-side neglect. The most common form of unilateral neglect is denial of illness. The basis of the disorder is cognitive rather than visual.

The hemispatial field is not the same as the visual field; it refers to the space to one side of the midline of the body, with a possible gradient across midline. Hemi-inattention is not only visual: it can affect olfaction, taste, touch, vibration, memory, and emotion. In an experiment using card placement to the left of, right of, and at the midline of the body, with identifying alphabetic characters at each end of a line drawn on the card, Heilman and Valenstein (1979) demonstrated that the neglect was not due to the line not being seen, but rather that there may be a defect in the orienting response (i.e., an attention–arousal defect), caused by disruption of the corticolimbic–reticular loop which is responsible for mediating arousal. Lesions that induce unilateral neglect produce an asymmetrical reduction of arousal, and the hypo-aroused hemisphere cannot prepare for action. As a result, any stimulus that comes into the hypo-aroused hemisphere is neglected.

One of the earliest and most serious claims for language processing in the right hemisphere was made by Eisenson (1962), who argued that right brain-damaged patients were deficient in vocabulary processing and sentence processing. From an earlier report of this research, we know that Eisenson studied 65 patients with right-brain damage, although we do not know whether there were tumor patients mixed in with stroke patients. We do not know whether he controlled for lesion size or lesion location in the right hemisphere. His patients were apparently not matched to control subjects on either an educational or an intelligence measure. However, even if we give him the benefit of the doubt on these matters, there are still some serious problems with the linguistics of the study. His recognition vocabulary test was adapted from the Institute of Educational Research Inventory's CAVD, which contains such words as *umbel* and *encomium*, words which we believe might cause a bit of difficulty in many normal, non-brain-damaged subjects. The problem with this kind of study, where the amount of right-hemisphere tissue damage is not determined, is that there can be no appropriate control group: left-hemisphere damaged nonaphasics may have less tissue loss; aphasics will clearly be inappropriate; and normals will not show the generalized deterioration of the right-hemisphere group. A further problem, and a rather interesting oversight, given what we know about the neglect phenomenon following right-hemisphere disease, is that after Eisenson presented each word in the test, he asked patients to point to one of four choices printed on a card. We do not know whether or not the visual field deficits (which some of his patients most certainly had) and the neglect (which some of his patients very likely had) were factored into the analysis of the results. Given the fact that his results show a very small difference between the right-hemisphere lesioned patients (10+ correct) and his control group (11+ correct), one is entitled to remain skeptical of this result. Eisenson's second claim, that right-hemisphere disease leads to an impairment in sentence processing, was based on data taken from the Minkus sentence test (from the Stanford–Binet Test, first published in 1942, in the *Journal of Applied Psychology*, in a paper by Robert Thorndike). In this test there are four sentences, each of which has a blank where a function word (grammatical formative) should be. The subject is asked to fill in the blank, that is, to supply the missing word. The problem is, only a few of the several possible words that will fit grammatically into the blanks are accepted for scoring purposes. Thus, for the sentence *We like to pop corn* _____ *to roast chestnuts over the fire,* only *and* and *or* are considered correct. If the subject had picked *but not,* it would be scored as an error. We do not know whether any of the right-hemisphere damaged patients gave answers which were in fact grammatically acceptable but did not happen to be permitted in the scoring of the test; however, we note that for one of the sentences, the disallowed words are actually of a higher frequency of occurrence in the language than the permitted ones. Until this study is replicated with proper neurological and linguistic controls, there is not much reason for accepting Eisenson's results.

Metaphorical Speech

Weinstein (1964) argued that the "right hemisphere is not a linguistically primitive shadow of the left hemisphere . . . it is especially involved in the relationship between language and perceptual and emotional processes" on which metaphorical speech is dependent. The left hemisphere appears dominant for phonological, sequential, syntactic, and referential functions of language, whereas the right hemisphere is more specialized for the experiential aspects of language. Disturbances of metaphorical speech are related to hemi-inattention in that there is a change in the interaction of the person with the environment and in the way the experience is perceived in the symbolic content of the physical world. Apparently an intact left hemisphere (right-hemisphere damage) does not necessarily insure adequate comprehension of linguistic messages, and an intact right hemisphere (left-hemisphere damage) does not guarantee adequate aesthetic sensitivity (Winner and Gardner, 1977.) In a task requiring both pictorial recognition of metaphors and verbal explanation of the same metaphor (e.g., *a heavy heart*), right-hemisphere damaged subjects chose the correct metaphorical pictorial match (a crying person) 43% of the time and a literal pictorial match (a large red heart being carried by a person who was staggering under its weight) 40% of the time, but were able to explain the metaphor linguistically in 85% of cases ("He's got many troubles."). Patients with left-hemisphere damage chose significantly more metaphorical pictures (58%) and rejected the literal pictures (choosing them only 18% of the time), and were more likely to offer a nonmetaphorical explanation. This dissociation between visual and linguistic interpretation would appear to suggest that right-hemisphere damaged subjects are not insensitive to metaphor, but as Weinstein asserts, they are insensitive to their environment and offer inappropriate emotional responses. Thus, their linguistic responses may be appropriate but they lack the ability to correctly identify situations by means of pictorial representations. (Neglect was controlled for in this study.)

In 1964, Weinstein demonstrated that patients with left-hemisphere damage had difficulty establishing the boundaries of phonetic (*love* for *glove*) and semantic (*cigarette* for *ashtray*) categories, whereas right-hemisphere lesioned patients experienced no such difficulties. Rather, they were disoriented for time and place (speaking about multiple Walter Reed Army Hospitals in different cities), confabulated about their illness (a patient who had become blind after a craniotomy for a brain tumor said he had been "hit on the head and rolled in a blind alley") or showed some form of neglect for the left side of the body either in delusional language (calling a paralyzed limb a "dummy"), or in metaphorical language (patients injured in car accidents described their own injuries in terms of what had happened to their car, such as "dented on top"), without being aware of their errors. This demonstrates an inability on the part of the patients to communicate according to preexisting systems of organization, including cultural roles. There is a change in the interaction of the patients with their environment. Unfortunately, as in Eisenson's case, Weinstein fails to mention the etiology of the lesions involved, so any conclusions must be accepted with reservation.

Semantics

A more precise investigation of the linguistic deficits of right-hemisphere brain-damaged persons was conducted by Lesser (1974) using syntactic, semantic, and phonological tests. Her subjects consisted of 15 left-hemisphere damaged patients, 15 right-hemisphere damaged patients, 9 persons with bilateral frontal leucotomies, and 15 non-brain-damaged controls. The syntax test consisted of 80 items illustrating pictures of two spoken sentences which differed in only one respect (word order, preposition, tense, plurality, etc.) in which the subject had to point to the sentence spoken by the examiner (e.g., "He walks" or "They walk."). The semantic test drew upon word association skills. Thirty stimulus words were illustrated, together with three words most frequently associated with it (e.g., *pencil, **paper**, pen, write*). The subject had to point to the picture named by the examiner. The phonological test used 20 stimuli words paired with 3 distractor words, which were the ones most commonly mistaken for it when spoken against a background of noise (e.g., *porch, torch, **scorch**, court*). The subject had to point to the picture named by the examiner. Lesser's results indicated that the right-hemisphere damaged patients were not impaired in their use of syntax or phonological discrimination, but had marked difficulty on the semantic test. Both the phonological and semantic tests utilized a format of picture verification, with four pictures placed in a square on each card. The subjects performed appropriately on the phonological test, so one can argue that their semantic deficit was not a result of unilateral neglect due to placement of stimuli, since both tests contained picture cards. She speculated that right-hemisphere damage interfered with the understanding of single words but did not interfere with left-hemisphere syntactic interpretation of sentences.

More recently, Gainotti, Caltagirone, and Miceli (1979) and Gainotti, Caltagirone, Miceli, and Masullo (1981) have investigated the semantic aspects of language in more detail. On an auditory comprehension test (1979) right-hemisphere damaged subjects made significantly more errors than controls. Of 110 subjects tested, 33 of them proved to exhibit a general mental deterioration, as shown by a battery of tests designed to reveal mental deficiencies. Thirty of the patients tested had tumors rather than cerebrovascular accident (CVA). It is not stated in the paper whether these patients are, in fact, the ones who were noted to have mental deterioration. The patients with deterioration accounted for most of the errors in the right-hemisphere damaged group. However, the importance of controlling for unilateral spatial neglect is well demonstrated in this study. When errors due to unilateral spatial neglect were ruled out (i.e., most of the errors occurred on items pictured on the left half of space) there was no significant difference between normal controls and right-hemisphere damaged subjects, either "deteriorated" or "nondeteriorated."

Gainotti, Caltagirone, Miceli, and Masullo (1981) used special care to minimize unilateral spatial neglect in a later series of semantic–lexical discrimination and phoneme discrimination tasks. The semantic–lexical discrimination tests consisted of an auditory language comprehension task and reading comprehension task. The stimuli were presented with instructions to match the stimulus word with the appropriately pictured item. The pictures were arrayed in a vertical column. Fifty subjects were tested, 36 with right-hemisphere CVA and 14 with right-hemisphere tumor. Eleven of the subjects proved to be mentally deteriorated as shown by the same battery of tests previously mentioned. Again, it is not stated in the paper whether the tumor-patient population proved to be the deteriorated subjects. The results indicated that the right-hemisphere damaged subjects scored significantly worse than normals on multiple choice tests of semantic discrimination and showed a tendency ($p = .077$) to perform worse than normal controls on the phoneme discrimination task, perhaps indicating of a general deficit. As before, most of the semantic errors were made by the "deteriorated" patients. However, in this study there was a significant difference between the "nondeteriorated" subjects and normal controls.

The differences between the results of the 1979 and the 1981 Gainotti studies are explained by the possibility that the 1981 nondeteriorated patients were actually under the influence of a mild, subclinical form of deterioration. We might suggest as an alternative explanation that unilateral spatial neglect is not easy to minimize, and that these researchers could not be sure that they were, in fact, minimizing hemi-inattention. As Gainotti, Caltagirone, Miceli, and Masullo (1981) conclude, it is possible that these subjects illustrate a high-level cognitive defect rather than a specific linguistic impairment. Archibald and Wepman (1968) noted that eight of their right-hemisphere damaged, right-handed patients who had language difficulties, also seemed to have a "general cognitive deficit." In their study there is a self-admitted lack of relevant neurological data from which to draw conclusions; but, again, the authors seemed only vaguely aware of the possibility of neglect, referring to their problems as "central visual involvement."

In a study using 34 right-hemisphere CVA subjects, Hier and Kaplan (1980) found mild syntactic and semantic deficits in some but not all of their patients. There was a high degree of correlation between verbal deficits and hemianopia–visuospatial deficits, which was not explained by "intellectual impairment" since all of the subjects had normal vocabulary scores. Those who exhibited deficits were impaired in their ability to explain the meaning of proverbs (e.g., *Don't cry over spilt milk*) and to comprehend logico-grammatical sentences (e.g., *Does lunch come before dinner?*). Subjects with posterior lesions did significantly worse on the logico-grammatical statements. Taken together, these observations suggest that verbal deficits are likely to occur only in association with other deficits and only after damage to certain areas of the right hemisphere.

Imagery Based Language

In another semantically based task (a false recognition paradigm using both visual and auditory modalities), Rausch (1981) found that patients with left temporal lobectomies (LTL) made significantly more false recognition errors than other subjects on semantically or acoustically related words during both auditory and visual tasks. This indicates an ability to encode verbal material initially, with a breakdown in information processing at a later stage. However, patients with right temporal lobectomies (RTL) did not differ from normal controls on the auditory task, but made significantly FEWER false recognition errors than controls on the visual task. It has been suggested that this dissociation between auditory and visual modalities is due to a breakdown in the encoding of the visual attributes of verbal material. The RTL patients do not appear to have a semantic–lexical memory deficit. They apparently made FEWER errors because they had initially encoded fewer stimuli as referents during the false recognition task, not because their memory proved to be any different from that of the controls.

Rausch suggests the damaged right hemisphere is more involved than the left hemisphere in the analysis of visual features of verbal material. Unfortunately she fails to mention whether she controlled for visuospatial neglect. Since this may be a confounding variable, we must accept her hypothesis with reservations.

However, data presented by Jones-Gotman and Milner (1978) appear to support her hypothesis. On a paired associates learning task using *concrete* (high imagery), high frequency words, patients with RTLs performed poorly, whereas they performed as well as normal controls on a task using *abstract* (low imagery) words. Their deficit appears to be one of image-mediated verbal learning. Apparently, RTL patients could not use visual imagery mnemonic devices as an aid to the recall of concrete words, an advantageous strategy the normal subjects were able to use to their benefit.

Milberg, Cummings, Goodglass, and Kaplan (1979) have reported a case history of a man with right-hemisphere frontotemporal damage that suggests a relationship between visuospatial and verbal ordering abilities contrary to previous assumptions of both being behaviorally distinct. The patient had difficulty performing tasks that required the use of positional information in space (picture arrangement) and revealed a breakdown of the ability to reproduce overlearned spatial–linear sequences (numbering the face of a clock). Verbal ordering abilities were also impaired. Sequential errors occurred during spontaneous writing (*girls* was written as *grils; in* was written as *ni*), and during recitation of the days of the week, months of the year, and letters of the alphabet.

Visuospatial processing is vulnerable to right-hemisphere damage. Milberg, Cummings, Goodglass, and Kaplan suggest that verbal sequential operations that depend on visuospatial processing may be affected secondarily to right-hemisphere damage.

The subject also demonstrated an interesting dissociation between arbitrary sequences and physically or logically constrained sequences. He had difficulty with arbitrarily ordered sequences (digit span recognition where the subject was only required to determine if two auditorally presented random number sequences were the same or not), but not with physically or logically constrained sequences (such as size relationships or quantity). This pattern of deficits proves interesting indeed. However, as the authors suggest, the patient was likely suffering from damage to more than the right hemisphere. He had remained in a coma for three months and suffered from hydrocephalus following an automobile accident in which he sustained a right temporal skull fracture.

Fried, Mateer, Ojemann, Wohns, and Fedio (1982) have also described an interesting phenomenon that appears to implicate the right hemisphere in visual mediation of language functions in the left hemisphere. During electrical stimulation of the posterior part of the right-hemisphere middle temporal lobe, patients demonstrated a consistent alteration in labeling emotional facial expressions pictured on cards (i.e., at one time the patient identified the face as "happy" and on a repeated trial labeled it as "angry"), yet previous tests of stimulation of the same area during verbal naming and recall failed to show similar effects. They proposed that this area of the cortex may play a role in the analysis of complex visual functions and therefore makes some contribution to linguistic function.

Visual imagery, as mediated by the right hemisphere, has reportedly been used as a strategy to teach a woman, with literal alexia (due to left-hemisphere parietotemporal cortical atrophy) to read (Carmon, Gordon, Bental, and Harness, 1977). Her inability to translate a sequence of printed letters into the corresponding sequence of spoken phonemes demonstrated a deficit in the sequential abilities of the left hemisphere. Retraining allowed her to process information pictorially by right-hemisphere processes, rather than analytically. However, she could only read the words she was trained for. This would suggest that the right hemisphere is capable only of holistic recognition of words. Zaidel (1977) reports that the right hemisphere lacks the capacity to decode visual information using grapheme–phoneme correspondence rules and may be capable of reading words by sight. This case history apparently supports that conclusion.

In deductive reasoning tasks employing congruent (e.g., *Bill is taller than John. Who is tallest?*) and incongruent (e.g., *A is taller than B. Who is shorter?*) problems, Caramazza, Gordon, Zurif, and DeLuca (1976) found that right-hemisphere damaged subjects appeared to have a deficit in visual imagery which affected verbal reasoning. The right-hemisphere damaged subjects were impaired on the incongruent form of the premise information (left-hemisphere damaged subjects, on the other hand, were impaired on all forms), suggesting that verbal reasoning of the sort required in a syllogistic task requires the formation of right-hemisphere based imagery at either a visual or general cognitive level. Apparently, right-hemisphere damaged subjects had an inability to manipulate an image of the premise information such that the linguistic input was not fully understood. Read (1981) similarly found that right-hemisphere damaged subjects exhibited significant difficulty with incongruent problems in a study that employed one additional term (e.g., *A is taller than B. B is taller than C. Who is shortest?*) thereby increasing the difficulty of the task. It is worth noting that 13 of 16 RTL subjects reported consistent use of imagery throughout the test whereas only 7 of 16 LTL subjects used such a strategy, but reportedly relied instead on verbal rehearsal. This may give us some information about the contribution of different areas of the right hemisphere, since Caramazza's patients presented with parietal damage, whereas Read's subjects with right temporal lobectomies had little difficulty. Read also found the LTL patients to have a marked overall impairment on both types of problems (as did Caramazza, Gordon, Zurif, and DeLuca, 1976). He suggests there may be a qualitative difference in image creation and representation between the hemispheres, depending upon the demands of the situation, and that the left hemisphere may create a form of "symbolic" imagery (the sort necessary for these tasks), whereas the right hemisphere may use perceptual imagery.

Wapner, Hamby, and Gardner (1981) also demonstrated that right-hemisphere damaged subjects exhibit difficulty in handling complex linguistic material. Their subjects had clear difficulty in integrating specific information and in drawing the proper point or moral from a story (e.g., *Keep trash in your own yard* rather than *Do unto others as you would have them do unto you*), tending instead to offer a literal response. This apparent difficulty in assessing the appropriateness of various facts, situations, and characterizations may indicate a deficit in handling complex ideational materials, rather than an actual deficit in the processing of complex linguistic material. Of interest is the fact that Wapner, Hamby, and Gardner noted deficits peculiar to site of lesion. Right-hemisphere anterior patients had a tendency to embellish and showed a lack of sensitivity to noncanonical elements. Lesións in central (temporoparietal) regions produced a tendency to commit errors with emotional material, especially if it was injury-related. Gardner, Brownell, Wapner, and Michelow (1981) have reported the same observation (i.e., that the right hemisphere has a role in modulating emotional responses). Patients with right-hemisphere anterior lesions exhibited extreme humor responses (abnormal in their perception of humor, perhaps due to a general euphoria) while those with right-hemisphere posterior lesions presented flat affect.

Prosody

Prosodic and pragmatic features of speech have not been frequently studied by researchers investigating the language skills of brain-damaged subjects. Ross and Mesulam (1979) studied the impaired ability to express emotion in two right-hemisphere lesioned patients. Their speech was characterized as aprosodic, accompanied by normal vocabulary, normal grammar, and normal articulation. One patient had a right posterior frontal lobe lesion and the other patient had an anterior parietal lobe lesion, suggesting a peri-Rolandic site in common. Following a review of ten cases of right-hemisphere damage, Ross (1981) proposed that the functional anatomic organization of the affective components of langue is localized in the right hemisphere in a manner that mirrors the organization of language in the left hemisphere. Ross's classification of the "aprosodias" by site of lesion is analogous to the classification of the aphasias by the Boston Veterans Administration (VA) sytem. He argued that his cases of aprosodia could be called global, motor, sensory, transcortical motor, transcortical sensory, and mixed transcortical aprosodia, stemming from lesions in areas of the right hemisphere homologous to the classical sites in the left hemisphere, which give rise to the corresponding aphasias. Although Ross's classification system makes an interesting claim about the right hemisphere, some caution is warranted before his results can be accepted. In many of his cases, the lesion loci were identified on computerized tomography (CT) scans made within a day or two of insult. It is well known that CT scans are unreliable within two months of a stroke, and it is therefore possible that the extent and location of the lesions in these cases were not accurate. One also must question the strategy of uncritically adopting an aphasia classification system and applying it to the prosodic aspects of language when (1) the aphasia classification system itself has been questioned and (2) no attempt was made to find out whether these prosodic disturbances appear in patients with lesions in other areas of the brain, such as other parts of the right hemisphere or of the left hemisphere. No analysis of the prosody of aphasic patients was offered in the Ross study. And finally, Ross' analysis of aprosodia itself is of questionable value since it makes no assessment of the known linguistic phonetic features of prosody: duration, pitch, intensity, and juncture. Replicating Ross' study will be difficult not only because there is so much room for interpreting just what aprosodia means, but because these features of prosody were not taken into account by Ross.

In addition to the lesions noted above, it is also possible that the basal ganglia play a role in the modulation of prosody. One of the case histories (Case 2) described by Ross and Mesulam (1979) included damage to the right-hemisphere basal ganglia. Parkinsonism, a disease of the basal ganglia, is also characterized by aprosodic speech.

The comprehension of prosody with respect to affective tone appears to present a somewhat clearer picture. Heilman, Scholes, and Watson (1975) have found that subjects with right-hemisphere parietal lesions are markedly impaired in identifying the affective components of language, but not the propositional components. Evidence from "same–different" judgments tasks (presented to patients with right-hemisphere parietal dysfunction and neglect) also demonstrated that these subjects were deficient in the ability to detect the emotional aspects of another individual's expressions and communication (Tucker, Watson, and Heilman, 1977). These subjects were also impaired in their ability to evoke affective tones in a posed (nonspontaneous acting) situation. This anatomic evidence lends support to the hypothesis that the right hemisphere is dominant for comprehending affective speech. Thus it appears that the right-hemisphere posterior sylvian area is necessary for the comprehension of affective speech, just as the left-hemisphere posterior area (Wernicke's area) is crucial for the comprehension of propositional speech.

Early Left-Hemisphere Damage: Hemispherectomy

It is well known that transfer of language functions to the right hemisphere is age-dependent; early and extensive left-hemisphere damage, prior to language acquisition, is very likely to cause language to switch to the right hemisphere. The question here is the adequacy of language functions under these circumstances. The evidence at present indicates that phonological and semantic systems transfer very well, with no discernible differences (Dennis and Whitaker, 1977; Kohn, 1980).

However, the acquisition of adequate syntactic abilities appears to be sensitive to several other dimensions of the medical history. Kohn (1980) studied 12 cases of children with epilepsy who were found to have right-hemisphere language lateralization (as measured by the Wada test). Only 3 of his subjects were determined to have intact comprehension of syntax, as measured by the Token Test and by the Active–Passive Test. The three cases had in common the facts that (1) left-hemisphere damage occurred prior to language acquisition, (2) they had been free of seizures for the first seven years of life, and (3) none of them had had frontal lobe involvement.

Dennis and Whitaker (1977) reported on three cases of Sterge–Weber–Dimitri syndrome, which resulted in the need to perform a hemispherectomy shortly after birth in an attempt to halt the progression of the disease. Two of the children were left hemidecorticates (i.e., had their left cortex removed) and later developed relatively normal right hemisphere language. IQ testing at the ages of 9–10 years revealed subtle syntactic deficits. Both subjects failed on two of the "criteria" discussed by Kohn (1980). Of course, one should always bear in mind that any pathology which causes such drastic measures may well have caused important structural changes in the remaining cortical and subcortical structures.

Hemispherectomy in late childhood or adult life will have a serious effect on cognitive functions (Whitaker and Ojemann, 1977). Right hemispherectomy patients are usually characterized by impaired visuospatial functioning and have intact speech and language, while left hemispherectomy patients utter expletives and automatic speech and have very limited propositional speech (Searleman, 1977). The linguistic capacities of the minor hemisphere are severely limited to more or less automatic speech (Selnes, 1974).

Selnes (1974) suggests there are several questions in the hemispherectomy literature which need to be addressed when considering cases of "late" hemispherectomy (i.e., after infancy):

1. To what degree is language lateralized in these patients?
2. Does the language observed represent a residual skill from an earlier developmental stage? That is, is the language capacity of the minor hemisphere subject to decay through disuse, such that the return of language may represent the restoration of a vestigial function? Or does the minor hemisphere share overlearned or automatic language with the dominant hemisphere?
3. What is the effect of surgical trauma itself on the intact hemisphere?

Aphasia

As has been previously reported (Weinstein, 1964; Lesser, 1974), right-hemisphere damaged patients experience no difficulty discriminating the boundaries of phonetic categories. Research with left-hemisphere damaged aphasic patients has supported this observation. In a consonant–vowel (CV) discrimination task, aphasics were sensitive to the voice feature but showed a decreased awareness for place, whereas right-hemisphere damaged patients processed both voice and place. This may suggest that the intact right hemisphere can process voice (since the damaged left hemisphere cannot) and the intact left hemisphere can process both voice and place (Perecman and Kellar, 1981). However, it may be that the damaged left hemisphere is impaired in the ability to process the place feature (which requires more restructuring of the acoustic waveform than the voice feature) and it can, therefore, process the simpler feature (voice) with residual abilities (Oscar-Berman, Zurif, and Blumstein, 1975).

Four cases of bilateral damage to the third frontal gyrus (F3) due to infarction are reported by Levine and Mohr (1979). Each of the four patients had some oral or written language output (Case 1: monotone; used single words or short phrases appropriately; slightly impaired comprehension. Case 2: lack of stress; articulation difficulties; used one or two word phrases; normal comprehension. Case 3: absent speech production; one or two word written output; slightly impaired comprehension with infrequent semantic errors. Case 4: usually answered questions with single word phrases; comprehension moderately impaired). Language output in these cases was most often limited to single words, demonstrating that appropriate syntactical manipulations were obviously impaired, while semantically appropriate single word answers were available to the patients. The authors suggest that the central region of the dominant hemisphere (including the inferior $\frac{2}{3}$ of the peri-Rolandic cortex and Rolandic operculum) may be more important to speech than the third frontal gyrus. This was demonstrated in one case where an additional lesion in F3 of the minor hemisphere had no effect beyond the Broca's aphasia caused by left-hemisphere damage (Case 4), but a lesion of the inferior peri-Rolandic region caused total muteness (Case 3). This last point is interesting; however, it involves a patient (Case 4) who is "strongly left-handed," so it may be inappropriate to classify and compare his lesions and deficits to those experienced by the remaining three right-handers. Of relevance to the issues discussed in this chapter is the authors' assertion that (contrary to what some other researchers have suggested) the minor (right) hemisphere third frontal gyrus cannot be mediating language output since it, too, is damaged, suggesting that language recovery depends on preserved areas of the dominant hemisphere, superior or posterior to Broca's area.

COMMISSUROTOMY

A fair percentage of the recent popularization of laterality originates from studies of the human split-brain patient. A perhaps not-unexpected consequence of popularization is a failure to recognize the limitations of research on split-brain patients. While it would be foolhardy to attempt to set the record straight in the media, it may be possible to convince a few colleagues to view this research with a more skeptical eye.

It should first be pointed out that the split-brain research in the scientific literature is actually a series of case history studies and not controlled, statistically validated, psychological experimentation. The case history (single subject research) has a very distinct and important role in the neurobehavioral sciences. From the case history one answers the question ''Is it possible for such-and-such to occur?''; from the case history one does *not* answer the question ''Does everyone exhibit such-and-such?'' In spite of this, it is not hard to find statements in the literature such as that the right hemisphere is better at recognizing shapes if they are not nameable, or, that the right hemisphere is incapable of syntactic processing, based upon the case history analysis of, frequently, one or two split-brain subjects. In point of fact one might very well be able to make such claims about the left and right hemispheres, but only in the context of the larger body of neuropsychological research—lesion studies, neurosurgical cases, and dichotic–tachistoscopic studies of non-brain-damaged subjects. Some of the data obtained from the split-brain patients fit in with the body of neuropsychological research and some does not; the part that does not is probably accounted for by the major fallacy in split-brain research, the failure to take into account the history and locus of extracallosal brain damage. Let us examine this in detail.

The split-brain operation is a neurosurgical procedure designed to limit the spread of epileptic seizures across the corpus callosum, the major (but not only) connection between the left and right hemispheres. In the operation, the skull is opened, one hemisphere (usually the nondominant) is pulled aside to expose the corpus callosum; a large number of arteries and veins which run between the two hemispheres are coagulated; and the corpus callosum (and occasionally one or more other interhemispheric commissures) is divided almost totally. The operation itself causes some brain damage, inevitably. Pulling aside one hemisphere (retraction) bruises the hemisphere along its mesial (inside, or middle) surface; coagulating the bridging arteries and veins causes the death of the tissue supplied by those arteries and veins. Presumably, retrograde axonal degeneration of the callosal fibers causes some changes in both the left and right hemispheres. However, these are not the significant extracallosal brain damage. The lesions responsible for the epilepsy are the principal extracallosal brain damage.

Neurosurgical remediation for epilepsy is not undertaken lightly. The treatment of choice now (and certainly the treatment of choice for the last decade) is the use of anticonvulsant drugs. Only those epileptic patients who cannot be adequately treated by drugs are candidates for surgery. For over 40 years there have been two accepted neurosurgical procedures for intractable epilepsy: (1) commissurotomy (the split-brain procedure) and (2) lobectomy (the removal of the damaged area causing the epilepsy). The latter has been the operation of choice if the epileptic focus is lateralized and limited in size; that is, if it is amenable to surgical removal, and if it is in an area of the brain that can be sacrificed with minimal postoperative impairments. An epileptic focus in the middle of the language cortex would clearly not qualify; one in the temporal pole does. More to the point, an epileptic focus that is too large will not qualify for resection of the damaged brain in most cases (although occasionally such patients undergo hemispherectomy). These patients might qualify for commissurotomy.

It is reasonable to conclude that the patients who underwent the split-brain procedure (commissurotomy) either had very large lesions causing the epilepsy, or had lesions in areas of brain that could not be safely removed, or both. It is also reasonable to conclude that these patients suffered from seizures, that is, they had some kind of lesion, for quite some time prior to the operation. In point of fact, the patients reported in the literature had suffered epilepsy for many years, most of them from early childhood.

One of the interesting properties of the human brain is its so-called plasticity. Particularly in the younger person, one effect of major brain damage is to cause functions to shift to another area of brain. For example, early and extensive lesions in the left frontal brain areas are likely to cause the speech production mechanisms to shift to the right hemisphere. Lesions in the anterior portions of the temporal lobe may cause some temporal lobe functions to shift up into the parietal lobe of the same hemisphere. It is an inescapable fact that the brains of the split-brain patients had the maximum opportunity to reorganize functionally, and we may assume that such reorganization occurred. Their lesions were large and involved functionally important tissue; they were of long-standing, often dating from early childhood; and, most importantly, none of the split-brain patients reported in the literature had even moderate language impairments (aphasia).

Let us now characterize the split-brain patient and discuss that patient as a research subject. Logically, the brain damage causing the epilepsy was either in the right hemisphere (RH), or the left hemisphere (LH) or both. The latter situation did obtain in some of these patients; clearly, if both hemispheres are seriously damaged, there is no possibility of testing normal lateralized functions. In actual fact, the extent of the brain damage in these patients usually prevented them from being tested at all and there are few reports on them in the split-brain literature. The other two types of patients pose an interesting problem. If the brain damage was in the RH, under the conditions described above it is clear that some RH functions would have transferred to the LH; if the brain damage was in the LH, the opposite would obtain, and some LH functions would have transferred to the RH. The brain-damaged hemisphere clearly cannot function at maximum capacity: it is impaired. The non-brain-damaged hemisphere (with respect to the epileptic lesion; the surgery may have also damaged it) poses an interesting problem. It cannot function like a normal person's hemisphere because it has acquired some new and additional functions due to plasticity. These functions may or may not have displaced those functions which the hemisphere would have naturally mediated had there been no brain damage. In short, the non-brain-damaged hemisphere cannot be compared to the same hemisphere in a normal subject. To place this argument in a more concrete setting, let us consider language functions. The evidence is clear from many lines of research that language functions usually, and normally, are mediated by the LH. If the epileptic lesion is in the RH, it (the usually nondominant hemisphere) obviously cannot subserve language functions normally. Therefore, any evaluation of language functions in the RH will fall short of what an unimpaired right hemisphere could do. If this type of epileptic lesion is in the LH however (the usually dominant hemisphere), some language functions will probably have transferred over to the RH. Therefore, any evaluation of language functions in the RH will exaggerate, or overestimate, what the normal RH does. The obvious conclusion is that one cannot assess RH language functions in the split-brain patient. But the same line of reasoning applies to any higher cortical function, either in the LH or in the RH; the nature of the split-brain patient is such that it is inappropriate to use such a patient to investigate lateralized functions in the human brain. Well beyond overgeneralizations, based on case histories of split-brain patients, that the right, or the left, hemispheres do certain things is a deeper fallacy; neither the RHs nor the LHs of the split-brain patients are comparable to the respective hemispheres of normal subjects.

REGIONAL CEREBRAL BLOOD FLOW

Increased metabolic activity, caused by certain tasks selectively activating local regions of the cortex, has been measured by means of a relatively well-established technique known as regional cerebral blood flow (rCBF) (Risberg, Halsey, Wills, and Wilson, 1975; Wood, 1980; Meyer, Sakai, Yamaguchi, Yamamoto, and Shaw, 1980).

Typically, measurements are made with the subject at rest in order to establish a baseline, followed by another measurement at least one hour later, while the subject is performing some specified task. Two parameters are generally used: (1) F1, the flow of rapidly perfused gray matter and (2) the initial slope index (ISI) as a control for local CBF changes measured with F1. Normal right-handed males exhibit a pronounced hemispheric symmetry of rCBF values while at rest (Risberg, 1980), indictive of a functional coupling of homologous cortical projection areas in the two hemispheres in the absence of stimulation. Halsey, Blauenstein, Wilson, and Wills (1980) demonstrated relative symmetry as well, when normal subjects were involved in speaking spontaneously about simple, banal topics, with principal activation noted at inferior frontal locations bilaterally. It was suggested that such activation mainly reflects bilateral cortical control for articulatory motor responses. However, hemispheric asymmetry may be obscured if factors such as sex, handedness, and individual differences in cognitive strategies are ignored (Gur and Reivich, 1980).

Localized flow changes during task-oriented behavior are indicative of changing metabolic activity and, by extrapolation, functional activation of neurons. Cognitive processing of verbal analogy tests causes larger flow increases on the left side, especially over Wernicke's area. A nonverbal perceptual task results in a flow increase in right-hemisphere frontal and parietal regions relative to the generalized bilateral flow increase due to cerebral activation (Risberg, Halsey, Wills, and Wilson, 1975). In a monaural verbal stimulation task, Maximilian (1981) found significant flow increases in left-hemisphere temporoparietal areas (compared to the analogous right-hemisphere area) regardless of which ear was receiving the stimuli, but also found significant flow increases in right-hemisphere frontotemporal regions during left-ear stimulation. He attributed the right-hemisphere activity to linguistic processing before transfer across the corpus callosum to the left hemisphere. Gur and Reivich (1980) have found similar left-hemisphere activation during a verbal task in 36 right-handed males, but a spatial (gestalt completion) test failed to produce a reliable increase in right-hemisphere flow. Apparently the spatial task was of the sort that could be solved by either left-hemisphere or right-hemisphere strategy, although better performance was associated with right-hemisphere activation. Meyer, Sakai, Yamaguchi, Yamamoto, and Shaw (1980) have identified two types of cerebral activation processes that are characterized by distinctive rCBF patterns:

1. Localized flow increases in cortical regions related to specific aspects of the mental task, such as that which Risberg, Halsey, Wills, and Wilson (1975) demonstrated in Wernicke's area.
2. A generalized activation of function necessary for conceptualization and performance of the task, including the brain stem–cerebellar region. This activation is probably mediated by the reticular activation system.

It has been suggested that this technique can provide clues to the reorganization of brain function following cerebral damage. Halsey, Blauenstein, Wilson, and Wills (1980) found that brain-damaged patients who attempted a task which was very difficult or impossible demonstrated a relatively bizarre activation pattern. Patients who attempted a task which could be accomplished with moderate or great difficulty demonstrated maximum focal activation elsewhere in the hemisphere than normally expected, or in the opposite hemisphere. Measurements of rCBF during a psychophysiological task given to a group of eight right-handed patients (with left-hemisphere damage who showed good recovery of speech within three months of their stroke) revealed a tendency for increased blood flow in most areas of both the LH and RH (Meyer, Sakai, Yamaguchi, Yamamoto, and Shaw, 1980). One of the few areas which did not show a flow increase was Broca's area. There were, however, sizeable increases in corresponding areas (posterior frontal and sylvian areas) of the RH, suggesting a possible transfer of motor speech functions from the disabled LH to the intact RH.

NEUROSURGICAL STUDIES

Kolb and Taylor (1981) studied 58 patients who had undergone a unilateral cortical excision for the relief of epilepsy; excisions were from the frontal, temporal, and parieto-occipital regions, in both the LH and the RH. One test was to match photographs from *Life* magazine with a standardized set of emotionally categorizable faces; the second test was to match the verbal categories of emotions expressed by the standardized set of faces to a sentence describing an event illustrated in a photograph from *Life* magazine. Kolb and Taylor found that right-hemisphere excisions impaired the photograph-matching test, regardless of lesion site, and that left-hemisphere excisions impaired the sentence-matching test, regardless of lesion site. In addition, they discovered that patients with right frontal lobe excisions frequently interrupted the testing session by spontaneously talking, whereas patients with left frontal excisions rarely did so. This finding contrasts with a previous study from the same group in which it was shown that patients with right or left frontal excisions exhibited a marked reduction in spontaneous facial expressions as compared to patients with excisions in temporal or parieto-occipital excisions. Since, at the moment, we do not have a linguistic–phonetic analysis of the prosodic features of the sentence-matching task, it is not really possible to determine whether Kolb and Taylor's (1981) findings are counterexamples to Tucker, Watson, and Heilman's (1977) or to Ross's (1981) classification of aprosodias.

One of the most direct methods of investigating the role of the cerebral cortex in language processing is the technique of electrical stimulation of the brain (ESB) during epilepsy surgery, pioneered by Wilder Penfield and his colleagues. Ojemann and Whitaker (1978a) presented 11 cases studied with ESB, including two in which the RH was studied. The advantages of ESB are that (1) the lesion locus (the epileptogenic focus) is relatively well demarcated, which means that much of the ESB mapping is done on normally healthy cortical tissue; (2) the ESB mapping itself is brief and thus does not allow the brain time to reorganize or adapt, and (3) the ESB mapping is very localized, typically subtending an area of half a centimeter in diameter. Although it is the case that subjects in these studies have brain damage, there is little reason to believe that left temporal lobe epilepsy causes language to shift to the RH, nor, for that matter, to believe that right temporal lobe epilepsy prevents language from locating in the RH. This is because all combinations of epileptic loci and language dominance are found in these subjects (right–right, right–left, left–left, and left–right), yet the distribution of language lateralization in these subjects is the same as in the general population as determined from aphasia research. Note the difference between these subjects and the commissurotomy subjects: the latter have extensive, nonoperable lesions that typically involve large areas of one hemisphere and sometimes both hemispheres. As has been shown repeatedly, the ESB mapping correlates quite well with the results of the sodium amytal test for language dominance (the Wada Test).

Of the two right-hemisphere patients reported by Ojemann and Whitaker (1978a), the first was right-hemisphere dominant for language and the second was not, as shown by Wada testing. The ESB produced an interference with naming in the first; there were no so-called 100% sites found (i.e., sites where stimulation caused the same result 100% of the

time), although this could have been due to the fact that clinical considerations limited the number of sites that could be sampled. The most interesting case is the second, a right-handed patient whose language dominance was clearly in the (typical) LH. Right-hemisphere ESB over 26
cortical sites produced no evidence of naming disturbance, rather clearly
indicating a minimal role in language processing by the RH when the LH
is dominant. Another case published by Ojemann and Whitaker (1978b)
involved a left-handed bilingual patient whose language was represented
in the RH and whose epileptic focus was also in the RH. In this case the
usual pattern of naming errors from ESB was demonstrated—for both
languages—although the two languages only partially overlapped across
the cortex. One other case, unpublished, involved a right-hemisphere operation. The patient was right-handed with normal left-hemisphere dominance for language. Ojemann mapped five different language and praxic
functions in the RH by ESB: naming, short-term verbal memory, reading,
sequential oral movements (by imitating a picture), and phoneme perception. ESB did not interfere with any of the functions in any cortical location—frontal, temporal, or parieto-occipital. Thus, there are two cases of
right-hemisphere language, left-handedness, and a right-hemisphere epileptic focus; these patients show the usual pattern of naming errors on
ESB. Two other cases of normal left-hemisphere language dominance,
right-handedness and a right-hemisphere epileptic focus showed no ESB
interference on a variety of language tasks. These latter two cases clearly
indicate that language lateralization can be quite complete, or, in other
words, they show that in some cases the RH does not substantially contribute to language processing (as measured by these tests, of course).

Andy and Bhatnagar (1982) reported electrical stimulation data for
three patients undergoing RH temporal lobe resections for medically intractable epilepsy. Confrontation naming was tested with standard black-
and-white slides of objects, which had a carrier phrase on top, in the manner of the Ojemann and Whitaker (1978a) studies (ESB was conducted
during slide presentation). The second task required responsive naming.
The patient first read functional information such as *What do we drive?* or
What do we cut with?, and then the patient was asked to name the object
being referred to obliquely. The third task required verbal completion.
The patient read a common phrase such as *Two and two are___*, or *Roses
are red___*. The patient was then asked to complete the phrase. Andy and
Bhatnagar noted both omission and misnaming–substitution errors during
RH stimulation, although they had found no 100% sites in the RH as Ojemann and Whitaker had identified during LH stimulations. In fact, patients in the Andy and Bhatnagar study typically made 50% or less naming
errors under ESB. In this preliminary version of their study, there is no

reference to control trials or error rates, nor is there any analysis of naming errors in terms of the three tasks they employed. As a result, their conclusions—that the RH has some language (naming) ability when the LH is dominant for language—must be considered tentative, particularly in view of the fact that their results directly contradict the Ojemann and Whitaker (1978a) study.

SODIUM AMYTAL (WADA TEST)

In preparation for craniotomies to be performed on conscious patients, as reported by Ojemann and Whitaker (1978a) and Andy and Bhatnagar (1982), it is typical to test the patient to determine which hemisphere mediates language. This is accomplished by the injection of sodium amytal, a fast-acting barbituate, into the internal carotid artery of one and then the other hemisphere, during which time language is tested. A number of patients have been assessed in this manner, and, as is the case with previously published data (Rasmussen and Milner, 1977), most patients' language functions are located in the LH. In over three dozen patients, we have seen only one example of bilateral language representation, and a few cases of unilateral right-hemisphere language. One of the latter cases was a bilingual patient (English and Spanish) who was left-handed. Both languages in this patient were located in the RH (Ojemann and Whitaker, 1978a). The same pattern has been observed in other bilingual cases: both or all languages are mediated by one hemisphere. To date eight bilingual patients have been studied (Rapport, Tan, and Whitaker, in press) with the amytal procedure; in all eight patients, all languages were in one hemisphere, in the LH in seven and in the RH in one. Language testing done during Wada Tests has been more restricted than that done during electrical stimulation of the brain; usually only naming has been assessed. Nonetheless, since naming is a relatively ubiquitous (spatially) language function, these studies have not lent much support to the view that the nondominant hemisphere has some language functions.

CONCLUSION

The present evidence for right-hemisphere language processing, when the left hemisphere is dominant for language, is equivocal. Certain facts are well established: the RH is capable of mediating language if the LH is damaged early in life. The evidence, such as from Dennis and Whitaker (1977), indicates that the RH is innately less successful than the LH as the language hemisphere, particularly in syntactic processing. It would appear that fairly extensive early LH damage must be sustained before language will switch to the RH, and even then there is the possibility of bilateral language representation, as in the case presented in Ojemann and Whitaker (1978a).

It is particularly·challenging to compare claims that RH lesions affect syntactic processing (Eisenson, 1962; Hier and Kaplan, 1981) with claims that the RH is not equal to the LH in syntactic processing ability (Dennis and Whitaker, 1977). Many factors must be considered in this discrepancy: developmental factors; differences between an early hemispherectomy and late focal lesions; or as-yet-unspecified inhibitory relationships between the hemispheres. Which particular factor or combination of factors remains to be seen. It certainly poses a challenging problem.

There is good evidence that the RH contributes to many factors of human behavior that play some role in communication. Affective behavior, especially as reflected in the appreciation of humor, imagery, visuospatial processing, and some aspects of attentional mechanisms, all have a RH contribution. Patients with denial of illness, or neglect, may select or use vocabulary items consistent with the behavioral impairment but, not surprisingly, inconsistent with normal word usage. In this sense, RH-lesioned patients may exhibit language that is more like that in a cognitive –affective disorder than in a linguistic disorder.

Once again one is impressed with the complexity and multidimensionality of language and communication. Whether the RH contributes to semantic processing, beyond imaginal–affective–metaphorical or humorous language use, is open to speculation. Eisenson's (1962) study has some faults, but, to our knowledge no direct replication has been attempted. Lesser (1974) and Gainotti, Caltagirone and Miceli (1979) and Gainotti, Caltagirone, Miceli and Masullo (1981) have data that support a semantic–vocabulary contribution of the RH, although as Gainotti et al. have shown, a general mental deterioration factor could play an important role in these data. Andy and Bhatnagar's (1982) data are particularly interesting in this regard, since the vocabulary items in their tests were all high-frequency, concrete, picturable words.

There is little doubt that enough data have accumulated to challenge the simple view that the LH is the language hemisphere and the RH does something else. The past two decades of research on RH language processing have, at the very least, produced some clear hypotheses, partially supported by "pilot studies." What seems to be called for now are some carefully designed studies that can separate out the linguistic features of language from the imaginal, affective, and cognitive features. The outcome will certainly be a more comprehensive understanding of the neurological substrates of communication.

REFERENCES

Andy, O. J., and Bhatnagar, S. (1982). Language in the right hemisphere. Paper presented at the International Neuropsychological Society Convention, Pittsburgh, Pa.

Archibald, Y. M. and Wepman, J. M. (1968). Language disturbance and nonverbal cognitive performance in eight patients following injury to the right hemisphere. *Brain, 91,* 117–127.

Caramazza, A., Gordon, J., Zurif, E. B., and DeLuca, D. (1976). Right hemisphere damage and verbal problem-solving behavior. *Brain and Language, 3,* 41–46.

Carmon, A., Gordon, H. W., Bental, E., and Harness, B. Z. (1977). Retraining in literal alexia: Substitution of a right hemisphere perceptual strategy for impaired left hemisphere processing. *Bulletin of the Los Angeles Neurological Societies, 42,* 41–50.

Czopf, J. (1972). Über die Rolle der nicht dominanten Hemisphäre in der Restitution der Sprache der Asphasischen. *Archiv fur Psychiatrie Nervenkranken, 216,* 162–171.

Dennis, M., and Whitaker, H. A. (1977). Hemispheric equipotentiality and language acquisition. In S. J. Segalowitz and F. Gruber (Eds.), *Language development and neurological theory.* New York: Academic Press.

Eisenson, J. (1962). Language and intellectual modifications associated with right cerebral damage. *Language and Speech, 5,* 49–53.

Fried, I., Mateer, C., Ojemann, G., Wohns, R., and Fedio, P. (1982). Organization of visuospatial functions in human cortex: Evidence from electrical stimulation. *Brain, 105*(2), 349–371.

Friedland, R. P., and Weinstein, E. A. (1977). Hemi-inattention and hemisphere specialization: Introduction and historical review. In E. A. Weinstein and R. P. Friedland (Eds.), *Advances in neurology,* Vol. 18. New York: Raven Press.

Gainotti, C., Caltagirone, C., and Miceli, G. (1979). Semantic disorders of auditory language comprehension in right brain-damaged patients. Journal of Psycholinguistic Research, *8*(1), 13–20.

Gainotti, G., Caltagirone, C., Miceli, G., and Masullo, C. (1981). Selective semantic-lexical impairment of language comprehension in right brain-damaged patients. *Brain and Language, 13,* 201–211.

Gardner, H., Brownell, H. H., Wapner, W., and Michelow, D. (1981). *Missing the point: The role of the right hemisphere in the processing of complex linguistic materials.* Unpublished manuscript.

Gur, R. C., and Reivich, M. (1980). Cognitive task effects on hemispheric blood flow in humans: Evidence for individual differences in hemispheric activation. *Brain and Language, 9,* 78–92.

Halsey, J. H., Jr., Blauenstein, U. W., Wilson, E. M., and Wills, E. L. (1980). Brain activation in the presence of brain damage. *Brain and Language, 9,* 47–60.

Heilman, K. M., Scholes, R., and Watson, R. T. (1975). Auditory affective agnosia. *Journal of Neurology, Neurosurgery, and Psychiatry, 38,* 69–72.

Heilman, K. M. and Valenstein, E. (1979). Mechanisms underlying hemispatial neglect. *Annals of Neurology, 5,* 166–170.

Heilman, K. M., and Van Den Abell, T. (1979). Right hemisphere dominance for mediating cerebral activation. *Neuropsychologia, 17,* 315–321.

Hier, D. B., and Kaplan, J. (1980). Verbal comprehension deficits after right hemisphere damage. *Applied Psycholinguistics, 1,* 279–294.

Jones-Gotman, M. K., and Milner, B. (1978). Right temporal lobe contribution to image-mediated verbal learning. *Neuropsychologia, 16,* 61–71.

Kohn, B. (1980). Right hemisphere speech representation and comprehension of syntax after left cerebral injury. *Brain and Language, 9,* 350–361.

Kolb, G., and Taylor, L. (1981). Affective behavior in patients with localized cortical excisions: Role of lesion site and side. *Science, 214,* 89–90.

Lesser, R. (1974). Verbal comprehension in aphasia: An English version of three Italian tests. *Cortex, 10,* 247–263.

Levine, D. N., and Mohr, J. P. (1979). Language after bilateral cerebral infarctions: Role of the minor hemisphere in speech. *Neurology, 29,* 927–938.

Maximilian, V. A. (1981). Cortical blood flow asymmetries during monaural verbal stimulation. *Brain and Language, 15,* 1–11.

Meyer, J. S., Sakai, F., Yamaguchi, F., Yamamoto, M., and Shaw, T. (1980). Regional changes in cerebral blood flow during standard behavioral activation in patients with disorders of speech and mentation compared to normal volunteers. *Brain and Language, 9,* 61–77.

Milberg, W., Cummings, J., Goodglass, H., and Kaplan, E. (1979). Case report: A global sequential processing disorder following head injury; A possible role for the right hemisphere in serial order behavior. *Journal of Clinical Neuropsychology, 1,* 213–225.

Ojemann, G., and Whitaker, H. A. (1978a). Language localization and variability. *Brain and Language, 6,* 239–260.

Ojemann, G., and Whitaker, H. A. (1978b). The bilingual brain. *Archives of Neurology, 35,* 409–412.

Oscar-Berman, M., Zurif, E. B., and Blumstein, S. (1975). Effects of unilateral brain damage on the processing of speech sounds. *Brain and Language, 2,* 345–355.

Perecman, E., and Kellar, L. (1981). The effect of voice and place among aphasic, nonaphasic right-damaged, and normal subjects on a metalinguistic task. *Brain and Language, 12,* 213–223.

Rapport, R., Tan, C. T., and Whitaker, H. A. (in press). Language functions and dysfunction among Chinese and English speaking polyglots. *Brain and Language,*

Rasmussen, T., and Milner, B. (1977). The role of early left-brain injury in determining lateralization of cerebral functions. In S. J. Dimond and D. A. Blizard (Eds.), *Evolution and lateralization of the brain. Annals of the New York Academy of Sciences, 299,*

Rausch, R. (1981). Lateralization of temporal lobe dysfunction and verbal encoding. *Brain and Language, 12,* 92–100.

Read, D. E. (1981). Solving deductive reasoning problems after unilateral temporal lobectomy. *Brain and Language, 12,* 116–127.

Risberg, J. (1980). Regional cerebral blood flow measurements by ^{133}Xe-Inhalation: Methodology and applications in neuropsychology and psychiatry. *Brain and Language, 9,* 9–34.

Risberg, J., Halsey, J. H., Wills, E. L., and Wilson, E. M. (1975). Hemispheric specialization in normal man studied by bilateral measurements of the regional cerebral blood flow—a study with the 133-Xe inhalation technique. *Brain, 98,* 511–524.

Ross, E. D. (1981). The aprosodias. *Archives of Neurology, 38,* 561–569.

Ross, E. D., and Mesulam, M. M. (1979). Dominant language functions of the right hemisphere? *Archives of Neurology, 36,* 144–148.

Searleman, A. (1977). A review of right hemisphere linguistic capacities. *Psychological Bulletin, 84* (3), 503–528.

Selnes, O. A. (1974). *Language functions of the nondominant hemisphere.* Unpublished manuscript.

Terzian, H. (1964). Behavioral and EEG effects of intracarotid sodium amytal injection. *Acta Neurochirurgica, 12,* 230–240.

Tucker, D. M., Watson, R. G., and Heilman, K. M. (1977). Discrimination and evocation of affectively intoned speech in patients with right parietal disease. *Neurology, 27,* 947–950.

Wapner, W., Hamby, S., and Gardner, H. (1981). The role of the right hemisphere in the apprehension of complex linguistic materials. *Brain and Language, 14,* 15–33.

Weinstein, E. A. (1964). Affections of speech with lesions of the non-dominant hemisphere. Res. Pub. Ass. Res. Nerv. Mental Dis., *42,* 220–225.

Weinstein, E. A., and Friedland, R. P. (1977). Behavioral disorders associated with hemi-inattention. In E. A. Weinstein and R. P. Friedland (Eds.), *Advances in neurology,* Vol. 18. New York: Raven Press.

Whitaker, H. A., and Ojemann, G. (1977). Lateralization of higher cortical functions: A critique. In S. J. Dimond and D. A. Blizard (Eds.), *Evolution and lateralization of the brain. Annals of the New York Academy of Sciences, 299,* 459–473.

Winner, E., and Gardner, H. (1977). The comprehension of metaphor in brain damaged patients. *Brain, 100,* 717–729.

Wood, F. (1980). Theoretical, methodological, and statistical implications of the inhalation rCBF technique for the study of brain–behavior relationships. *Brain and Language, 9,* 1–8.

Zaidel, E. (1977). Unilateral auditory language comprehension on the Token Test following cerebral commissurotomy and hemispherectomy. *Neuropsychologia, 15,* 1–18.

Zangwill, O. L. (1967). Speech and the minor hemisphere. *Acta Neurologica et Psychiatrica Belgica, 67,* 1013–1020.

PART II

What Should Be the Brain Base for Language?

INTRODUCTION

The question posed by the title leads to the core of the dilemmas of neuropsychology in general and neurolinguistics in particular. What parts of the brain are most involved in language processing? What type of neurophysiological unit is appropriate for the matching? What is it that constitutes evidence for the involvement of particular neurophysiological tissue? Despite occasional philosophical inadequacies of scientific neuropsychology (Uttal, 1978, Ch. 5), useful advances have been achieved. For example, electrical stimulation of the brain has provided some specific information concerning good places to start the search (Penfield and Roberts, 1958; Ojemann and Whitaker, 1978). Studies of tissue damage in aphasia reinforce these data (Kertesz, Lesk, and McCabe, 1977) as do experiments with normal subjects (Bryden, 1982; Desmedt, 1977).

In this section, Mateer and Ojemann marshal evidence for left thalamic involvement in language processing, and Mateer considers the evidence for a basis in praxis for the asymmetric cortical representation of language. The functional basis for the involvement of brain tissue in each case can, of course, vary. Thus, the task in all the studies reviewed is complicated by the difficulty of knowing what to look for, as well as what to look at. Tissue asymmetries per se in the cerebral cortex across hemispheres naturally form a place to continue the search (see Witelson, this volume), yet we still lack conclusive evidence that any such asymmetries reflect functional asymmetries.

These difficulties, though by no means making progress impossible, are reflected in an examination of the logical requirements for a neurolinguistic theory, that is, for the placement of language knowledge, experience, and production in a brain model. It is very premature to evaluate critically along these lines the localization issues with respect to language. However, an outline of such a critical appraisal of a neuropsychological theory of visual experience is given in this part by Anderson and Leong.

REFERENCES

Bryden, M. P. (1982). *Laterality: Functional asymmetry in the intact brain.* New York: Academic Press.

Desmedt, J. E. (Ed.), (1977). *Recent developments in the psychobiology of language: The cerebral evoked potential approach.* London: Oxford Univ. Press.

Kertesz, A., Lesk, D., and McCabe, P. (1977). Isotope localization of infarcts in aphasia. *Archives of Neurology, 34,* 590–601.

Ojemann, G. A., and Whitaker, H. A. (1978). Language localization and variability. *Brain and Language, 6,* 239–260.

Penfield, W., and Roberts, L. (1959). *Speech and brain mechanisms.* Princeton, N.J.: Princeton Univ. Press.

Uttal, W. R. (1978). *The psychobiology of mind.* Hillsdale, N.J.: Lawrence Erlbaum Associates.

CHAPTER 5

Bumps on the Brain: Right–Left Anatomic Asymmetry as a Key to Functional Lateralization*

Sandra F. Witelson

INTRODUCTION

Functional differentiation of the cerebral cortex clearly exists. The localization of different cognitive functions in the anteroposterior plane of the human brain, first noted in the nineteenth century and exemplified by the different functions of the frontal and posterior language regions, has been easy to conceptualize in anatomical terms. Such regions are situated differently in the brain and accordingly are adjacent to different primary cortical regions and at the least must have different intercortical connections. Moreover, cytoarchitectonic studies (e.g., Brodmann, 1908; Campbell, 1905; von Economo and Koskinas, 1925) subsequently showed that different regions within a hemisphere involve different types of cortex,

* Preparation of this paper and some of the research reported within were supported in part by U.S. NINCDS Contract #N01-NS-6-2344 and Ontario Mental Health Foundation Grant #803. Some of the results were presented at the 13th Annual Winter Conference on Brain Research, Keystone, Colorado, Jan. 1980 and at the 19th Annual meeting of the Academy of Aphasia, London, Ont., Oct. 1981.

which vary in laminar and cellular parameters. Consequently, this dimension of functional differentiation of the cortex was readily conceptualized as having neuroanatomical underpinnings.

Functional differentiation of right and left cortex, however, although documented prior to the anteroposterior specialization, has been more difficult to conceptualize anatomically. The bisection of the brain along the anteroposterior axis results in right and left homologous halves. Thus, any particular region and its intercortical connections are similar for each side. Moreover, the cytoarchitectonic studies did not report any right–left differences in types of histological regions. In fact, such differences were not investigated and still remain to be done. Possible anatomical substrates of lateral differentiation of the cortex remained an enigma for a long time. However, recent neuroanatomical discoveries and developments point to some possible morphological substrates.

This chapter will include a description and a review of some of the consistently observed right–left neuroanatomical asymmetries. The question will also be addressed whether these anatomical asymmetries are in fact a morphological substrate of cerebral dominance. These issues will be considered in conjunction with the implications that such neuroanatomical substrates may have for elucidating aspects of functional specialization and mechanisms of brain–behavior relationships. The optimistic view will be presented here that the study of variation in neuroanatomical asymmetry may not only reveal a possible biological basis of cerebral dominance, but may also serve to unlock some currently unresolved issues in functional asymmetry.

NATURE OF HEMISPHERE SPECIALIZATION

It appears useful to discuss briefly the nature of hemisphere functional specialization. The existence of the phenomenon of functional lateralization is obvious. However, the precise nature of just what cognitive processes are lateralized remains elusive. The extensive research in this area in the past few decades has brought current conceptualizations of functional specialization a long way from the initial description of cerebral dominance as involving only the left hemisphere and a silent right hemisphere and, subsequently, as a dichotomy described simply as "verbal versus nonverbal" or "auditory versus visual."

The unique functions of the left hemisphere and the functional differences between the hemispheres are still considered to be closely related to language functions, but somewhat less directly than previously thought. A reasonable working hypothesis, given the information available to date, is that the hemispheres are specialized for two different types or modes of information processing, and it is the type of processing required by the task and/or chosen by the individual that is the determining factor in lateralization, and not the type of stimulus or test involved. The left hemisphere appears specialized, that is, more involved or superior, in processing information by analyzing the stimuli as discrete items in reference to their temporal arrangement. In this scheme, linguistic skills are conceptualized as being mediated predominantly by the left hemisphere because they are heavily dependent on this analytic–temporal mode of information processing. The programming for the production of sequences of voluntary movements (praxis) which Liepmann (1908) and more recently Geschwind (1975) and Kimura (1976) have indicated to be mainly dependent on the left hemisphere, may also be conceptualized within this scheme of left-hemisphere specialization for processing items in series.

The right hemisphere is now also clearly documented to be dominant or specialized in its own right, and not just the "nonverbal" hemisphere. Its relatively specialized function may be described as one in which information is processed so that stimuli are synthesized or unified into a holistic percept and in which the temporal aspects of the stimuli are superceded. The perception of spatial relationships, regardless of the sensory modality involved, appears to depend mainly on this type of cognitive processing.

This description of hemisphere functional specialization has the advantage that it not only accounts for most of the empirical data of functional lateralization in normal and brain-damaged adults, but it is also compatible with the existence of hemisphere specialization which has been demonstrated in the young preverbal child (Witelson, 1977a, 1982) and to a lesser extent in nonhuman species (e.g., Harnad, Doty, Goldstein, Jaynes, and Krauthamer, 1977; Witelson, 1977b).

UNRESOLVED ISSUES

As the general pattern and nature of cerebral dominance became clearer, other complicating factors emerged, such as individual differences in cerebral laterality. It has long been known that there is variation in the direction of functional lateralization, as, for example, in right-handers versus left-handers. It has also become evident that there is interindividual variation in the degree of laterality. Again, using hand-preference groups as an example, left-handers appear to be a heterogeneous group. While the majority seem to have an overall pattern similar to right-handers, some appear to have a reversed pattern of asymmetry (e.g.,

Satz, 1980; Witelson, 1980). However, some left-handers, possibly those having a history of familial sinistrality (Hécaen and Sauguet, 1971; Hécaen, De Agostini and Monzon-Montes, 1981) or those with inverted hand posture in writing (Levy and Reid, 1976), appear to have less lateralization of function, that is, a greater degree of bihemispheric representation of function compared to other left-handers and to right-handed individuals.

With bihemispheric representation of a function, the issue is further complicated by the possibility of two mechanisms. Does "bihemispheric" representation involve duplicated representation of a cognitive function in both hemispheres (a "twin" situation), or does it involve divided representation of a cognitive function with some aspects lateralized to one hemisphere and other aspects within the same general mode represented in the other hemisphere (a "subdivision of labor" situation)? The latter possibility is an example of intraindividual variation discussed later in this chapter.

Gender may also be a factor in individual differences in hemisphere specialization. Both males and females show the same direction of laterality, but some studies suggest that the degree of functional uniqueness of each hemisphere may be less in females, as indicated in studies of normal individuals (e.g., Witelson, 1976) and brain-damaged patients (McGlone, 1980). Atypical brain organization also appears to be associated with atypical patterns of cerebral dominance as in the case of Turner's Syndrome (XO) (e.g., Swallow, 1980; Waber, 1979). In contrast to handedness differences, the sex difference in lateralization appears to be one only of degree, not of direction; and accordingly may have a different biological basis, perhaps neurochemical rather than morphological (Witelson, 1980). Some cognitive disorders, such as development dyslexia, may have as one neurological factor an atypical pattern of lateralization of function, such as bihemispheric representation of "holistic" functions (Witelson, 1977c).

The picture becomes more complicated with the study of homogeneous groups of normal individuals. Even among strongly right-handed normal males, experimental data raise the possibility of some variation in cerebral lateralization. Approximately 10% of such samples may not show an expected perceptual asymmetry, for example, a clear right-ear superiority on linguistic dichotic tasks. The usual interpretation of such results is either that they reflect different cognitive strategies used by these subjects on this test, or that they reflect the imperfectness of this lateral perceptual task as an index of the dominant speech hemisphere. In the former case, different strategies may have either an experiential or biological basis and this does not resolve the question of possible variation in brain organization. In the latter case, if the perceptual tests do misclassify some individuals, then the question remains as to how one decides when to interpret the perceptual asymmetry as valid, as is done in the case of varying results for hand-preference groups, and when to question observed perceptual asymmetry.

The hypothesis raised here is whether some right-handers may have a different pattern of brain organization than the majority of dextrals. Data from other lines of research (to be indicated later) support such a possibility. The hypothesis emerges that right-handers may be phenotypically similar, but may be a heterogeneous group in relation to functional lateralization and genotypically heterogeneous in this respect. This situation would be consistent with Annett's (1972) model of the inheritance of handedness, in which it is proposed that individuals not having the genetic factor for right-hand preference may manifest either right- or left-hand preference.

The elucidation of functional lateralization is further complicated by intraindividual variation in laterality. The implicit assumption exists that the pattern of lateralization for one cognitive skill holds for all related skills which are considered to require the same mode of information processing. This may be an oversimplication. For example, Rasmussen and Milner (1977) have reported some cases in which speech production and language comprehension were not subserved by the same hemisphere (albeit after early brain damage and possible neural reorganization).

Another example of possible intraindividual variation concerns the imperfect correlation between the lateralization of aphasic and apraxic symptoms. Aphasia is not a good predictor of the occurrence of apraxia, particularly in left-handers. As indicated above, praxis (or motor learning and programming) appears to be mainly dependent on the left hemisphere, as is language. This appears to be the case in right-handers. However, in some left-handers it may be the case that the right hemisphere plays the major role in motor learning and praxis, when it is not the major hemisphere for speech and language functions (Heilman, Coyle, Gonyea, and Geschwind, 1973; Geschwind, 1975).

In addition, correlations between perceptual asymmetry on different indices of hemisphere specialization supposedly based on similar cognitive functions have been found to be far from perfect (e.g., Bryden, 1965; Witelson, 1977d). For example, the correlation obtained between ear and visual field asymmetry on tasks involving verbal processing is low.

Is the low correlation a result of low validity of the tests, is cerebral dominance just one factor affecting performance on these tests, or may different subskills within one type of information processing be lateralized differently in some individuals?

It is clear that there are some individual differences in patterns of hemisphere specialization. The questions follow: Which are real? How may they be validly measured? What is the basis for such variation? Is there a genetically determined biological basis underlying the individual differences?

LIMITATIONS OF NEUROCOGNITIVE METHODS

It is difficult to ascertain precisely the apparent, different patterns of cerebral dominance. The current methods of studying lateralization of function provide some information, but by their nature, cannot provide a complete picture. The study of patients with well-defined cortical lesions or ablations and associated cognitive deficits can indicate clearly what role that damaged region subserves, or more precisely, what the brain can and cannot do without that region. But such studies cannot yield the negative data, that is, whether similar cognitive deficits would be caused by damage to homologous regions in the other hemisphere. Thus the possibility of some degree of bihemispheric representation is difficult to rule out with the study of brain-damaged patients. Examination of severity of deficit and extent of recovery can provide only hypotheses about the other hemisphere's previous or current role in mediating the cognitive function under study.

The study of patients with interhemispheric commissurotomy can provide information about the level of each type of cognitive function of which each separated hemisphere is capable (e.g., Zaidel, 1976). However, a hemisphere's capacity when removed from the influence of the other hemisphere may give an unnatural picture. Furthermore, in split-brain patients, there usually is a history of early brain damage underlying the epilepsy, and thus the possible associated effects of structural plasticity and resultant functional reorganization on lateralization of function. In most split-brain patients, this is a confounding variable in the determination of subsequent direction and degree of the pattern of lateralization.

The use of hand preference as the independent index of cerebral dominance is greatly limited by its known heterogeneity in brain organization, particularly with sinistrals, and also by the ambiguity of whether the observed hand preference was modified by environmental influences.

Research using the various lateral perceptual tasks devised in different modalities (e.g., dichotic, dichoptic, dichhaptic) has yielded the clearest data relevant to the issue of patterns and degree of cerebral laterality (e.g., Levy and Reid, 1978). However, there are some limitations here too. The tasks may well include some confounding variables which affect the behavioral asymmetry, for example, directional scanning in visual tasks, praxis in dichhaptic tests. The most valid score needed to reflect the magnitude of right–left perceptual asymmetry is a difficult issue to resolve (e.g., Bryden and Sprott, 1981). Moreover, as indicated above, there is the major problem in ascertaining the validity of the tests as indices of functional lateralization. When the expected lateral field effect is not obtained in every individual of a sample studied, it has been nearly impossible to discern whether the tasks involve unknown cognitive factors affecting behavior, whether these individuals are using different cognitive strategies, whether the subjects have different experiential biases, or whether they are indeed different biologically. It is hoped that some of the new technological advances, including cerebral blood flow measurement and positron emission tomography in conjunction with neurocognitive performance, may help elucidate some of these issues. The study of neuroanatomical asymmetry also may provide a key to some of these issues.

Neuroanatomical asymmetry is a new aspect of study in hemisphere specialization. The field was reopened by Geschwind and Levitsky (1968) when they reported that a gross right–left morphological asymmetry is present in the size of a cortical association area in the temporal lobes. This work stimulated much research, which in turn has raised many questions and opened up new avenues of study.

There are two main issues associated with neuroanatomical asymmetry relevant to the present discussion. The first is whether the documented neuroanatomical asymmetry is a morphological substrate for the functional uniqueness of each hemisphere. If so, is this the case at a gross anatomical and/or histological level? The second concerns the possibility that the study of neuroanatomical asymmetry in relation to functional asymmetry may help elucidate some of the issues of functional lateralization raised above, such as individual differences, the validity of perceptual indices of brain lateralization, and the genetics of laterality.

NEUROANATOMICAL ASYMMETRY

In respect to the first issue, I shall briefly review the current evidence concerning various morphological asymmetries between the hemispheres and how they may relate to cerebral dominance. Consider first the temporal plane or planum temporale (the posterior region of the superior surface of the temporal lobe), as it has been the most extensively studied area to date, and involves an area whose role in cognition is relatively well documented.

Right–left asymmetries on the lateral surface of the human brain in the perisylvian region were beginning to be documented before the turn of the century by anatomists such as Cunningham (1892), Eberstaller (1890), and later by Shellshear (1937) and Connolly (1950). These authors noted right–left differences in the length and angulation of the Sylvian (lateral) fissure, with the left being longer and more horizontal in direction than the right (see Geschwind, 1974; Rubens, Mahowald, and Hutton, 1976; Witelson, 1977b for reviews). In addition to this external asymmetry, it was also noted early on by Pfeifer (1921, 1936), and von Economo and Horn (1930) that within the Sylvian fissure, the transverse gyri (Heschl gyri) and the expanse of cortical surface (planum temporale) posterior to the Heschl gyrus (known to include the primary auditory cortex) showed marked variation in gross morphology, with the left planum often being larger (see Figure 1). Unfortunately, this observation was not included in von Economo's (1929) English volume.

In spite of these observations, no hypothesis of the possible relationship of neuroanatomical variation to the newly observed functional asymmetry was made by the early neuroanatomists. This may have been partly attributable to the fact that perisylvian asymmetries had also been observed to exist to some extent in chimpanzees (e.g., Fischer, 1921). Since the concept of cerebral dominance was then synonymous with that of speech lateralization, the association between an anatomical feature, also present in nonverbal animals, and functional lateralization may have seemed untenable.

When the possibility of an association was finally raised decades later, it was suggested that the two asymmetries were likely not related, as the perisylvian anatomical asymmetries appeared too insignificant to account for the marked functional asymmetries (von Bonin, 1962). Eventually, the hypothesis of an association was considered (e.g., Geschwind and Levitsky, 1968; von Bonin, 1981).

Following the work of Geschwind and Levitsky (1968) who first measured the exposed planum temporale in a large number of specimens, several other studies examined asymmetry in the magnitude of the planum, both in respect to lateral length and surface area. Most of these studies have been reviewed in detail previously (e.g., LeMay, 1976; Witelson, 1977b). A brain specimen showing typical gross asymmetry in the planum temporale is shown in Figure 2.

A current summary of the studies of planum asymmetry in humans, both adults and infants, is presented in Table I. All studies found the left

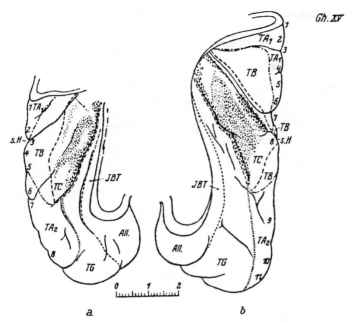

FIGURE 1.   Sketch of Brain XV shows the superior surface of the temporal lobes with the left hemisphere presented on the right side. The sulci are represented by solid lines. In this specimen each hemisphere has only one Heschl transverse gyrus and sulcus, the latter marked as "s.H." The marked gross asymmetry is obvious with the posterior end of the Sylvian fossa extending more posteriorly in the left temporal lobe. The larger extent of the left planum temporale behind "s.H." is also clear. The various cytoarchitectonic regions, such as TC, TB, and TA_1, are marked. They too may be seen as asymmetrical in extent between right and left sides (TA_1: left = 1.1 cm², right = 0.5 cm²) (SFW). [From von Economo and Horn, 1930: Illus. (Abbildung) 7, a and b.]

planum to be larger, by about a third to almost double the size of the right planum. In absolute units this difference is at least 1 cm in length and 1.5 cm² in area, a difference easily observable by gross visual inspection. The left planum is also larger in the majority of specimens, in about 70% of cases over all studies. The consistency between studies attests to the reliability of this asymmetrical feature.

Other neuroanatomical asymmetries have been observed, but with less documentation. Asymmetries exist in the ventricular system, in the vascular pattern, and in the breadth and alignment of the frontal and posterior regions of the hemispheres. In general, these asymmetries have not proven as reliable, and are more difficult to interpret in terms of function

Table I

ANATOMIC ASYMMETRY IN THE PLANUM TEMPORALE IN HUMANS

Study	Specimens		Measure (cm)	Mean size			Distribution (%)		
	No.	Age		Left	Right	L/R	L > R	Equal[b]	R > L
Adult									
Geschwind and Levitsky, 1968	100		Length	3.6	2.7	1.3	65	24	11
Teszner, 1972; Teszner, Tzavaras, Gruner, and Hécaen, 1972	100		Area (via wax molds)	—	—	—	64	26 (nearly)[b]	10
Witelson and Pallie, 1973	16		Length "nonbiased"[a]	2.5	1.5	1.7	69	0 (2 mm^2)	31
			Area "nonbiased"[a]	4.8	3.0	1.6	69	0 (2 mm^2)	31
Wada, Clarke, and Hamm, 1975	100		Area (planimetric units)	37.0	18.4	2.0	82	8 (5 units)	10
Rubens, Mahowald, and Hutton, 1976	21		Length	3.1	1.8	1.7	67	—	—
Kopp, Michel, Carrier, Biron, and Duvillard, 1977	83		Area (via aluminum foil mold)	5.9	4.3	1.4	77	1.3	21.7
Infant									
Teszner et al., 1972	1 specimen and 7 iconographs	7–9 mo gestation	Length	—	—	—	75	12.5 (nearly)	12.5
Witelson and Pallie, 1973	14	1 day–3 mo postnatal	Length "nonbiased"[a]	1.7	0.9	1.9	79	7 (2 mm)	14
			Area "nonbiased"[a]	1.9	1.1	1.7	64	36 (2 mm^2)	0
Wada, et al., 1975	100	7 mo gestation–18 mo postnatal	Area (planimetric units)	20.7	11.7	1.8	56	32 (5 units)	12
Chi, Dooling, and Gilles, 1977	207 celloidin-sectioned brains	10–44 wk gestation	Length via visual inspection	—	—	—	54	28	18

[a] These figures differ slightly from those in the 1973 report in that "nonbiased" measures (planum defined as posterior to second Heschl sulcus if present) were not reported for length, and "same" was defined as having identical scores

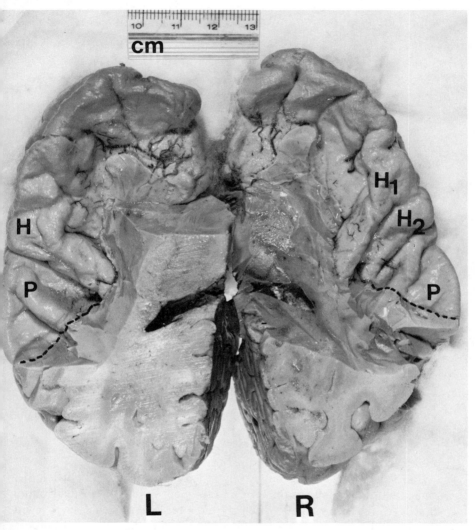

FIGURE 2. Superior view of the right and left temporal lobes of Brain 101. R, right; L, left; H_1, first Heschl gyrus; H_2, second Heschl gyrus; P, planum. (Left planum = 4.9 cm²; right planum = 4.0 cm².)

because they do not include regions clearly demarcated as functional units, as in the case of the planum. These anatomical asymmetries have been reviewed in detail elsewhere (LeMay, 1976; Witelson, 1977b). They will be discussed below in relation to data relavant to their association to cerebral dominance.

PROBLEMS IN RELATING PLANUM ASYMMETRY AND FUNCTIONAL LATERALIZATION

The planum is the superior surface of the posterior part of the first temporal gyrus which is continuous with the supramarginal gyrus and the angular gyrus, regions known to be relevant for language comprehension and praxis. The planum is clearly part of Wernicke's posterior language region (e.g., Bogen and Bogen, 1976). Its function, coupled with its greater expanse on the left side, readily leads to the hypothesis that this anatomical feature is a substrate of cerebral dominance. One source of skepticism toward this hypothesis concerns the assumption that a larger region is associated with dominant function. However, in the domain of cortical localization there is a precedence that more complex functioning is subserved by larger regions. For example, in the primary sensory areas, the macular region in vision has a greater area of representation than do the peripheral fields; in somesthesis the hand is represented by a larger region than is the trunk. At a speculative level, such larger regions may allow for advantages in neural circuitry. If comparable cell packing density exists between sides, therefore more cells would be present in the side with the greater region. Should cell packing differ between sides, then the larger side would either have more neurons per unit volume, or have fewer neurons per unit volume and more intercellular space for greater dendritic development and different synaptic density, either of which may confer functional advantages. Any specific histological hypothesis at this point would be premature.

Another problem undermining support of an association between anatomical and functional asymmetries is the poor match of the distribution figures for these asymmetries. If the temporal lobe morphological asymmetry is in fact a substrate of cerebral dominance, then one might predict that the frequency distributions of speech lateralization and of the side of the greater planum would be similar. These distributions are not as close as one might expect.

The distribution of speech and language lateralization in the population may be best represented by the results of intracarotid sodium amytal testing. I suggest this because the incidence of aphasia associated with unilateral lesions only indicates if the damaged hemisphere is involved to some degree in language, but it does not necessarily indicate the possible contribution by the other hemisphere. Thus, the estimate that 95% of the population has speech represented in the left hemisphere on the basis of the incidence of aphasia associated with left-sided damage (e.g., Zangwill, 1967) may indicate the proportion of individuals with some left-hemisphere representation, but may not reflect the proportion of individuals with only or at least predominantly left-hemisphere specialization for linguistic functions. The results of amytal testing would appear to indicate more precisely whether speech is represented in the left, right or both hemispheres. On the basis of such data, for individuals with no history of early brain damage, approximately 95% of right-handers have speech lateralized only in the left hemisphere, as do approximately 70% of left-handers (Milner, 1974; Rasmussen and Milner, 1977).

If one assumes that consistent right-handers constitute about 70% of the population (Annett, 1972), then approximately 90% of the total population (95 of 70% plus 70 of 30%) might be expected to have speech functions only or mainly lateralized in the left hemisphere.

The left planum is larger on the average in 70% of specimens, with the results of individual reports varying from approximately 55 to 80% as shown in Table I. Thus there is considerable discrepancy from the 90% estimate for the incidence of left-speech lateralization.

There are limitations with the available data for both the anatomical and functional parameters. The speech lateralization data available from amytal testing are based on a very limited sample of speech and language functions and only on individuals having varying brain pathology (Milner, 1975).

The true distribution of planum asymmetry is also difficult to discern for several reasons. The anatomical boundaries of the planum are relatively clear compared to those of other gyri, for example, the angular gyrus, but there still is sufficient ambiguity to present difficulty. The posterior border of the planum is defined as the posterior end of the Sylvian fossa at the temporoparietal junction (called the Sylvian point). If there is branching in the posterior segment of the Sylvian fissure, it is often difficult to discern from only lateral inspection of the brain, which ramus it is that may include the posterior surface of the planum. If the rami are equally long and deep, then it is impossible to be sure which surface to use for gross planum measurement, even after dissection. Perhaps only three-dimensional study of the histology of the region may resolve this issue.

The sharper upward angulation of the Sylvian fissure in some hemispheres, usually on the right, complicates the issue further. This asymmetry may lead to a systematic bias in dissection to expose the planum and in planum measurement. Different studies have used different dissection methods to expose the planum. In some studies (e.g., Witelson and Pallie, 1973), the brain was cut at a level dorsal to the Sylvian fissure and any posterior rami. The end of the main ramus of the Sylvian fissure was determined and exposed by gradually chipping away inferior parietal tissue until the full planum was visible. The method of cutting through the Sylvian fissure in its main horizontal axis to the end of the brain, as some other studies have done, may in some cases inadvertantly remove the posterior part of the planum which may be superior or possibly inferior to the main axis of the fissure.

The anterior border of the planum is even more difficult to define. Here studies vary in the definition of the border. The difficulty hinges (1) on whether the planum is defined as being posterior to the first or to the second transverse Heschl sulcus, and opinions vary on this, and (2) on the anatomical problem of how to define which of the numerous sulci in the region may be the first and possibly the second Heschl sulci. These issues too are still not resolved and may require cytoarchitectonic delineation. Von Economo and Horn (1930) clearly expressed this anatomical ambiguity in some of their specimens by labeling various possible Heschl sulci with question marks (see Figure 3).

Another possible source of error in planum measurement is whether the two-dimensional surface of the planum is a sufficiently valid measure, or whether some three-dimensional measure is required. There is also the issue of whether the planum is one functional unit. It certainly is heterogeneous histologically. Finally, there is the statistical issue of what magnitude of anatomical asymmetry constitutes a difference. All these factors may introduce error into the distribution figures.

In summary, the direction of the mean difference is the same for anatomical and functional asymmetry, but the distribution figures are discrepant. Until there is more direct evidence, it is difficult to state with confidence that the two asymmetries are associated and may not just be individual examples of asymmetry as are other lateralized phenomena, such as the position of the heart, the larger right claw (used functionally as the crusher claw) in the stone crab (Cheung, 1976), or the honey bee's dance system involving flight to the right side of the sun in situations of ambiguity (Brines and Gould, 1979).

One type of data that could address this issue is the study of the correlation between planum asymmetry and indices of functional lateralization, such as amytal testing or perceptual asymmetry. An association would be supported if proportionately more instances of non-left-functional lateralization were found in those cases not having a larger left planum. It would also be convincing to ascertain whether there is intraindividual covaria-

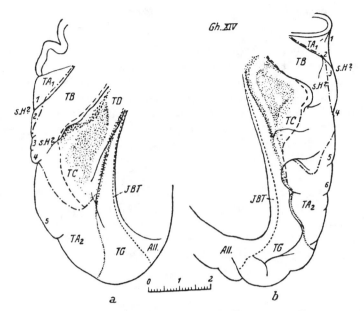

FIGURE 3. Sketch of Brain XIV shows the superior surface of the temporal lobes, presented with the left hemisphere on the right side. The sulci are represented by solid lines. In this case (the subject was known to be ambidextrous) there are two relatively transverse sulci on each side, all labelled "s.H.?" The decision as to which may be Heschl sulci would affect whether there are two Heschl gyri on both sides, and associated very small and similar plana on each side; or whether there is only one Heschl gyrus on each side, thus with larger plana on both sides; or whether there is an asymmetry in number of Heschl gyri between sides and a resultant marked planum asymmetry (SFW). [From von Economo and Horn, 1930: Illus. (Abbildung) 4, a and b.]

tion on functional and anatomical asymmetries. Such studies have not yet been reported. Such research would require neurocognitive testing of individuals for whom subsequent postmortem analysis of the brain would be available, which clearly provides logistical complexity. I will return to this issue later.

One easier method involves correlation of planum asymmetry with hand preference, which can be obtained retrospectively. As indicated above, unfortunately hand preference is only grossly correlated with speech lateralization. However, on the basis of the available data concerning hand preference and language lateralization, one could expect some group differences between dextrals and sinistrals. One might predict a greater incidence of a larger right planum, or less anatomical asymmetry among left-handers than among right-handers.

To date, retrospective handedness information was obtained only in one study (Wada, Clarke, and Hamm, 1975), but even in this case the authors suggested that the handedness data were too ambiguous to analyze. There are, however, a few individual case reports which do support an association. Tuge and Ochiai (1978) observed a larger right planum in a left-handed woman, who also was a gifted pianist and composer. Galaburda and Kemper (1979) observed plana of equal size in a left-handed male, but who also was dyslexic and epileptic. The case of von Economo and Horn, as described in Figure 3, was non-right-handed and the left planum temporale was not larger.

Clearly, the required data are not yet available. The only way available as yet to measure the planum is by direct observation after brain dissection, which makes postmortems necessary and thus correlative studies administratively difficult. Eventually computerized tomography (CT) of the brain may be able to reflect aspects of the morphology of the planum in vivo, but this is not yet technically possible (LeMay, 1977, 1980).

ASYMMETRIES IN VIVO
AND FUNCTIONAL LATERALIZATION

Several other parietal, frontal, and occipital asymmetries have been documented, some of which have the advantage of being obtainable in vivo. This obviously allows more readily for correlation of anatomical asymmetry with functional asymmetry. For some of these measures there are data already available concerning correlation with functional asymmetry. There is a disadvantage, however, with most of these anatomical measures in that it is not clear which specific gyri are included in these measured regions and, accordingly, it is not clear what cognitive functions are associated with these regions. In addition, the reliability of these asymmetries is not as high as in the case of the planum temporale.

One asymmetry in the temporoparietal region is the degree of angulation formed by the arches of the posterior branches of the middle cerebral artery as it courses posteriorly through the Sylvian fossa and emerges laterally at the end of the fissure. This angulation is visible in carotid arteriograms. A larger arterial angle has been noted to occur more frequently on the right than on the left side (LeMay and Culebras, 1972). This may reflect the pattern of a sharper rise of the Sylvian fissure, an associated smaller downward expansion of the parietal operculum, and a higher Sylvian point on the right side. This asymmetry in position of Sylvian points was observed directly in postmortem study of a small sample of specimens. The arterial asymmetry was further studied in relation to handedness and a correlation was observed. The typical pattern described above was found for right-handers, but the asymmetry was attenuated in left-handers, with the majority showing nearly equal angulation (Hochberg and LeMay, 1975). A more recent study has shown the arterial asymmetry to be significantly correlated with speech lateralization determined by amytal testing (Ratcliff, Dila, Taylor, and Milner, 1980).

Another *in vivo* measure involves the use of CT scans in which the frontal and occipital breadths of the hemispheres are measured 5 mm from the poles (LeMay, 1977; LeMay and Kido, 1978). It was observed that the typical pattern of greater frontal expanse on the right side and greater occipital expanse on the left side was somewhat attenuated in a group of left-handers. A subgroup consisting of familial left-handers, considered to be those most likely to have bilateral speech lateralization (e.g., Hécaen and Sauget, 1971), were found to differ most from the right-handed group. A study of direct volume measurement of the frontal and occipital lobes on the right and left sides (although measured well behind 5 mm from the poles) supports the asymmetry noted in CT scans, both for adults and infants; however, the asymmetry appears less clear for the frontal region (Weinberger, Luchins, Morihisa, and Wyatt, 1982).

Other *in vivo* measures, including length of the posterior horns of the lateral ventricles (via ventriculograms) in adults (McRae, Branch, and Milner, 1968) and in children (Strauss and Fitz, 1980), and vascularization patterns (Carmon and Gombos, 1970; Carmon, Harishann, Lowinger, and Lavy, 1972; Di Chiro, 1962) have shown right–left asymmetry and some correlation with functional lateralization via hand-preference and dichotic listening scores. These individual studies are reviewed in detail elsewhere (Witelson, 1977b, 1980).

Whether these various anatomical asymmetries are associated with planum asymmetry remains to be demonstrated. Some may be quite independent of each other. Sylvian fissure length and arterial arch may be correlated with planum size. The greater occipital breadth and greater posterior protrusion (petalia) on the left than right side may be related to the longer posterior horn of the left lateral ventricle, but may be unrelated to planum asymmetry. Preliminary study of the association of frontal and occipital breadth asymmetry via CT scans and planum asymmetry observed postmortem does not yet provide support of a correlation (Pieniadz and Naeser, 1981).

The cumulative results of this research support a correlation between neuroanatomical asymmetry and functional lateralization. They raise the likelihood that neuroanatomical asymmetries may be a substrate of hemisphere specialization.

PLANUM ASYMMETRY
AND SPEECH LATERALIZATION

The questions of a correlation between patterns of hemisphere specialization and planum asymmetry, other gross brain measurements, and
quantitative neurohistological analyses of the temporal lobe region are
being addressed in some of my current research. In essence, the procedure involves the psychological testing of seriously ill individuals and correlation of these data with the results of anatomical study of the brain
specimens of these individuals obtained via clinical postmortems. An extensive neurocognitive test battery is administered which includes indices
of right- and left-hemisphere specialization (such as handedness tests and
dichotic and dichhaptic tests) and measures of level of ability on various
lateralized cognitive skills such as language, spatial orientation, and
music. The neuroanatomical study includes measures of weight, volume,
radiographic measures of ventricular regions, arterial caliber measurements, extensive direct caliper measurements of various cortical ex-
panses on the lateral and medial aspects of the hemispheres, measures of
specific gyri and interhemispheric fiber tracts, detailed linear and area
measurement of the surface of the temporal lobe (including the planum
and adjoining Heschl gyri), and histological analyses of this temporal region. Specific procedural details are given elsewhere (Witelson, 1981).

I will summarize here some preliminary results on a first subgroup of 12
cases for a subset of gross anatomical measures of the planum temporale
and for some of the functional lateralization measures, including hand
preference, finger tapping rate, and ear asymmetry on linguistic and musical dichotic stimulation tests.

Of these 12 cases, 9 were strongly right-handed individuals on the basis
of Annett's (1970) hand preference questionnaire and reported to have always had dextral preference on the basis of a developmental, social, and
medical questionnaire. Of the 9 right-handed cases, 7 had a left planum
that was considerably larger in area than the right; two cases did not
(Brains 103 and 4). Brain 103 had almost identical right and left plana,
both in size and shape. Brain 4 had a larger right planum.

Brain 103 was from the only right-handed individual with almost identical ear scores, showing slightly greater left-ear scores on a dichotic consonant–vowel (CV) test on two different administrations. Hand difference
on the tapping test was in favor of the right hand but was less than normal.
Ear asymmetry on the dichotic melody test was the smallest for all cases
tested.

Brain 4 was also from an individual who showed little ear asymmetry on the two CV dichotic test sessions, with a slight bias towards the right ear. This subject showed almost no hand difference in tapping rate and was the only case in which the rate of tapping with the right index finger in the condition of concurrent verbalization showed the unusual pattern of virtually no interference effect (e.g., Lomas and Kimura, 1976; Peters, 1977). All the other right-handed individuals who received dichotic and tapping tests ($n = 4$ cases) showed typical behavioral asymmetry both in direction and degree.

The psychological test data of these two cases are compatible with non-left-hemisphere representation of speech. This inferred pattern of functional laterality may be related to the atypical anatomical pattern of the left planum not being larger.

The remaining three cases were of left-handers. Given the heterogeneous nature of left-handers in relation to speech lateralization, one would not expect all cases of left-hand preference to show reversed anatomical asymmetry. One might predict that more cases of atypical planum asymmetry would occur in a subgroup of left-handers than right-handers. Unfortunately the group is still too small for statistical group comparisons. However, as above, analysis of correlation within cases may be considered. One case (Brain 104) showed a clearly larger left planum. Interpretation of the dichotic scores in this case is confounded by the presence of a lesion in the left Heschl gyrus. However, this individual likely had speech representation mainly in the left hemisphere, as right frontal epilepsy involving left arm and facial paralysis was not associated with aphasia. He wrote with a noninverted hand posture. The second left-hander (Brain 9), also a noninverted writer, had a borderline larger left planum and showed minimal ear asymmetry with slight right-ear advantage on both administrations of the CV dichotic test. Tapping rate was higher with the left (dominant) hand by a typical difference. Left-hand tapping in the concurrent verbalization condition showed a marked drop in rate, compatible with some speech functions in the right hemisphere. The third left-hander (Brain 1) had plana of equal magnitude. Unfortunately no dichotic tests were administered nor were any neurological symptoms noted as lateralizing signs. Only hand preference and familial handedness were obtained. Of the 12 cases, this was one of only two cases with a positive family history of left-hand preference (the other was a right-hander with typical laterality patterns). In Case 1, the individual's father was left-handed (noninverted posture) and she herself wrote with an inverted posture. The factor of familial sinistrality has been associated with a greater degree of bihemispheric representation of speech in some left-handers rather than with left-sided speech (Hécaen and Sauguet, 1971). Inverted hand posture has also been suggested as a parameter associated with greater bihemispheric representation of speech, in this case as opposed to predominant right-sided speech (Levy and Reid, 1976).

The suggestion that one of three left-handers has predominantly left-hemisphere speech representation and that the other two possibly have bihemispheric speech is compatible with the known heterogeneity of brain organization of sinistrals.

In regards to the right-handers, however, at first consideration the suggestion that two of nine strongly right-handed individuals with no history of early brain damage may have bihemispheric or right-sided speech lateralization might seem highly improbable. However, the results of amytal testing indicate that approximately 4% of right-handers without clinical evidence of early left-hemisphere damage appear to have right-hemisphere lateralization for speech (Rasmussen and Milner, 1977). From the neurological literature, there have been individual reports of crossed aphasia in dextrals (e.g., Wechsler, 1976). In addition, reviews of large groups have indicated that dysphasic symptoms may follow right-hemisphere lesions in right-handers in about 10% of cases (e.g., Zangwill, 1967). A recent review (Hécaen, De Agostini, and Monzon-Montes, 1981) reported that approximately 8% (calculated from Table 9 of Hécaen et al., 1981) of right-handers showed aphasic symptoms with right-sided lesions, particularly females with familial sinistrality. From a third source, Annett (1975) calculated that on the basis of her theory of the inheritance and manifestation of hand preference in relation to speech lateralization, the percent of dextrals with right-hemisphere speech may be as high as 9%.

On the basis of these reports, it may not be so surprising to find non-left-hemisphere speech representation in a minority of right-handers. The right-handed cases (103 and 4) may be just such individuals as those atypical right-handers described above. In my series, the right-handed cases 103 and 4 and left-handed cases 9 and 1 may be similar to each other genotypically in regards to the lack of the factor for right-hand preference and for a factor for left-speech lateralization (following Annett, 1975). The lack of clear anatomical asymmetry may be part of the manifestation of such genetic information.

In summary, for the 12 specimens studied, the anatomical and functional asymmetry data appear to be highly correlated, supportive of the hypothesis that the anatomical asymmetry may be a substrate of hemisphere functional specialization. Clearly more cases of both right-handers and left-handers with varying laterality patterns need to be analyzed with statistical methods.

Whether the individual variation in gross anatomical asymmetry is reflected in histological asymmetry is clearly a next research step. Histological analysis is even more pertinent to assess if function proves to be correlated with gross anatomy, since one would expect function to be more closely related to anatomical parameters of cells than of gyri.

Histological study of the planum region to date has focused on the parcellation of different cytoarchitectonic regions. Marked variation was noted early in the topography of the different cytoarchitectonic regions between brains as well as between the right and left hemispheres in a series of seven brains (von Economo and Horn, 1930). Unfortunately the results were not quantified or summarized for the series of brains studied, nor were the results analyzed for any consistent right–left differences. More recently, in a series of four brains, Galaburda, Sanides, and Geschwind (1978) have reported that the extent of the most posterior part of the planum, their Area Tpt, comparable to Area TA_1 (von Economo) or Area 22 (Brodmann, 1909), was larger in the left side, with a positive correlation between asymmetry in extent of this region and planum asymmetry.

A different but complimentary histological question is being addressed in my current research with the collaboration of Dr. M. Colonnier. The basic question is whether the quantitative parameters per unit volume of tissue are similar between right and left sides for homologous tissue. For example, the numbers of different types of neurons in different laminae under 1 mm^2 of cortical surface are being calculated. It appears that there are quantifiable right–left differences, but whether they are related to gross anatomical or functional asymmetry patterns is not yet clear.

IMPLICATIONS OF A NEUROANATOMICAL SUBSTRATE OF CEREBRAL DOMINANCE

The demonstration that there may be neuroanatomical substrates of hemispheric specialization which are present from before birth has theoretical and clinical implications.

Such an association between anatomical and functional factors indicates a prewired neurobiological precursor of functional lateralization, one of the relatively unique organization factors in the human brain. Such neural specificity may be comparable to that of the visual cortex (Hubel and Wiesel, 1979), involving preprogrammed morphological cellular arrangement and associated functional specificity.

However, such congenital neural specificity may still be susceptible to environmental influences, such as the lack of relevant species-predictable stimulation. Early coordinated visual stimulation to both eyes is required so that the neural specificity in the visual system is not modified (Hirsch and Spinelli, 1970). Early auditory stimulation with speech sounds may be needed for the genetic pattern of cerebral dominance to be sustained (see Witelson, 1977a, p. 269). A possible example of a modified pattern of brain organization may be the case of the girl Genie who suffered linguistic, other cognitive and emotional deprivation from infancy and appeared to have bihemispheric speech representation (Fromkin, Krashen, Curtiss, Rigler, and Rigler, 1974).

Such prewired neural specificity does not necessarily preclude functional plasticity or reorganization based on structural plasticity—a characteristic particularly operative in the immature brain (Goldman-Rakic, 1980). The observation of the existence of hemisphere specialization in children in the same age range for whom transferability of speech functions to the undamaged hemisphere has been reported has often been considered a paradoxical situation. However, potential structural and functional plasticity does not preclude functioning specialization of the hemispheres. They may coexist as orthogonal factors. This is true for other specialized cortical regions, such as the motor cortex. It has preprogrammed specific functions, but if damaged, there is the capacity for some reorganization, which is greater earlier in life. Similarly, hemisphere specialization may be present anatomically and functionally, but some reorganization may occur under atypical situations, particularly in the immature brain. Accordingly, greater recovery earlier in life may be conceptualized as a result of structural plasticity, not as a result of a lack of specialization (Witelson, 1982).

If a preprogrammed anatomical substrate proves to underlie functional lateralization, it suggests that there may be limitations to interhemispheric plasticity or equipotentiality. The severe and permanent language and spatial deficits which often follow brain damage sustained at maturity attest to hemisphere specialization, but also to limited equipotentiality. Similarly, cases of hemispherectomy in which considerable language and spatial functions are present, regardless of which hemisphere remains, attest to the functional plasticity of the young brain; however, the less obvious but nevertheless demonstrable cognitive deficiencies which are present, depending on which hemisphere was removed (Kohn and Dennis, 1974), attest to the prewired specificity of the two hemispheres and to the limits of interhemispheric equipotentiality, even after very early neurological insult.

If hemisphere specialization has a basis in neuroanatomical asymmetry, then the individual differences observed in hemisphere specialization may have a biological basis. Neuroanatomical study may help elucidate different subgroups of right-handers, generally considered (perhaps incorrectly) to be a homogeneous group in respect to brain organization. This would have theoretical implications in that the atypical "crossed" aphasics would be seen as a different type of right-hander, and the finding of right-handers showing right-hemisphere speech on amytal testing would not be interpreted as necessarily a result of neural reorganization subsequent to brain damage. At a more speculative level, some aspects of cognitive functioning, such as preferred cognitive strategy (analytic–temporal or synthetic–spatial) may have a biological basis. For example, the difference between musicians who may prefer the analytic processing of melodies, rather than synthetic–holistic processing as used by the general population (Bever and Chiarello, 1974), may have neuroanatomical underpinnings (Witelson, 1980). The differences between human societies that developed oral languages and orthographies differing in phonetic, syntactic, and visual–holistic demands may have neurobiological differences.

The question of the inheritance of cerebral dominance has been difficult to resolve. The main problem has been to determine with certainty the pattern of functional lateralization of individuals and their family members. The pattern of specialization determined by the study of deficits following brain damage is clearly not a likely method for study of this issue. Neither is handedness, which is known to be heterogeneous in respect to brain organization and may be subject to environmental influences. The study of neuroanatomical asymmetry may be useful here. If it is correlated with functional lateralization, then this gross morphological feature, unaffected by environment, may serve as a valid index of genotype of brain lateralization for speech. Such research, however, would depend on subsequent technological advances of *in vivo* measures, such as computerized tomography for brain scanning.

At a more practical level, the study of the correlation between anatomical and functional asymmetry may help determine the validity of the lateral perceptual tests used as indices of hemisphere specialization in neurologically intact individuals. To date they are considered valid as group measures, but sufficient numbers of right-handers perform contrary to expectation to render suspect the tests' level of confidence for individual classification. It may be that study of the correlation of anatomical asymmetry with functional asymmetry will reveal that the perceptual asymmetry is a more precise index of functional lateralization than is hand preference, which is currently used as the baseline. Kimura (1961) suggested that dichotic listening correlated with amytal test results better than did hand preference. "Atypical" perceptual asymmetry may reflect a true individual difference not detectable on the basis of the cruder measure of hand preference.

Magnitude of anatomical asymmetry may have relevance for degree of recovery of function. Left-handers as a group show less cognitive deficits following brain damage than do right handers. It is unlikely that the basic neural properties underlying structural plasticity are different in left-handers. However, sinistrals have greater bihemispheric representation of functions which may be related to less anatomic asymmetry. This anatomical parameter may be the basis of the lesser severity or greater recovery of deficits in sinistrals. The same factor may apply in general to variation in deficits and recovery subsequent to brain damage.

Finally, clinical disorders in which atypical hemisphere specialization is suspected may now be considered in terms of some specific structural precursors as etiological factors. For example, developmental dyslexia, the disorder of relative difficulty in learning written language skills from the start, has been suggested to be associated with an atypical pattern of brain organization (e.g., Dalby and Gibson, 1981; Witelson, 1977c). The question may follow whether the neuroanatomical morphology is atypical in such individuals or whether the correlation between anatomical and functional asymmetry is atypical. One case study provides support for such speculation (Galaburda and Kemper, 1979). Atypical hemisphere specialization has also been implicated in childhood autism (Hier, LeMay, and Rosenberger, 1979) and schizophrenia (Luchins, Weinberger, and Wyatt, 1979). Knowledge of the etiology of a disorder can only serve to facilitate the development of effective treatment programs.

REFERENCES

Annett, M. (1970). A classification of hand preference by association analysis. *British Journal of Psychology, 61,* 303–321.
Annett, M. (1972). The distribution of manual asymmetry. *British Journal of Psychology, 63,* 343–358.
Annett, M. (1975). Hand preference and the laterality of cerebral speech. *Cortex, 11,* 305–328.
Bever, T. G., and Chiarello, R. I. (1974). Cerebral dominance in musicians and nonmusicians. *Science, 185,* 537–539.
Bogen, J. E., and Bogen, G. M. (1976). Wernicke's Region—Where is it? *Annals of the New York Academy of Sciences, 280,* 834–843.
Brines, M. L., and Gould, J. L. (1979). Bees have rules. *Science, 206,* 571–573.
Brodman, K. (1909). *Vergleichende Lokalisationslehre der Grosshirnrinde in ihren Principien dargestellt auf Grund der Zellenbauer.* Leipzig: Barth.
Bryden, M. P. (1965). Tachistoscopic recognition, handedness, and cerebral dominance. *Neuropsychologia, 3,* 1–8.
Bryden, M. P., and Sprott, D. A. (1981). Statistical determination of degree of laterality. *Neuropsychologia, 19,* 571–581.
Campbell, A. W. (1905). *Histological studies on the localization of cerebral function.* Cambridge: University Press.

Carmon, A., and Gombos, G. M. (1970). A physiological vascular correlate of hand preference: Possible implications with respect to hemispheric cerebral dominance. *Neuropsychologia, 8,* 119–128.

Carmon, A., Harishanu, Y., Lowinger, E., and Lavy, S. (1972). Asymmetries in hemisphere
blood volume and cerebral dominance. *Behavioral Biology, 7,* 853–859.

Cheung, T. S. (1976). A biostatistical study of the functional consistency in the reversed
claws of the adult male stone crabs. *Crustaceana, 31,* 137–144.

Chi, J. G., Dooling, E. C., and Gilles, F. H. (1977). Left–right asymmetries of the temporal
speech areas of the human fetus. *Archives of Neurology, 34,* 346–348.

Connolly, C. (1950). *External morphology of the primate brain.* Springfield, Ill.: Thomas.

Cunningham, D. J. (1892). *Contribution to the surface anatomy of the cerebral hemispheres.*
Dublin: Royal Irish Academy.

Dalby, J. T., and Gibson, D. (1981). Functional cerebral lateralization in subtypes of disabled readers. *Brain and Language, 14,* 34–48.

Di Chiro, G. (1962). Angiographic patterns of cerebral convexity veins and superficial dural
sinuses. *American Journal of Roentgenology, Radium Therapy, and Nuclear Medicine,
87,* 308–321.

Eberstaller, O. (1890). *Das Stirnhirn.* Wien and Leipzig: Urban and Schwarzenberg.

Fischer, E. (1921). Über die variationen der hirnfurchen des schimpansen. *Anatomischer
Anzeiger, 54* (Suppl), 48–54.

Fromkin, V. A., Krashen, S., Curtiss, S., Rigler, D., and Rigler, M. (1974). The development of language in Genie: A case of language acquisition beyond the "critical period."
Brain and Language, 1, 81–107.

Galaburda, A. M., and Kemper, T. L. (1979). Cytoarchitectonic abnormalities in developmental dyslexia: A case study. *Annals of Neurology, 6,* 94–100.

Galaburda, A. M., Sanides, F., and Geschwind, N. (1978). Cytoarchitectonic left–right
asymmetries in the temporal speech region. *Archives of Neurology, 35,* 812–817.

Geschwind, N. (1974). The anatomical basis of hemispheric differentiation. In S. J. Dimond
and J. G. Beaumont (Eds.), *Hemisphere function in the human brain.* London: Paul
Elek.

Geschwind, N. (1975). The apraxias: Neural mechanisms of disorders of learned movement.
American Scientist, 63, 188–195.

Geschwind, N., and Levitsky, W. (1968). Human brain: Left–right asymmetries in temporal
speech region. *Science, 161,* 186–187.

Goldman-Rakic, P. S. (1980). Morphological consequences of prenatal injury to the primate
brain. *Progress in Brain Research, 53,* 3–19.

Harnad, S., Doty, R. W., Goldstein, L., Jaynes, J., and Krauthamer, G. (Eds.). (1977). *Lateralization in the nervous system.* New York: Academic Press.

Hécaen, H., De Agostini, M., and Monzon-Montes, A. (1981). Cerebral organization in left-
handers. *Brain and Language, 12,* 261–284.

Hécaen, H., and Sauguet, J. (1971). Cerebral dominance in left-handed subjects. *Cortex, 7,*
19–48.

Heilman, K. M., Coyle, J. M., Gonyea, E. F., and Geschwind, N. (1973). Apraxia and
agraphia in a left hander. *Brain, 96,* 21–28.

Hier, D. B., LeMay, M., and Rosenberger, P. B. (1979). Autism and unfavorable left–right
asymmetries of the brain. *Journal of Autism and Developmental Disorders, 9,* 153–159.

Hirsch, H. V. B., and Spinelli, D. N. (1970). Visual experience modifies distribution of horizontally and vertically oriented receptive fields in casts. *Science, 168,* 869–887.

Hochberg, F. H., and LeMay, M. (1975). Arteriographic correlates of handedness. *Neurology, 25,* 218–222.

Hubel, D. H., and Wiesel, T. N. (1979). Brain mechanisms of vision. *Scientific American,* *241,* 150–162.

Kimura, D. (1961). Cerebral dominance and the perception of verbal stimuli. *Canadian Journal of Psychology, 15,* 166–171.

Kimura, D. (1976). The neural basis of language qua gesture. In H. Avakian-Whitaker and H. A. Whitaker (Eds.), *Studies in neurolinguistics,* Vol. 2. New York: Academic Press.

Kohn, B., and Dennis, M. (1974). Patterns of hemispheric specialization after hemidecortication for infantile hemiplegia. In M. Kinsbourne and W. L. Smith (Eds.), *Hemispheric disconnection and cerebral function.* Springfield, Ill.: Thomas.

Kopp, N., Michel, F., Carrier, H., Biron, A., and Duvillard, P. (1977). Étude de certaines asymétries hémisphériques du cerveau humain. *Journal of the Neurological Sciences, 34,* 349–363.

LeMay, M. (1976). Morphological cerebral asymmetries of modern man, fossil man, and nonhuman primates. *Annals of the New York Academy of Sciences, 280,* 349–366.

LeMay, M. (1977). Asymmetries of the skull and handedness. *Journal of the Neurological Sciences, 32,* 243–253.

LeMay, M. (1980). Personal communication.

LeMay, M., and Culebras, A. (1972). Human brain morphologic differences in the hemispheres demonstrable by carotid arteriography. *New England Journal of Medicine, 287,* 168–170.

LeMay, M., and Kido, D. K. (1978). Asymmetries of the cerebral hemispheres on computed tomograms. *Journal of Computer Assisted Tomography, 2,* 471–476.

Levy, J., and Reid, M. (1976). Variations in writing posture and cerebral organization. *Science, 194,* 337–339.

Levy, J., and Reid, M. (1978). Variations in cerebral organization as a function of handedness, hand posture in writing, and sex. *Journal of Experimental Psychology: General, 107,* 119–144.

Liepmann, H. (1908). *Drei Aufsätze aus dem Apraxiegebeit.* Berlin: Karger.

Lomas, J., and Kimura, D. (1976). Intrahemispheric interaction between speaking and sequential manual activity. *Neuropsychologia, 14,* 23–33.

Luchins, D. J., Weinberger, D. R., and Wyatt, R. J. (1979). Schizophrenia: Evidence of a subgroup with reversed cerebral asymmetry. *Archives of General Psychiatry, 36,* 1309–1311.

McGlone, J. (1980). Sex differences in human brain asymmetry: A critical survey. *The Behavioral and Brain Sciences, 3,* 215–263.

McRae, D. L., Branch, C. L., and Milner, B. (1968). The occipital horns and cerebral dominance. *Neurology, 18,* 95–98.

Milner, B. (1974). Hemispheric specialization: Scope and limits. In F. O. Schmitt and F. G. Worden (Eds.), *The Neurosciences: Third Study Program.* Cambridge, Mass.: MIT Press.

Milner, B. (1975). Psychological aspects of focal epilepsy and its neurosurgical management. *Advances in Neurology, 8,* 299–321.

Peters, M. (1977). Simultaneous performance of two motor activities: The factor of timing. *Neuropsychologia, 15,* 461–465.

Pfeifer, R. A. (1921). Die Lokalisation der Tonskala innerhalb der kortikalen Hörsphäre des Menschen. *Monatsschrift für Psychiatrie und Neurologie, 50,* 99–108.

Pfeifer, R. A. (1936). Pathologie der Hörstrahlung und der corticalen Hörspäre. In O. Bumke and O. Foerster (Eds.), *Handbuch der Neurologie,* Vol. 6. Berlin: Springer.

Pieniadz, J. M., and Naeser, M. A. (1981). Correlation between CT scan hemispheric asymmetries and morphological brain asymmetries of the same cases at postmortem. Presented at the 19th Annual meeting of the Academy of Aphasia. London, Ontario: October, 1981.

Rasmussen, T., and Milner, B. (1977). The role of early left-brain injury in determining lateralization of cerebral speech functions. *Annals of the New York Academy of Sciences, 299*, 328–354.

Ratcliff, G., Dila, C., Taylor, L., and Milner, B. (1980). The morphological asymmetry of the hemispheres and cerebral dominance for speech: A possible relationship. *Brain and Language, 11,* 87–98.

Rubens, A. B., Mahowald, M. W., and Hutton, J. T. (1976). Asymmetry of the lateral (sylvian) fissures in man. *Neurology, 26,* 620–624.

Satz, P. (1980). Incidence of aphasia in left-handers: A test of some hypothetical models of cerebral speech organization. In J. Herron (Ed.), *Neuropsychology of left-handedness.* New York: Academic Press.

Shellshear, J. L. (1937). The brain of the Aboriginal Australian: A study in cerebral morphology. *Philosophical Transactions of the Royal Society of London, Ser. B. 227,* 293–409.

Strauss, E., and Fitz, C. (1980). Occipital horn asymmetry in children. *Annals of Neurology, 8,* 437–439.

Swallow, J. A. (1980). *The influence of biological sex on cognition and hemisphere specialization: A study of Turner Syndrome.* Unpublished M.Sc. Thesis, McMaster University, Hamilton, Ontario.

Teszner, D. (1972). *Étude anatomique de l'asymetrie droite-gauche du planum temporale sur 100 cerveaux d'adults.* Unpublished doctoral dissertation, Université de Paris.

Teszner, D. A., Tzavaras, A., Gruner, J., and Hécaen, H. (1972). L'asymetrie droite-gauche du planum temporale: A propos de l'étude anatomique de 100 cerveaux. *Revue Neurologique, 126,* 444–449.

Tuge, H., and Ochiai, H. (1978). *Further investigation on the brain of a pianist based upon microscopic observations with reference to cerebral laterality.* Contribution from The Brain Institute, Japan Psychiatric and Therapeutic Center, Machida, Tokyo.

von Bonin, G. (1962). Anatomical asymmetries of the cerebral hemispheres. In V. B. Mountcastle (Ed.), *Interhemispheric Relations and Cerebral Dominance.* Baltimore: Johns Hopkins Press.

von Bonin, G. (1981). Personal communication.

von Economo, C. (1929). *The cytoarchitectonics of the human cerebral cortex.* New York: Oxford Univ. Press.

von Economo, C., and Horn, L. (1930). Über Windungsrelief, Maße und Rindenarchitektonik der Supratemporalfläche, Ihre individuellen und Ihre Seitenunterschiede. *Zeitschrift für die gesamte Neurologie und Psychiatrie, 130,* 678–757.

von Economo, C., and Koskinas, G. N. (1925). *Die Cytoarchitecktonik der Hirnrinde des erwachsenen Menschen.* Berlin: Springer.

Waber, D. P. (1979). Neuropsychological aspects of Turner's Syndrome. *Developmental Medicine and Child Neurology, 21,* 58–69.

Wada, J. A., Clarke, R., and Hamm, A. (1975). Cerebral hemisphere asymmetry in humans. *Archives of Neurology, 32,* 239–246.

Weinberger, D. R., Luchins, D. J., Morihisa, J., and Wyatt, R. J. (1982). Asymmetrical volumes of the right and left frontal and occipital regions of the human brain. *Neurology, 11,* 97–100.

Wechsler, A. F. (1976). Crossed aphasia in an illiterate dextral. *Brain and Language, 3,* 164–172.

Witelson, S. F. (1976). Sex and the single hemisphere: Specialization of the right hemisphere for spatial processing. *Science, 193,* 425–427.

Witelson, S. F. (1977a). Early hemisphere specialization and interhemisphere plasticity: An empirical and theoretical review. In S. Segalowitz and F. Gruber (Eds.), *Language development and neurological theory*. New York: Academic Press.

Witelson, S. F. (1977b). Anatomic asymmetry in the temporal lobes: Its documentation, phylogenesis, and relationship to functional asymmetry. *Annals of the New York Academy of Sciences, 299*, 328–354.

Witelson, S. F. (1977c). Developmental dyslexia: Two right hemispheres and none left. *Science, 195*, 309–311.

Witelson, S. F. (1977d). Neural and cognitive correlates of developmental dyslexia: Age and sex differences. In C. Shagass, S. Gerson, and A. Friedhoff (Eds.), *Psychopathology and brain dysfunction*. New York: Raven Press.

Witelson, S. F. (1980). Neuroanatomical asymmetry in left handers: A review and implications for functional asymmetry. In J. Herron (Ed.), *The neuropsychology of left-handers*. New York: Academic Press.

Witelson, S. F. (1981). Neuroanatomical asymmetry in the human temporal lobes and related psychological characteristics. *9th Semi-Annual Progress Report*, U. S. NINCDS, Contract Number N01-NS-6-2344, September, 1981.

Witelson, S. F. (1982). Hemisphere specialization from birth. *International Journal of Neuroscience, 17*, 54–55.

Witelson, S. F., and Pallie, W. (1973). Left hemisphere specialization for language in the newborn: Neuroanatomical evidence of asymmetry. *Brain, 96*, 641–646.

Zaidel, E. (1976). Unilateral auditory language comprehension on the token test following cerebral commissurotomy and hemispherectomy. *Neuropsychologia, 15*, 1–17.

Zangwill, O. L. (1967). Speech and the minor hemisphere. *Acta Neurologica et Psychiatrica Belgica, 67*, 1013–1020.

CHAPTER 6

Motor and Perceptual Functions of the Left Hemisphere and Their Interaction*

Catherine A. Mateer

INTRODUCTION

It is well recognized that the cerebral hemispheres are organized asymmetrically for functional behavior. The left hemisphere in almost all right-handed individuals is specialized for language, the right for nonverbal functions. Most of the literature regarding the nature and function of the "dominant" left hemisphere stresses the representational or sign–referent characteristics of "language." In the past decade, however, a number of lines of research have suggested a variety of nonlinguistic behaviors that are better performed by the left hemisphere. Though in themselves nonlinguistic, these specialized motor and perceptual abilities may underlie and subserve the left hemisphere's capacity for linguistic behavior. In

* Supported by NIH Grant NS 180517111 and by NIH Teacher–Investigator Development Award 1-K07 NS 00505, both awarded by the National Institute of Neurological and Communicative Disorders and Stroke, PHS/DHHS. The author is an affiliate of the Child Development and Mental Retardation Center of the University of Washington.

145

this chapter I review some of the evidence for specialized motor functions of the left hemisphere and summarize the auditory perceptual abilities in which it excels. Finally, I present evidence which supports a specialized, tightly knit interdependence of input–output capabilities in the dominant hemisphere which may serve as the fundamental basis on which languge develops and depends.

AUDITORY PERCEPTUAL FUNCTIONS
OF THE LEFT HEMISPHERE

Cerebral localization of language comprehension derived from lesion studies has been used to argue for the existence of neural mechanisms specialized for the reception of language. Many studies of the neural basis of language have thus dealt with such high level referential receptive language functions as word and sentence comprehension and their dissolution. Unfortunately, the characteristics of the neural mechanisms that underlie such complex capabilities, involving at least phonemic, morphemic, and syntactic analysis, are extremely difficult to specify. In addition, oral language is highly redundant, only a portion of the signal being necessary for transmission of meaning. An alternative approach, the investigation of lower-level language and nonlanguage perceptual functions, has proved very fruitful in delineating what the fundamental perceptual capabilities of the left hemisphere may be on which the higher level analyses ultimately depend.

Aside from the receptive language disorders following left-hemisphere lesions, the most consistent demonstrations that the dominant left hemisphere is important for language reception have been derived from dichotic studies. If pairs of contrasting digits or words are presented simultaneously to right and left ears, those presented to the right ear are usually more accurately reported (Kimura, 1961). The effect is normally attributed to the functional prepotency of the contralateral pathway from the right ear to the language-dominant left hemisphere. The interpretation of the right-ear advantage (REA) for verbal or speech stimuli is complemented by studies demonstrating that the ear advantage was reversed for nonspeech materials (melodies, tonal signals, environmental noise) (Kimura, 1964; Chaney and Webster, 1965; Curry, 1967). A left-ear advantage for words and digits presented dichotically, found in subjects known to have right-hemisphere language dominance (Kimura, 1961) served as additional support for involvement of the language dominant hemisphere in processing verbal material.

Further studies, however, clearly demonstrated that a right-ear advantage (REA) in left-hemisphere language dominant subjects was not dependent on stimuli being meaningful. Such meaningless stimuli as nonsense syllables and backward speech yield a consistent right-ear effect (Shankweiler and Studdert-Kennedy, 1967; Curry, 1967; Curry and Rutherford, 1967; Kimura, 1967; Kimura and Folb, 1968; Darwin, 1969; Haggard, 1969). In addition, if the speech signal is pulled apart, the perception of some components appears to be relatively more dependent on the processing capabilities of the dominant hemisphere. Not all classes of speech sounds produce equivalent REAs when dichotically presented. The strongest significant REAs have been found for stop consonants, a weaker REA for liquids (/r/, /l/). Vowels have typically failed to result in any significant ear advantage (Studdert-Kennedy and Shankweiler, 1970; Cutting, 1973).

In a study of patients with lateralized lesions, patients with left-hemisphere lesions were significantly impaired relative to right-hemisphere damaged patients and control subjects in the discrimination of stop consonants. Although both left- and right-hemisphere damaged patients were impaired on vowel discrimination relative to controls, there was no significant difference in vowel discrimination between left- and right-hemisphere damaged groups (Yeni-Komshian and Rao, 1980). Thus, the perception of consonants, particularly stop consonants, appears to be more left-hemisphere dependent than for other classes of phonemes.

Although stop consonants and vowels differ according to phonetic class, Tallal and Newcombe (1978) have pointed out that they also differ in the rate of change of the acoustic cues which characterize their spectra. They suggest that a critical factor underlying left-hemisphere processing of certain stimuli may relate to specific aspects of the acoustic rather than the linguistic nature of the stimuli. A number of studies of normal patients, using both dichotic and electrophysiologic techniques, have demonstrated that some nonverbal acoustic processing, particularly of signals characterized by rapidly changing temporal cues, occurs in the left rather than the right hemisphere (Efron, 1963; Lackner and Teuber, 1973; Cutting, 1974; Molfese, 1978; Tallal and Newcombe, 1978; Divenji and Efron, 1979; Mills and Rollman, 1979). Several of these authors have suggested that the left hemisphere does not appear to discriminate between phonetic and nonphonetic transitions—those that are linguistically meaningful versus those that are not. Perhaps the most compelling piece of evidence in support of an acoustic rather than a strictly linguistic processing propensity in the left hemisphere is that reported by Schwartz and Tallal (1980). Dichotically presented phonemically similar CV syllables, differing only in a temporal acoustic cue (the rate of change of the formant transition), appeared to be processed differently by the hemispheres. When the rate of acoustic change within syllables was reduced, while keeping phonemic characteristics constant, the characteristic asymmetry in processing of the speech stimuli, that is, a right-ear advantage, was significantly decreased. Tallal and her colleagues suggest that it is a deficit in perceiving rapidly changing sequential information, in detecting transitional elements, that underlies the receptive language impairments of some aphasic adults and language-disabled children (Tallal and Piercy, 1974; Tallal and Newcombe, 1978; Stark and Tallal, 1978, 1979). The superiority of the left hemisphere for processing of linguistic or verbal stimuli may reflect, at least in part, left hemispheric superiority in processing sequences of auditory stimuli, particularly those incorporating the very rapidly changing acoustic events that characterize much of speech stimuli.

Given the evolutionary relationship of human and other primates, it is possible that the analysis of communicative auditory information would also be lateralized to one hemisphere. Beecher, Petersen, Zoloth, Moody and Stebbins (1979), using a monaural task, reported that five Japanese macaques showed a significant REA (left hemisphere processing) for aspects of conspecific vocalizations that were dependent on peak-relevant processing. Corresponding pitch-relevant processing requirements yielded a LEA in two animals tested. Since the peak-relevant task involved a temporal discrimination and the pitch-relevant task a pitch discrimination, demonstration of a REA on the temporal task in this species of nonhuman primates is remarkably consistent with the human data indicating left hemisphere involvement in fast temporal discriminations. Employment of neurally lateralized mechanisms, presumably in the left hemisphere, for the analysis of conspecific communication sounds requiring temporal distinctions may be a property not only of the human but of the nonhuman primate nervous system.

Motor Functions of the Left Hemisphere

Many decades ago Liepmann (1913) advanced the notion that the left hemisphere was particularly important in the control of certain types of practiced motor activity, that it was fundamentally the "hemisphere of action." He suggested that aphasic and apraxic disorders observed after left-hemisphere damage were intrinsically related to dysfunction of underlying motor control mechanisms rather than to disorders of verbal mediation. Kimura and her colleagues in a series of experiments and articles have expanded upon and developed a compelling case for these early views (Kimura, 1973; Kimura, 1976; Mateer and Kimura, 1976; Kimura, 1978, 1979).

EVIDENCE FOR LATERALIZED
MANUAL MOTOR CONTROL

The most obvious evidence for lateralized motor control is, of course, the higher incidence of right-handedness than of left-handedness (right handedness being a characteristic directly related to the probability of left-hemispheric language representation). The most salient feature of left-hemisphere damage aside from aphasia is limb apraxia. Limb apraxia is usually defined as an inability to use or demonstrate use of objects or to otherwise perform skilled practiced acts in the absence of significant paresis or sensory loss. The deficit, however, is not only apparent during familiar, practiced motions but can be demonstrated in a variety of nonmeaningful, unpracticed motor tasks. It is often observed bilaterally, with movements of the ipsilateral left hand impaired in addition to those of the contralateral right hand. Patients with left-hemisphere damage were impaired bilaterally on the imitation of meaningless manual sequences (Kimura and Archibald, 1974) and in the acquisition and subsequent performance of a task requiring sequential changes in the configuration of the hand and upper limb musculature (Kimura, 1978). Impairments in finger spelling and signing in deaf individuals after left-hemisphere damage have traditionally been interpreted as another instance of left-hemisphere specialization for linguistic sign referential behavior. However, in a case study of such a deaf aphasic, Kimura, Battison, and Lubert (1976) demonstrated that imitation of meaningless sequences of hand postures similar to those used in signing was disturbed bilaterally. Using this case in conjunction with a review of neurologically based manual signing disorders, Kimura (1981) reinterpreted the deficits in signing, following left hemisphere damage, as reflecting, at least in part, impairments in producing complex sequences of limb movement. The left hemisphere motor "dominance" is not reflected in fine digital (finger) movement. On measures of digital dexterity such as individual finger flexion, adoption of unusual hand postures and finger tapping speed, the contralateral deficit is as great as or greater than with right-hemisphere lesions as with left-hemisphere lesions (Kimura, 1979). Rather, the left hemisphere appears to have a specialized capacity for control of certain kinds of complex, sequenced motor performance requiring changes in upper limb position and orientation.

Finally, there are several lines of evidence that support an overlap in the neural mechanism underlying speech and right-hand control. During speaking, more free movements or gestures, those which do not result in self-touching, are made with the right arm than with the left (Kimura, 1973; 1976). Studies of the interaction between speaking and manual activity have demonstrated that speech depresses performance on a variety of manual tasks, the effect seen largely on the right hand (Hicks, Provenza and Rybstein, 1975; Kinsbourne and Cook, 1971; Lomas, 1980; Lomas and Kimura, 1976; McFarland and Ashton, 1975; 1978).

EVIDENCE FOR LATERALIZED ORAL MOVEMENT CONTROL

Description of Terminology

The overwhelming evidence for lateralized oral movement control, for both verbal and nonverbal movement, is derived from the relationship of aphasia and oral apraxia to left-hemisphere damage. Before presenting evidence for an overlap in the control of verbal and nonverbal oral movement, some of the terminological problems in reviewing the literature of aphasia need to be addressed. A large number of labels have been applied to the description and interpretation of speech disturbances. A patient may have difficulty producing even a single phoneme or may demonstrate many errors in phoneme selection during speech attempts. Such articulatory or phonemic disorders of speech production have been variously termed verbal apraxia (Canter, 1967), apraxia of speech (Johns and Darley, 1970), phonetic disintegration (Alajouanine and Lhermitte, 1964), anarthria (Liepmann, 1913), cortical dysarthria (Bay, 1965), and literal paraphasia (Goodglass and Kaplan, 1972). Patients may also demonstrate many errors in the selection and production of words or phrases. This second category includes anomia, verbal paraphasia (substitution of a semantically closely related word, for example, *mother* for *father*), and grammatic or syntactic errors (correct words produced out of order or words produced without appropriate grammatic inflection).

Although a classification into two different types of aphasia is often proposed, the criteria for distinguishing between articulatory and higher-level, linguistically-based errors are rarely given or not easily applied. For example, Darley, Aronsen, and Brown (1975) have strongly urged the separation of apraxia of speech from aphasia. They define apraxia of speech as "an impairment in the capacity to program the positioning of the speech musculature and the sequencing of movements for the production of phonemes," aphasia as "an impairment in symbolic processing." The criteria used for inclusion in the apraxia-of-speech category, however, would be met by most if not all aphasic patients. Thus, expression in most classifications of aphasia contains initial phoneme errors on some occasions, more phoneme substitutions and omissions than distortions, and "islands of error-free speech" (Darley, 1970).

Darley, Aronson & Brown (1975) have also proposed that apraxia of speech can be distinguished from aphasia on the grounds that apraxia is not a word-finding difficulty. Unfortunately, the decision on this point is often very inferential, for example, the patient who produces *tup* for *cup* is said not to have word-finding difficulty, whereas the patient who produces *spag* for *cup* does have such difficulty. Without knowing something about the reliability of a particular misnaming or whether the patient can perhaps write it, though not be able to speak it, there is little basis for such inferences. Since assumptions regarding such a separation of apraxic and aphasic speech are often not based on empirical evidence, it seems more reasonable at present to consider the whole range of expressive speech disruption under the term aphasia.

Broca's aphasia (also termed expressive aphasia, aphemia, motor aphasia, and anarthria) is often characterized by laborious articulation even on a single phoneme level, with a severe reduction in the flow of speech. In other types of aphasia, speech may seem fluently and effortlessly articulated but produced with many errors. This dimension of fluency has been shown to discriminate usefully between what have classically been called Broca's (expressive) and Wernicke's (receptive) aphasias (Goodglass, Quadfasel, and Timberlake, 1964; Kerschensteiner, Poeck, and Brunner, 1972; Mateer and Kimura, 1976).

Fluency is usually defined by the mean length of verbal responses (in syllables or words) in a picture-description task or other open-ended task. The criterion is easily applied and usable across examiners, a characteristic notably lacking in most aphasia classification systems. The fluent–nonfluent classification will be used in general discussions throughout this chapter, but reviews of specific studies will employ the investigator's terminology and, where possible, a description of its use within that study.

Nonverbal Oral Movement Impairments

The association of aphasia with an inability to imitate nonlinguistic oral–facial movement in the absence of significant oral weakness was first documented by Jackson (1878) in his classical paper on nonprotrusion of the tongue. The terms "faciolingual apraxia" (Woltman, 1923), "facial apraxia" (Nathan, 1947), "nonverbal apraxia of the oral mechanism" (Eisenson, 1962) and most commonly "oral apraxia" (DeRenzi, Pieczuro, and Vignolo, 1966) have been used to describe an inability to perform voluntary movements with muscles of the tongue, lips, and jaw, although automatic movement of these same muscles is preserved.

Oral–facial apraxia, traditionally demonstrated by requesting imitation of single oral postures such as lip protrusion or tongue lateralization, has most often been associated with nonfluent aphasia (Nathan, 1947; Alajouanine, 1956; DeRenzi, Pieczuro, and Vignolo, 1966; Poeck and Kerschensteiner, 1975). The inability to produce isolated oral movements appears quite compatible with the effortful and impaired production of even isolated speech sounds, often characteristic of nonfluent aphasia. In contrast, fluent aphasics often have seemingly effortless speech although it is usually marked by many paraphasic errors. They rarely have difficulty with production of single oral movements, the traditional test of oral praxis. As a result, the impairments in speech production demonstrated in fluent aphasia have been either interpreted as linguistic errors of a semantic or phonological nature, or related to an auditory deficit in monitoring their own speech or that which they may be attempting to imitate.

In contrast to this view, Mateer and Kimura (1976) reported that most or all aphasic patients, even those with predominantly receptive disorders and fluent speech output, demonstrate impairments on a task that involves the imitation of a sequence of nonverbal oral movements. Patients with unilateral cerebral vascular damage were required to perform a number of oral motor tasks, both verbal and nonverbal. Nonfluent aphasics were impaired in the imitation of single oral movements, such as tongue protrusion or mouth opening, as had previously been reported. However, on the imitation of a series of three such nonverbal oral movements, not only nonfluent aphasics, but also fluent aphasics were impaired. Despite up to three available trials on the same movement sequence, only two aphasics were able to produce correctly even one of the five oral movement sequences. This impairment was not explicable on the basis of visual memory or oral–tactile perceptual deficits in the fluent aphasics (Mateer, 1976). The task did not depend on auditory input for either stimulus presentation or for response monitoring. The requirement for impairment appeared to be use of an oral motor task which required successive changes in the oral facial configuration. The degrees of impairment on these sequential nonverbal oral motor tasks and on the verbal tasks were well correlated and the nature of errors bore striking similarities across tasks (Mateer, 1978). Patients with right-hemisphere damage did not differ from controls on either the acquisition of nonverbal oral movement sequences or verbal repetition tasks. The findings suggest that deficits in coordinating oral movements are fundamental to most aphasic impairments, the meaningfulness of the responses not being a critical factor in the appearance of the defect.

Further evidence of asymmetric hemispheric involvement in production of nonverbal sequential oral motor tasks has been observed by Mateer and Dodrill (unpublished observations) during intracarotid injection of sodium amytal. Administration of this drug, an ultra-quick-acting barbiturate, temporarily inactivates one cerebral hemisphere for between five and ten minutes. It is used clinically in neurosurgical candidates to determine language lateralization and hemispheric involvement in memory processes. The study was carried out during the amytal procedure in eight right-handed patients, all of whom were determined by the procedure to be left-hemisphere dominant for language. Photographs depicting three oral–facial postures were presented for imitation, regularly interspersed with presentation of naming and reading tasks. Following left perfusion, all eight patients demonstrated a variable period of time ranging from 30 seconds to approximately three minutes of generally decreased responsiveness. Neither speech attempts nor attempts at imitating the oral postures were observed. Resumption of these behaviors began within 15 seconds of each other. Unsuccessful adoption of single postures was observed during the period in which even single-word naming responses were incorrect. Usually unrelated perseverative consonant–vowel combinations were observed. Even after single-word naming responses became intelligible, though not necessarily errorless, severe disruption in the imitation of sequential oral–facial postures was observed. Isolated movements were often accurate, but the rate and order of sequencing were disturbed. Other errors included perseveration on postures, omission of particular postures, and production of unrelated movements. During this period of disrupted oral movement sequencing, speech responses on the more demanding connected speech task involving sentence reading were characterized by marked difficulty with both articulatory and grammatical features. Articulatory alterations included phoneme substitutions, omissions, and reversals. Right perfusions yielded no disturbances in oral movement sequencing, except for a mild asymmetry of movement related to the induced contralateral hemiparesis. Thus nonverbal oral movement control appears to be dependent on the left hemisphere and highly related to speech production.

TRADITIONAL LOCALIZATION
OF ORAL MOTOR CONTROL

Beyond its lateralization to the dominant left hemisphere, what is known about the intrahemispheric localization of oral praxis? Using standard lesion localization techniques, oral apraxia has traditionally been related to dysfunction of the premotor and/or precentral face area of the left hemisphere (Bay, 1965; Nathan, 1947). In a study relating impairments of oral movement production to lesion sites using CT scans, Tognola and Vignolo (1980) report that critical areas necessary for the correct performance of single oral gestures are located in the left hemisphere and include the inferior frontal cortex and adjacent parts of the first temporal convolution and the anterior insula.

All of these studies, however, used impairment of single oral movements as the criterion for oral apraxia. They do not identify which areas may be involved in the production of a sequence of oral movements. Impaired sequential oral movement production, in the context of unimpaired production of the component movements, was most often found in fluent aphasics in the Mateer and Kimura (1976) study. Because it has been shown that fluent aphasics usually have damage to the posterior perisylvian areas (Benson, 1967; Naesar and Hayward, 1978; Mazzochi and Vignolo, 1979), so that one might expect more posterior involvement in the sequential movement tasks.

Localization of Oral Motor Control
by Cortical Stimulation Mapping

Further determination of the intrahemispheric cortical localization of sequential nonverbal oral movement production was carried out by Ojemann and Mateer (1979) using the technique of electrical cortical stimulation mapping in awake patients undergoing craniotomy in the course of neurosurgical procedures. Mapping to electrical stimulation, during which specific brain sites undergo transient, focal disruption, affords a degree of resolution unavailable with most other methods of functional localization (Ojemann, 1978). The cortical mapping studies have been carried out in eight patients prior to resection of a left anterior temporal lobe epileptic focus. A number of language and language-related functions were mapped. These include naming, reading of sentences which required generation of appropriate syntactic verb structures for completion, short-term verbal memory, mimicry of oral facial movements, and phonemic discrimination.

Naming is measured in conjunction with achromatic slide presentations depicting a carrier phrase such as *This is a* and a line drawing of an object. Reading is measured with slide presentations of typed sentences which are presented to be read after the naming slide. Following the reading slide which acts as a distractor, the patient sees a slide on which is printed *Recall*. This serves as a cue to recall the last object named as a measure of short-term verbal memory. Stimulation of discrete cortical sites is presented during pseudorandomly occurring trials of each task. Nonstimulation trials serve as the control baseline by comparison to which the significance of stimulation at each site, and thus its involvement in the function, is determined.

Motor mimicry is measured in a task in which the patient is instructed to mimic postures displayed on a slide. In one series of slides, the same oral-facial posture is repeated three times; in another, three different postures are shown, which the patient is to follow in sequence. There is no memory component to this task: the slide model is displayed during the entire time that the mimicry occurs. Stimulation occurs during the presentation of randomly selected items of each type for each site. Responses are recorded on videotape for offline analysis, blind to whether stimulation has occurred.

Identification of phonemes is measured in a separate test, a modification of the Stitt consonant identification task. The stop consonants, /p/, /b/, /t/, /d/, /k/, and /g/ are imbedded in the carrier phrase /ae_ma/. In this task, stimulation is applied only during the two seconds in which the consonant is presented, with a two-second response period without stimulation so the ability to detect the imbedded consonant is not confounded with effects on motor output mechanisms. These procedures are outlined in detail elsewhere (Ojemann, 1979; Ojemann and Mateer, 1979a,b).

The results obtained with stimulation on the motor mimicry task and their relationship to other language functions provide some insight into underlying left-hemisphere organization. Stimulation during the motor mimicry test of sites in the posterior inferior frontal lobe, in an area of cortex just in front of face motor area in five of the eight patients, altered the ability to produce even single oral–facial movements. Stimulation at these same sites also alters performance on all tasks requiring speech output. During naming, there is an arrest of all speech, including inability to read the phrase *This is a*. Similarly, during reading there is often an arrest of output, though this verbal memory task, there is commonly no response when the current is applied at the time of recall. Application of the current at the time the patients first saw the item, or during the time that the memory must be stored, produces no effects. Thus, these sites appear to be part of the final motor pathway for oral–facial movements, a pathway that extends forward from the face motor area into the posterior inferior frontal cortex. These locations correspond well with localization of single oral movement impairments derived from lesion studies. Stimulation in the homologous area of the nondominant hemisphere does not produce any language disturbance or change in motor mimicry, although stimulation of motor cortex itself in the nondominant hemisphere alters the ability to produce single facial movements and evokes an arrest of speech (Penfield and Roberts, 1959; Ojemann and Whitaker, 1978). Thus, the area involved in direct control of face movements appears to be appreciably larger in the dominant, rather than in nondominant hemisphere, extending forward into the inferior frontal cortex.

An association is apparent between the ability to mimic oral-facial movements and to identify phonemes. At three of the six final motor pathway sites (50%), identified in five patients, there was an evoked disturbance in identification of phonemes, even though the technique of the phoneme identification task is such that there is no current applied at the time of output. Defects in phoneme identification have also been reported after resection of dominant hemisphere face motor cortex (Darwin, Taylor and Milner, 1975), though the patients did not demonstrate any detectable aphasia.

Turning to the sequential part of the motor task, stimulation of some sites altered only the ability to mimic a sequence of oral-facial movements, while single repeated movements were intact. In the eight-patient series described above, only sequential movements were altered at 18 sites in the perisylvian cortex of the dominant hemisphere, in portions of inferior frontal, superior temporal, and parietal lobes. It is this area of perisylvian cortex surrounding the final motor pathway which likely subserves the lateralized control of sequential facial movements as described by Mateer and Kimura (1976).

That such cortex also plays a significant role in language is evident by the overlap with changes in naming or reading. In the eight-patient series, 72% of the sites with sequential motor change show naming or reading changes; this accounts for 56% of the sites where stimulation altered naming and/or reading. The areas involved in spontaneous brain lesions that produce persisting motor aphasias, as described by Mohr (1976), encompass this same perisylvian cortex that stimulation mapping relates to sequential motor movements. This cortex is one of the crucial substrates for the generation of language and is likely damaged in most patients with persisting aphasias.

An even more striking association than that between the final common pathway sites and phoneme identification sites is present between the sequential motor sites and those where phonemic identification is altered. Seventy-eight percent of sites showing sequential motor changes also showed changes in phoneme identification; this accounts for 82% of the sites with phoneme identification changes in those patients. The overlap in sequential motor and phonemic identification function is significant at the .01 level. This identifies an area of cortex common to the decoding of speech sounds and the sequencing of oral motor movements. In the following section, we will review support from a variety of sources of a tight

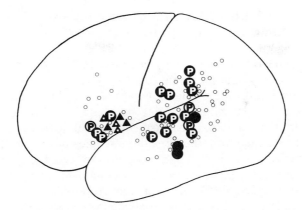

FIGURE 1. Composite of stimulation mapping results on the mimicry of nonverbal oral–facial postures and identification of phonemes in eight patients. All sampled sites are indicated by a circle or a triangle. The triangles indicate sites where repeated production of the same movement was altered. Sites where repeated production of a single movement was intact, but production of sequences of three different movements was impaired, are indicated by ●. Evoked alterations in phoneme identification are indicated by ▲ at repeated movement sites and by large P's at all other sites. The overlap between sites where oral movement production and phonemic identification were altered is significant at the .01 level.

interactive linkage between motor and perceptual mechanisms, a sequential motor–perceptual subsystem lateralized to left hemisphere. (Details of this organization are diagrammed in Figure 1.)

Motor–Perceptual Interactions in the Left Hemisphere

Evidence has been presented which suggests specialized nonlinguistic functions which are left-hemisphere dependent. First, the left hemisphere is involved in certain aspects of auditory processing, particularly of stimuli, including speech stimuli, which are characterized by their rapidly changing acoustic spectra. Second, the left hemisphere is involved in the control of sequential movements of the oral and brachial musculature involving successive changes in position. The perisylvian sequential motor–phonemic identification system identified by stimulation mapping has properties of mediating both motor sequencing and speech decoding. How does such a system fit with the classical division of language cortex into expressive and receptive zones? Is it compatible with other behavioral and neurophysiological data and with current theories regarding central organization of speech and language?

Some Difficulties with Classical Views
of Language Organization

Human language has traditionally been localized to two regions of the perisylvian cortex of the dominant, usually left, hemisphere: an inferior frontal subdivision for the production of speech and a parietal–temporal region for the understanding of speech. There are, however, obvious problems with this dichotomous organization. First, there appears to be considerable evidence from the behavioral studies of an overlap between measures of speech expression and reception. Indeed, many studies have reported in general a high degree of correlation between the measures (Schuell and Jenkins, 1959; Kimura and Archibald, 1974; Karis and Horenstein, 1976). Patients with severely decreased expression are significantly more likely to have severely reduced comprehension, irrespective of overall lesion size. Second, distinction into disorders of language production or language understanding is rarely, if ever, absolute. Rather, both expressive and receptive deficits are present in virtually all aphasic patients, although one or the other may predominate.

PERCEPTUAL STUDIES

Recently, a number of studies have addressed aspects of speech perception and language comprehension in the anteriorly and posteriorly based aphasias. Broca's aphasics, historically so classified on the basis of effortful production but good understanding of speech, have been shown to demonstrate a significant and almost constant impairment in verbal comprehension on the sensitive nonredundant Token Test (DeRenzi and Vignolo, 1963). Basso, Casati, and Vignolo (1977) found that poor resolution of voicing perception was positively related to nonfluency and impaired phonetic output. Studies by Carpenter and Rutherford (1973), and Blumstein, Baker, and Goodglass (1977) have shown that despite sometimes relatively good comprehension of meaningful speech, discrimination of phonemes is often impaired in nonfluent aphasics. Wernicke's aphasics with poorer comprehension of language typically showed milder speech perception impairments. However, impairments in speech perception at some level are characteristic of most aphasic groups, even those that evidence quite good auditory language comprehension.

The lack of direct correspondence between phonemic discrimination and language comprehension (Gainotti, Caltagirone and Ibba, 1975) might be accounted for on the basis of other cues for meaning (i.e., intonation, expectancy, redundancy) associated with connected speech and/or other requirements for comprehension of language. Alternatively, it may be that once language has been acquired, larger acoustic units (i.e., words or phrases) come to be associated with meaning, so that analysis on a phonemic level is not a requirement for comprehension. Given a series of tasks designed to increase systematically the demands for semantic processing of auditorally presented words, Wernicke's aphasics demonstrate an increase in the number of errors made and also produce more semantically based than phonologically based errors (Baker, Blumstein and Goodglass, 1979).

The stimulation mapping data (Ojemann and Mateer, 1979a,b) suggest that a broad range of perisylvian cortex in frontal, temporal, and parietal lobes may play a role in phonemic identification. This task requires not only the discrimination of physical attributes relating to phonetic categories, but also the ability to label the phonetic category. Blumstein, Cooper, Zurif, and Caramazza (1977) suggest that Wernicke's patients in particular may have difficulty with the latter labeling task. Since the task used during stimulation mapping did not allow us to separate these aspects, it may be that stimulation at the different sites disrupted the perceptual requirements and the labeling requirements of the task differentially, but the broad representation of this function and its strong association with disruption of sequential motor output remains.

PRODUCTION STUDIES

The effortful and impaired speech production in most Broca's aphasics is well recognized. Patients are classified as having Wernicke's or receptive aphasia predominantly on the basis of impaired comprehension. However, their oral expression, although usually fluent, is invariably impaired to some degree. Nearly all aphasics produce phonological errors as part of their expressive disorder. The phonological errors include phoneme substitutions, omissions, and additions, as well as errors in phonemic sequencing. Neither the distribution of errors by phonological type nor the phonological distance between the target phoneme and the substituted phoneme differentiate between Broca's, Wernicke's, and conduction aphasics, although all errors occur more frequently in the Broca group (Blumstein, 1973).

On closer inspection, however, the Broca's group demonstrates phonetic (acoustic) errors in addition to phonemic (class or shape) errors. Unlike controls or Wernicke's aphasics, the productions of Broca's aphasics are not characterized by clear dissociations in the voice onset time characteristics of the alveolar stop consonants /t/ and /d/ (Blumstein, Cooper, Goodglass, Statlender, and Gottlieb, 1980). In other studies of apraxic speech, such alterations in fundamental production parameters as voice onset time and segment duration (Freeman, Sands, and Harris, 1978) and inconsistencies in the production of oral and nasal consonants (Itoh, Susanama, and Hirose, 1978) have been shown. The production deficits in anterior aphasics appear to reflect a disruption of oral motor coordination on a very fundamental level. This fundamental impairment in motoric aspects of articulation is consistent with the findings of difficulty with even single nonverbal oral movement production observed in nonfluent aphasics (Mateer and Kimura, 1976) and with cortical stimulation in the left precentral region (Ojemann and Mateer, 1979b).

The lack of phonetic errors in the posterior aphasics has been taken to suggest that the speech output deficit in these patients is one in which the phonological form of the target word is programmed incorrectly and thus an incorrect though correctly articulated phoneme is produced (Blumstein, 1981). This interpretation depends on invoking a linguistic explanation, that is, a higher-level encoding deficit related to a phonological system. However, the data with regard to deficits in sequential oral movement production in these patients would suggest that oral motor sequencing on even a nonverbal level is impaired. That is, the selection and adoption of oral postures and the production of transitions between postures is altered for both speech and nonspeech targets in association with posterior temporal and parietal lobe dysfunction.

The lesion data thus support widespread involvement in the perisylvian cortex of both speech perception, and verbal and nonverbal oral movement production. Alterations in both features to some degree are found in association with most if not all aphasic syndromes. The overlap in these functions seen with the focally disruptive technique of stimulation mapping is thus not inconsistent with observations of aphasic performance in patients with spontaneous lesions.

NEUROPHYSIOLOGICAL STUDIES SUGGESTING AN OVERLAP IN PERCEPTION AND PRODUCTION

The stimulation mapping evidence for a common cortex for motor sequencing and phonemic identification in the perisylvian region has been described. Another piece of neurophysiological evidence in support of overlapping functional cortical areas for oral movement or speech production and perception comes from studies which measure regional cerebral blood flow. Cerebral blood flow is a variable which appears to be closely coupled with metabolic tissue function, and it can be measured in awake patients as they engage in a variety of cognitive activities. During overt speech production there is activation in several cortical regions of the left dominant hemisphere (Roland, Skinhoj, Lassen, and Larsen,

1980; Lassen and Larsen, 1980). Although cortical landmarks cannot be directly observed with this essentially noninvasive procedure, the regions activated during speech appear to include the supplementary motor area, the somotosensory motor and premotor face areas including that cortex situated in the classical Broca's area, and the auditory cortex including posterior aspects of both superior and middle temporal gyri. Of interest is that Broca's area, classically thought to be involved in expressive language functions, is activated together with primary auditory areas when a normal subject either listens to speech. These data suggest that during auditory analysis, there is evoked activity in an area thought to be primarily part of a motor response pathway.

BEHAVIORAL STUDIES

There is also evidence from behavioral studies in normal subjects which supports a tightly linked association between oral motor responses and perceptual functions. Sussman and MacNeilage (1971) have utilized a pursuit auditory tracking task to investigate hemispheric dominance for sensorimotor control of speech-related movements. Subjects were better able to match a continuously varying target tone with a tone controlled by movements of the tongue, mandible, or lips when the cursor tone, the tone under speech articulator movement control, was presented to the right rather than the left ear. This right ear–cursor advantage was hypothesized to result from the presence in the left hemisphere of a specialized auditory sensorimotor integration mechanism which is specialized for integrating kinesthetic and tactile feedback resulting from articulatory movements with auditory concomitants of those same movements.

Porter and Lubker (1980) have provided further support of an acoustic–gestural linkage with regard to verbal performance using a "shadowing" task. Subjects were asked to listen to random series of CV or VCV syllables and repeat them as rapidly as possible. The exceedingly rapid shadowing which approached the lower latency limit for the motor gestures themselves suggested that initiation of the response occurred prior to the perception of the phonetic segments. They concluded that early phases of speech analysis yield input directly convertible to information required for speech production.

THEORIES REGARDING
THE MOTOR–PERCEPTUAL LINK

It is not a new idea that listeners may generate internally an articulatory gesture which corresponds to an acoustic pattern which they hear. Liberman, Cooper, Shankweiler, and Studdert-Kennedy (1967) suggested that such a common mechanism was involved in speech perception. Their "motor theory of speech perception" is based on two phenomena; acoustic invariance and categorical perception. First, each phoneme does not have a unique acoustic representation in the speech signal. Two very different formant transitions can both be heard as the same phoneme (e.g., /d/) if the phoneme is presented in a different consonant–vowel context. Second, perception of phonemes does not follow continuously varying acoustical cues, but rather shows the quality of categorical perception. If for example the lag between the first and second formant, (a cue to "voicing") is varied along a continuum, the resultant stimuli are not perceived continuously but as belonging to one of two categories. Speech perception appeared to depend on more than the constitution of the acoustic signal. Liberman, Cooper, Shankweiler, and Studdert-Kennedy (1967) suggested that the brain may generate a motor model of speech as part of the decoding process, that perception is accomplished by reference to a speaker's "knowledge" of articulatory maneuvers and their acoustic results.

Despite its appeal in dealing with the phenomenon of categorical perception for speech sounds, the motor theory of speech perception has recently come up against findings that severely challenge it. First, some of the perceptual phenomena thought to be unique to speech sound processing are not limited to speech. Second, the perception of human speech sounds, at least as described by the phenomena of categorical perception and perceptual constancy do not appear to be unique to adult speakers or even to *Homo sapiens*. Data gathered from prelinguistic human infants (Eimas, 1974, Kuhl, 1979) and from several animals including primates (Morset & Snowdon, 1975) and chinchillas (Kuhl, and Miller, 1975) also yield categorical perception curves for speech stimuli. Since neither infants nor animals produce these categories or distinctions, it has been proposed that something about the mammalian auditory system facilitates the categorization, that some of the perceptual discontinuities that underlie speech sound contrasts are the natural result of mammalian auditory physiology. Kuhl (1979) suggests that the limited repertoire of contrastive speech sounds used by humans was originally "selected" in the evolutionary scheme of things precisely because they exploited the perceptual discontinuities of, and thus were ideally suited to, the auditory system. The acoustic correlates of certain phonetic features that show quantal properties in production are precisely those features that show categorical-like perception confirming a close relationship between attributes of production and distinctions made in perception (Stevens, 1972). Evidence from infant studies points to a loss of discriminative abilities for certain speech sounds over time if the distinctions are not utilized in the environment. Over the same period, increasing differentiation is a hall-mark of the development of productive language skills. In the context of this discussion the infant language learner develops articulatory maneu-vers that mirror both the inherent perceptual discontinuities and the dis-tinctions utilized in the linguistic environment. The adult's system reflects these associations behaviorally, and the perisylvian cortex, where stimu-lation mapping shows changes in both motor sequencing and phoneme discrimination (Ojemann and Mateer, 1979a,b) would appear to represent an appropriate anatomic substrate for such a process.

Perceptual invariance presents the problem such that in two signals with very different formant transitions (*ud/id*), the consonant is identified as /d/. There is also the problem of motor invariance. The actual move-ments associated with the two productions of /d/ are very different. It has been proposed that the invariance, if it exists, may be found in the ideal-ized target positions of the end point vocal tract configurations (Lind-blom, 1963; Stevens and House, 1963; Halle and Stevens, 1962). In Mac-Neilage's (1970) version of a target-based theory, the speaker possesses an internalized space coordinate system that specifies invariant spatial "targets" for the articulators to achieve. Motor commands are generated as target-directed speech movements. These target positions of the oral musculature are associated with particular phoneme productions, while the actual movement pattern depends on the preceding and following pho-neme targets. Critical to this view are appropriate levels of motor control capability to obtain the targets' spatial configurations and representation at some level of the association between the changing spatial configura-tions of the oral structures and the acoustic output with which they are associated.

The nature of motor control for speech is not well understood. Debates continue as to whether it primarily involves a closed loop system depen-dent on ongoing feedback (tactile, proprioceptive, auditory) or an open loop system of central commands. Some degree of hearing is obviously requisite to the development and long-term maintenance of normal speech production (Chase, 1967; Hardy, 1970; Ringel, 1970; Abbs and Hughes, 1975). Once speech has been developed, however, attempts to disrupt speech production by altering speakers' auditory feedback of their own production appear to be only minimally effective. With amplification, at-tenuation, or selective filtering of their speech, speakers decrease or in-crease vocal intensity or display different resonance characteristics (Sie-gel and Pick, 1974; Lane and Tranel, 1971; Garber, 1976). With elimination of auditory feedback, speakers tend to increase intensity and prolong voicing, but suffer no loss of intelligibility (Lane and Tranel, 1971). Although intelligibility may be maintained, however, it has been shown that even very short duration (200 msec) interruptions in feedback can alter carefully measured lip and jaw movements (Barlow and Abbs, 1978). Thus a strong ongoing perceptual–motor linkage can be identified in normal speakers.

In recent years a number of electrophysiological and behavioral studies have demonstrated that various animals have auditory detectors "tuned" to signals that are of interest to them. Single cells in the auditory system which are matched to the formant frequencies of species-specific vocalizations have been identified in crickets, bullfrogs, and squirrel monkeys (Capranica, 1965; Frishkop and Goldstein, 1963; Wollberg and Newman, 1972). The perceptual apparatuses of these animals appear to be structured in terms of the acoustic parameters to which their sound-producing apparatuses are tuned. It has been hypothesized on the basis of speech sound adaptation studies that humans possess feature detectors in their auditory systems, which are tuned to the acoustic correlates of phonetic features of speech (Eimas and Corbit, 1973). The electrical stimulation mapping studies (Ojemann and Mateer, 1979b) have identified a cortical region in the left hemisphere where mechanisms for both verbal and nonverbal oral movement production and phonemic identification overlap. It may reflect the existence of a coherent evolutionary process in which the human brain evolved unique mechanisms structured in terms of the matched requirements for the development and maintenance of speech production and perception.

SUMMARY AND CONCLUSION

A review of both speech perception and production capacities preferentially dependent on the left hemisphere suggests that in neither area is the hemispheric specialization dependent on symbolic or representational function. First, the left hemisphere appears specialized for the perception of sequences of auditory stimuli, particularly those characterized by rapidly changing acoustic spectra. This kind of analysis is necessary for the phonetic decoding of speech, though not necessarily for the comprehension of larger units of speech. Second, evidence suggests that the left hemisphere contains a system for the accurate internal representation of moving body parts, important for the control of changes in the position of both the oral and brachial musculature. Both of these fundamental features of left-hemisphere function appear to utilize the left dominant perisylvian region. Achievement of single oral movements or postures can apparently be mediated by a more anterior system on the left, with coordination of these individual movements into a series being critically dependent on the posterior system. The perceptual system appears to overlap with both systems.

Evidence for a strong overlap in oral motor control and perceptual function is derived both from behavioral assessment of patients who has sustained lesions, and studies with normal subjects. Neurophysiological evidence for the interdependence and its localization in perisylvian cortex comes from both cortical electrical stimulation studies and cerebral blood flow studies.

The tightly knit organization of fundamental oral-motor control capabilities and acoustic analysis in the dominant perisylvian region are hypothesized to reflect a phylogenetically developing system on which the development and maintenance of oral communication depend.

REFERENCES

Abbs, J. H., and Hughs, O. M. (1975). Motor equivalence coordination in labial-mandibular system. *Journal of the Acoustical Society of America, 58,* 539.

Alajouanine, T. (1956). Verbal realization in aphasia. *Brain, 79,* 1–28.

Alajouanine, T., and Lhermitte, F. (1964). Aphasia and physiology of speech. *Journal of Nervous and Mental Disease, 42,* 204–219.

Baker, E., Blumstein, S. E., and Goodglass, H. (1979). Phonological versus semantic factors in the auditory comprehension of aphasics. New York: International Neuropsychology Society.

Barlow, S. M., and Abbs, J. H. (1978). Some evidence of auditory feedback contributions to the ongoing control of speech production. *ASHA: A Journal of the American Speech and Hearing Association, 20,* 728.

Basso, A., Casati, G., and Vignolo, L. (1977). Phonemic identification defect in aphasia. *Cortex, 13,* 85–95.

Bay, E. (1965). The concepts of agnosia, apraxia and aphasia after a history of a hundred years. *Journal of Mount Sinai Hospital, 32,* 637–650.

Beecher, M. D., Petersen, M. R., Zoloth, S. R., Moody, D. B. and Stebbins, W. C. (1979). Perception of conspecific vocalizations by Japanese macaques. *Brain, Behavior and Evolution, 16,* 443–460.

Benson, D. F. (1967). Fluency in aphasia: Correlation with radioactive scan localization. *Cortex, 3,* 373–394.

Blumstein, S. E. (1973). *A phonological investigation of aphasic speech.* The Hague: Mouton.

Blumstein, S. E. (1981). Neurolinguistic disorders: Language-brain relationships. In S. B. Filskov and T. J. Boll (Eds.), *Handbook of Clinical Neuropsychology.* New York: Wiley.

Blumstein, S. E., Baker, E., and Goodglass, H. (1977). Phonological factors in auditory comprehension in aphasia. *Neuropsychologia, 15,* 19–30.

Blumstein, S. E., Cooper, W. E., Goodglass, H., Statlender, S., and Gottlieb, J. (1980). Production deficits in aphasia: A voice-onset time analysis. *Brain and Language, 9,* 153–170.

Blumstein, S., Cooper, W., Zurif, E., and Caramazza, A. (1977). The perception and production of voice-onset time in aphasia. *Neuropsychologia, 15,* 371–383.

Canter, C. J. (1967). Neuromotor pathologies of speech. *American Journal of Medicine, 46,* 659–666.

Capranica, R. R. (1965). Sound communication with bullfrog, *Rana catesbeiana. American Zoologist, 5,* 693–695.

Carpenter, A. L., and Rutherford, D. R. (1973). Acoustic cue discrimination in adult aphasia. *Journal of Speech and Hearing Research, 16,* 534–542.

Chaney, R. B., and Webster, J. C. (1965). Information in certain multidimensional acoustic signals. Report No. 1339, *USN Electronics Laboratory Reports,* San Diego.

Chase, R. A. (1967). Abnormalities in motor control secondary to congenital sensory deficits. In J. F. Bosma (Ed.), *Symposium on oral sensation and perception.* Springfield, Ill.: Thomas.

Curry, F. K. W. (1967). A comparison of left-handed and right-handed subjects on verbal and non-verbal dichotic listening tasks. *Cortex, 3,* 343–352.

Curry, F. K. W., and Rutherford, D. R. (1967). Recognition and recall of dichotically presented verbal stimuli by right- and left-handed persons. *Neuropsychologia, 5,* 119–125.

Cutting, J. E. (1973). A parallel between degree at encodedness and the ear advantage: Evidence from an ear monitoring task. *Journal of the Acoustical Society of America, 52,* 368–376.

Cutting, J. E. (1974). Two left-hemisphere mechanisms in speech perception. *Perception and Psychophysics, 16,* 601–612.

Darley, F. L., Aronson, A. E. and Brown, J. R. (1975). Apraxia of speech: Impairment of motor speech programming. *Motor Speech Disorders,* Philadelphia, W. B. Saunders.

Darwin, C. (1969). Laterality effects in recall of steady-stage and transient speech sounds. *Journal of the Acoustical Society of America, 46,* 114–121.

Darwin, C., Taylor, L., and Milner, B. (1975). Proceedings of the Seventeenth International Symposium of Neuropsychology. *Neuropsychologia, 13,* 132.

DeRenzi, E., Pieczuro, A., and Vignolo, L. A. (1966). Oral apraxia and aphasia. *Cortex, 2,* 50–73.

DeRenzi, E., and Vignolo, L. A. (1963). The Token Test: A sensitive test to detect receptive disturbance in aphasics. *Brain, 85,* 665–678.

Divenji, P. L., and Efron, R. (1979). Spectral versus temporal features in dichotic listening. *Brain and Language, 7,* 375–386.

Efron, R. (1963). Temporal perception, aphasia and déjà vu. *Brain, 86,* 403–424.

Eimas, P. (1974). Linguistic processing of speech by young infants. In R. L. Schiefelbusch and L. L. Lloyd (Eds.), Language Perspectives: Acquisition, Retardation, Intervention. Baltimore: University Park Press.

Eimas, P. and Corbit, J. (1973). Selective adaptation of linguistic feature detectors. *Cognitive Psychology, 4,* 99–109.

Eisenson, J. (1962). Aphasia. In L. Levin (Ed.), *Voice and Speech Disorders: Medical Aspects.* Springfield, Ill.: Thomas.

Freeman, F. J., Sands, E. S., and Harris, K. S. (1978). Temporal coordination of phonation and articulation in a case of verbal apraxia: A voice onset time study. *Brain and Language, 6,* 106–111.

Frishkop, L. S., and Goldstein, M. H. (1963). Responses to acoustic stimuli from single units in eighth nerve of bullfrog. *Journal of the Acoustical Society of America, 35,* 1219–1224.

Gainotti, G., Caltagirone, C., and Ibba, A. (1975). Semantic and phonemic aspects of auditory language comprehension in aphasia. *Linguistics, 154,* 15–29.

Garber, S. (1976). The effects of feedback filtering on nasality. Paper presented at American Speech and Hearing Association Convention, Houston, Texas, 1976.

Goodglass, H., and Kaplan, E. (1972). *The assessment of aphasia and related disorders.* Philadelphia: Lea and Febinger.

Goodglass, H., Quadfasel, F. A. and Timberlake, W. H. (1964). Phrase length and the type of severity of aphasia. *Cortex, 1,* 133–153.

Haggard, M. P. (1969). Perception of semivowels and laterals. *Journal of the Acoustical Society of America, 46,* 115–121.

Halle, M., and Stevens, K. N. (1962). Speech recognition: A model and a program for research. *IRE Transactions on Information Theory,* **IT-8,** 155–159.

Hardy, J. C. (1970). Development of neuromuscular systems underlying speech production. In *Speech of the Dentofacial Complex: The State of the Art.* Proceedings of the Workshop: ASHA Report 5. Washington, D.C.: American Speech and Hearing Association.

Hicks, R. E., Provenza, F. J., and Rybstein, E. R. (1975). Generalized and lateralized effects of concurrent verbal rehearsal upon performance of sequential movements of the fingers by the left and right hands. *Acta Psychologica, 39,* 119–130.

Jackson, J. H. (1878, 1932). Remarks on non-protrusion of the tongue in some cases of aphasia. In J. Taylor *Selected Writings of John Hughlings Jackson,* Vol. 2. London: Hodder and Stoughton, 1932.

Itoh, M., Susanuma, S., and Hirose, H. (1978). Articulatory dynamics in a patient with apraxia of speech. Paper presented at meetings of the Academy of Aphasia, Chicago.

Johns, D. F., and Darley, F. L. (1970). Phonemic variability in apraxia of speech. *Journal of Speech and Hearing Research, 13,* 556–583.

Karis, R., and Horenstein, S. (1976). Localization of speech parameters by brain scan. *Neurology, 26,* 226–230.

Kerschensteiner, M., Poeck, K., and Brunner, E. (1972). The fluency–nonfluency dimension in the classification of aphasic speech. *Cortex, 8,* 233–247.

Kimura, D. (1961). Some effects of temporal lobe damage in auditory perception. *Canadian Journal of Psychology, 15,* 156–165.

Kimura, D. (1964). Left–right differences in the perception of melodies. *Quarterly Journal of Experimental Psychology, 16,* 335–338.

Kimura, D. (1967). Functional asymmetry of the brain in dichotic listening. *Cortex, 3,* 163–178.

Kimura, D. (1973). Manual activity during speaking: I. Right-handers. *Neuropsychologia, II,* 45–50.

Kimura, D. (1976). The neural basis of language qua gesture. *Studies in Neurolinguistics, 2,* 145–156.

Kimura, D. (1978). Acquisition of a motor skill after left hemisphere damage. *Brain, 100,* 527–542.

Kimura, D. (1981). Neural mechanisms in manual signing. *Sign Language Studies, 33,* 291–312.

Kimura, D. (1979). Neuromotor mechanisms in the evolution of human communication. In H. D. Steklin and M. J. Raleigh (Eds.), *Neurobiology of social communication in primates.* New York: Academic Press. 197–219.

Kimura, D. and Archibald, Y. (1974). Motor functions of the left hemisphere. *Brain, 97,* 337–350.

Kimura, D., Battison, R., and Lubert, B. (1976). Impairment of nonlinguistic hand movements in a deaf aphasic. *Brain and Language, 3,* 556–571.

Kimura, D., and Folb, S. (1968). Neural processing of backwards-speech sounds. *Science, 161,* 395–396.

Kinsbourne, M., and Cook, J. (1971). Generalized and lateralized effects of concurrent verbalization on a unimanual skill. *Quarterly Journal of Experimental Psychology, 23,* 341–345.

Kuhl, P. K. (1979). Models and mechanisms in speech perception: Specks comparisons provide further contributions. *Brain Behavior and Evolution, 16,* 374–408.

Kuhl, P. K. and Miller, J. D. (1975). Speech perception by the chinchilla: voice-voiceless distinction in alveolar plosive consonants. *Science, 190,* 69–72.

Lackner, J. R., and Teuber, H. L. (1973). Alterations in auditory fusion thresholds after cerebral injury in man. *Neuropsychologia, II,* 409–415.

Lane, H. L., and Tranel, B. (1971). The Lombard sign and the role of hearing in speech. *Journal of Speech and Hearing Research, 14,* 677–709.

Lassen, N. A., and Larsen, B. (1980). Cortical activity in the left and right hemispheres during language-related brain functions. *Phonetica, 37,* 27–37.

Liberman, A. M., Cooper, F. S., Shankweiler, D. P., and Studdert-Kennedy, M. (1967). Perception of the speech code. *Psychological Review, 74,* 431–461.

Liepmann, H. (1913). Motor aphasia, anarthria and apraxia. *Translations of the 17th International Congress of Medicine (London),* Section XI, Part 2, 97–106.

Lindblom, B. (1963). Spectrographic study of vowel reduction. *Journal of the Acoustical Society of America, 35,* 1773–1779.

Lomas, J. (1980). Competition within the left hemisphere between speaking and unimanual tasks performed without visual guidance. *Neuropsychologia, 18,* 141–149.

Lomas, J., and Kimura, D. (1976). Intrahemispheric interaction between speaking and sequential manual activity. *Neuropsychologia, 14,* 23–33.

McFarland, F., and Ashton, R. (1975). The lateralized effects of concurrent cognitive activity on a unimanual skill. *Cortex, II,* 283–290.

McFarland, F., and Ashton, R. (1978). The influence of brain lateralization of function on a manual skill. *Cortex, 14,* 102–111.

MacNeilage, P. F. (1970). The motor control of serial ordering of speech. *Psychological Review, 77,* 182–196.

Mateer, C. (1976). Patterns of oral-facial sensory impairment after unilateral hemispheric damage. University of Western Ontario Research Bulletin No. 393, London, Canada.

Mateer, C. (1978). Impairments of nonverbal oral movements after left hemisphere damage: A follow-up analysis of errors. *Brain Language, 6,* 334–341.

Mateer, C. and Dodrill, C. (Unpublished observations). Overlapping lateralization of nonverbal oral motor control and speech dominance: Evidence from sodium amytal procedures.

Mateer, C., and Kimura, D. (1976). Impairment of nonverbal oral movements in aphasia. *Brain and Language, 4,* 262–276.

Mazzochi, F., and Vignolo, L. A. (1979). Localization of lesions in aphasia: Clinical CT scan correlations in stroke patients. *Cortex, 15,* 627–654.

Mills, L., and Rollman, G. (1979). Left hemisphere selectivity for processing duration in normal subjects. *Brain and Language, 1,* 320–335.

Mohr, J. (1976). Broca's area and Broca's aphasia. *Studies in Neurolinguistics, 1,* 201–236.

Molfese, D. L. (1978). Left and right hemisphere involvement in speech perception: Electrophysiological correlates. *Perception and Psychophysics, 23,* 237–243.

Morse, P. A. and Snowdon, C. T. (1975). An investigation of categorical speech discrimination by rhesus monkeys. *Perception and Psychophysics, 17,* 9–16.

Naesar, M. A., and Hayward, R. W. (1978). Lesion localization in aphasia with cranial-computed tomography and Boston Diagnostic Aphasia Exam. *Neurology, 28,* 545–551.

Nathan, P. W. (1947). Facial apraxia and apraxic dysarthia. *Brain, 70,* 449–478.

Ojemann, G. A. (1978). Organization of short-term verbal memory in language area of human cortex: Evidence from electrical stimulation. *Brain and Language, 5,* 331–348.

Ojemann, G. A. (1979). Individual variability in cortical localization of language. *Journal of Neurosurgery, 50,* 164–169.

Ojemann, G. A., and Mateer, C., (1979a). Cortical and subcortical organization of human communication: Evidence from stimulation studies. In H. D. Steklis and M. J. Raleigh, (Eds.), *Neurobiology of social communication in primates.* New York: Academic Press.

Ojemann, G. A., and Mateer, C. T. (1979b). Human language cortex: Localization of memory, syntax and sequential motor-phoneme identification systems. *Science, 205,* 1401–1403.

Ojemann, G. A., and Whitaker, H. (1978). Language localization and variability. *Brain and Language, 6,* 239–260.

Penfield, W., and Roberts, L. (1959). *Speech and brain mechanisms.* Princeton, N.J.: Princeton Univ. Press.

Poeck, K., and Kerschensteiner, M. (1975). Analysis of the sequential motor events in oral apraxia. In K. J. Zulch, O. Creutzbeldt, and B. C. Galbraith (Eds.), *Cerebral localization.* Berlin: Springer-Verlag.

Porter, R. J., and Lubker, J. F. (1980). Rapid reproduction of vowel–vowel sequences. Evidence for a fast and direct acoustic–motoric linkage in speech. *Journal of Speech and Hearing Research, 23,* 593–602.

Ringel, R. L. (1970). Oral sensation and perception. *American Speech and Hearing Association Reports, 5,* 188–206.

Roland, P. E., Shinhoj, E., Lassen, N. A., and Larsen, B. (1980). Different cortical areas in man in organization of voluntary movements in extrapersonal space. *Journal of neurophysiology, 43,* 137–150.

Schuell, H. M., and Jenkins, J. J. (1959). The nature of language deficit in aphasia. *Psychological Review, 66,* 45–67.

Schwartz, J., and Tallal, P. (1980). Rate of acoustic change may underlie hemispheric specialization for speech perception. *Science, 207,* 1380–1381.

Shankweiler, D., and Studdert-Kennedy, M. (1967). Identification of consonants and vowels presented to left and right ears. *Quarterly Journal of Experimental Psychology, 19,* 59–63.

Siegel, G. M., and Pick, H. L., Jr. (1974). Auditory feedback in the regulation of the voice. *Journal of the Acoustical Society of America, 56,* 1618–1624.

Stark, R., and Tallal, P. (1978). Effect of acoustic-cue redundancy of perception of stop–vowel syllables by language-delayed children. *Journal of the Acoustical Society of America, 64(S1),* S50–S51.

Stark, R., and Tallal, P. (1979). Analysis of stop consonant production errors in developmentally dysphasic children. *Journal of the Acoustical Society of America, 66,* 1703–1712.

Stevens, K. N. (1972). The quantal nature of speech: evidence from articulatory-acoustic data. In David and Deres (Eds.), *Human Communication: A Unified View.* New York: McGraw-Hill.

Stevens, K. N., and House, A. S. (1963). Perturbation of vowel articulations by consonantal context. *Journal of Speech and Hearing Research, 6,* 111–128.

Studdert-Kennedy, M., and Shankweiler, D. (1970). Hemispheric specialization for speech perception. *Journal of the Acoustical Society of America, 48,* 579–594.

Sussman, H. M., and MacNeilage, P. F. (1971). The laterality effect in lingual–auditory tracking. *Journal of the Acoustical Society of America, 49,* 1874–1880.

Tallal, P., and Newcombe, F. (1978). Impairment of auditory perception and language comprehension in dysphasia. *Brain and Language, 5,* 13–24.

Tallal, P., and Piercy, M. (1974). Developmental aphasia: Rate of auditory processing and selective impairment of consonant perception. *Neuropsychologia, 12,* 82–83.

Tognola, G., and Vignolo, L. A. (1980). Brain lesions associated with oral apraxia in stroke patients: A clinico-neuroradiological investigation with the CT scan. *Neuropsychologia, 18,* 257–272.

Wollberg, Z., and Newman, J. D. (1972). Auditory cortex of squirrel monkey—response patterns of single cells to species-specific vocalizations. *Science, 175,* 212–214.

Woltman, H. (1923). Review of clinical and anatomic studies of apraxia with special reference to papers of R. Brun. *Archives of Neurology and Psychiatry, 10,* 344–358.

Yeni-Komshian, G. H., and Rao, P. (1980). Speech perception in right and left CVA patients. Paper presented at International Neuropsychological Society Meetings, San Francisco.

CHAPTER 7

Thalamic Mechanisms in Language and Memory*

Catherine A. Mateer and George A. Ojemann

INTRODUCTION

This chapter discusses changes in speech, language, memory, and motor control associated with altered function of the thalamus. Suggestions that the thalamus and nearby basal ganglia might have a specific language role were put forth very early in the history of aphasiology. Pierre Marie's (1906) anarthric zone included portions of both these structures. Language deficits were reported by Dejerine and Roussy (1906) as part of their early description of the effects of focal thalamic lesions. Despite

* Portions of this research were supported by NIH Research Grant NS 1805 17111, awarded by the National Institute of Neurological and Communicative Disorders and Stroke, PHS/DHHS. C.A.M. is the recipient of NIH Teacher Investigator Development Award 1KO7 NS 00505. Both authors are affiliates of the Child Development and Mental Retardation Center at the University of Washington, Seattle, Washington, 98195.

these early observations, the prevailing view for the next fifty years held that the thalamus and basal ganglia, though important in maintaining overall alertness, did not play a specific role in the generation of language (Neilson, 1946). In the last two decades, however, there has been a substantial interest in the role of these structures in language.

Several factors contributed to this renewed interest. The widespread application of angiography to identify more discretely the location of spontaneous intercerebral hemorrhages during life provided the basis for more detailed descriptions of clinical syndromes associated with different intracerebral hemorrhages. It was noted that left but not right thalamic hemorrhages were commonly associated with a language disturbance (Fisher, 1959). Penfield and Roberts (1959) hypothesized that the thalamus was an important link between anterior and posterior cortical language areas. With the advent of computerized tomographic (CT) scanning, greater numbers of cases with spontaneous thalamic lesions of all kinds (particularly hematomas) can be identified during life, so that detailed assessment of language function can be obtained (Mohr, Watters, and Duncan, 1975).

At the same time, stereotaxtic surgery of the thalamus became widely practiced, most often as treatment for dyskinesias, particularly parkinsonism. Following placement of lesions in the left ventral lateral thalamus, it came as something of a surprise that a significant number of patients demonstrated clinically evident dysnomias (in excess of 40% in one series [Selby, 1967]). The detailed evaluation of language function after these discrete surgical lesions in the thalamus and adjacent structures, both acutely and over the long term, has provided additional insight into subcortical language mechanisms (Bell, 1968). Further insight into these mechanisms has also come from the intraoperative testing of language functions during focal electrical stimulation of thalamic nuclei during stereotaxic operations, using techniques similar to those previously developed for identification of cortical areas involved in language functions. (Ojemann, Fedio, and Van Buren, 1968).

The general consensus from both spontaneous and surgically induced lesion studies is that there is lateralization of language functions at the level of the thalamus, usually to the left side, and that those thalamic language functions differ from language functions subserved by any cortical area. In the model we have developed to synthesize the lesion and stimulation data, the left and right thalami function asymmetrically to gate access to the cortex by means of a "specific alerting response" which is related to material-specific input. The support for and nature of this function will be discussed below.

ANATOMICAL REVIEW

The two thalami form the great bulk of the diencephalon. The left and right thalami appear as asymmetric egg-shaped nuclear masses lying obliquely across the cranial end of the brain stem. Continuous with the midbrain, each thalamus is bounded laterally by the internal capsule and medially and superiorly by the third ventricle.

Depending upon the classification system used, each thalamus contains up to 25 separate nuclear masses. Several of these nuclei have a specific area of cortex with which they are interconnected. The cortical connections are always bidirectional, both from the thalamus to the cortex, and from the cortex back to the same area of thalamus. The thalamocortical connections are generally facilitory. Activation of a minute portion of the thalamus activates the corresponding and much larger portion of the cerebral cortex.

Three nuclei in the lateral portion of the thalamus, the ventral anterior (VA) the ventral lateral (VL) and the ventral posterior (VP) have projections to the premotor, motor, and sensory areas of the cortex respectively. At least three other nuclei have connections with association areas of the cortex: the dorsomedial nucleus (DM) with the entire prefrontal cortex, the intralaminar nucleus with the orbital frontal cortex, and the pulvinar nucleus (P) with the posterior parietotemporal area. Phylogenetically the thalamus enlarges as the cortex develops. Most of this thalamic enlargement is in pulvinar, and more dorsal areas of the lateral thalamus. Those dorsal parts of the human lateral thalamus do not show the clear anatomic subdivisions of the ventral lateral thalamic area. The enlargement of these areas in primate development is related to an additional later migration of precursor neurons into these thalamic areas—neurons that also contribute to the development of the basal ganglia (Rakic, 1972).

ALTERATIONS IN LANGUAGE AND MEMORY AFTER THALAMIC LESIONS: EVIDENCE FROM SPONTANEOUSLY OCCURRING LESIONS

Naturally occurring lesions (restricted to the thalamus) which have been associated with language deficits include tumors, hematomas, and infarcts of the left thalamus (Arseni, 1958; Bugiani, Conforto, and Sacco, 1969; Cappa and Vignolo, 1979; Ciemins, 1970; Cheek and Taveras, 1966; Fazio, Sacco, and Bugiani, 1973; Fisher, 1959; Horenstein, Chung, and Brenner, 1978; Luria, 1977; McKissock and Paine, 1958; Mohr, Watters,

and Duncan, 1975; Penfield and Roberts, 1959; Reynolds, Harris, Ojemann, and Turner, 1978; Reynolds, Turner, Harris, and Ojemann, 1979;
Samarel, Wright, Sergay, and Tyler, 1976). Other sources for information
on the thalamic role in language are the rare cases with spontaneous
neuronal degeneration primarily, although seldom solely, in the thalamus
(Brown, 1979). The case material is variable with respect to etiology and
to the strength of the anatomic localization. For all types of case material,
there are the additional problems of the adequacy of the language assessment and the lack of longitudinal evaluations. (A recent longitudinal case
study by Glosser, Kaplan and LoVerme (1982) represents a welcome exception.) Despite these limitations, several consistent observations associated with thalamic involvement have emerged from these reports with
regard to speech and language behavior.

The most frequent speech-related change cited during the early postin-
sult stage is aphonia or muteness. It is particularly evident in cases of bilateral ventrolateral thalamic lesions (Bertrand, Martinez, Hardy, Molina-
Negro, and Velasco, 1973). In the course of recovery following this early
stage, a frequently noted general characteristic is wide fluctuation in performance over intervals as short as minutes. This fluctuation is apparent
both for general levels of activity, with the patient moving between deep
somnolence and normal alertness, and for specific behaviors such as
speech output, with the patient ranging from being mute to producing log-
orrhea (Mohr, Watters, and Duncan, 1975). Although this situation is
most apparent in the early stages after a lesion, aspects of it often persist.
This degree of variability is not typically seen following cortical lesions.
When patients with thalamic lesions produce language, it is commonly
fluent. Understanding of speech is usually relatively intact, as is repetition
(an important distinction from the usual pattern of conduction aphasia
after damage to the nearby temporal lobe stem). The fluent speech generally shows a substantial degree of dysnomia, with frequent perseveration. Unlike dysnomic errors seen with cortical lesions, naming errors
seen with thalamic lesions are often characterized by wildly extraneous
intrusions (Luria, 1977). The naming errors frequently have no obvious
phonemic or semantic paraphasic qualities. They are perfectly good
words that are simply irrelevant to the topic at hand. A striking example is
a case reported by one of the authors, of a patient who repeatedly called
pictures of simple objects "affirmative action" (Ojemann, 1976). Deficits
are usually more marked in spoken than in written language. Reading, in
the few cases of thalamic lesion in which it has been studied, has been
relatively intact (Cappa and Vignolo, 1979; Glasser, Kaplan, and Lo-
Verme, 1982).

Many features of this "syndrome" of thalamic aphasia were evident in a recent patient of ours. This patient, a 59-year-old female, was admitted with severe headache, and had sustained a left posterior communicating artery aneurysm bleed with hemorrhage into the left subarachnoid and intraventricular spaces. Following increased obtundation, a left frontal ventriculostomy was performed. Following this procedure she was characterized as alert, oriented, socially appropriate, and without receptive or expressive aphasia. Five days later she underwent a left frontal craniotomy with clipping of the aneurysm. Immediately after the operation she demonstrated decreased responsiveness and right hemiparesis. Two days later cerebral blood flow studies were normal but there was CT scan evidence of an infarction only in the left anterior thalamus and in the anterior quarter of the posterior limb of the internal capsule (Figure 1). This area was distant from the aneurysm and the operative manipulation, and likely represented an infarct in the distribution of the anterior choroidal artery. During the first three days after the operation, there was gradual recovery of alertness, though intermittent episodes of lethargy and somnolence were noted throughout her recovery. When awake, she began responding variably to simple commands but was essentially mute for the first two to three postoperative weeks. There was no spontaneous speech and no speech could be elicited. Little spontaneous movement was observed and she did not orient herself to visual stimuli or to speech with either postural change or with eye movements. In the third postoperative week she began to respond to simple questions with single word responses. Verbalizations increased over the next three weeks, at which time she demonstrated use of simple and occasionally complex sentences. Performance on language tests was highly variable not only day to day, but minute to minute.

Two months following surgery, the right upper and lower limbs were back to almost full strength while language impairments persisted. Despite a continued reduction in spontaneous speech, her speech was fluent with intact prosody, articulation, and grammar. The most striking abnormality was a marked anomia, evident on over 50% of items to be named. The semantic content of verbal responses was severely restricted because of the limited number of content words used. In most cases, neologisms or unrelated real words were substituted for content words in a perseverative fashion (e.g., persapil–glass; characters–penny, watch–pencil). Repetition of words and sentences was intact. Comprehension was relatively less impaired, though deficits were apparent with long and/or linguistically complex material (Revised Token Test 55/64). Despite intact immediate memory span, short-term postdistractional memory for verbal (but not nonverbal) material was severely impaired (Wechsler Memory Scale MQ-73). Oral reading was unimpaired and reading comprehension, which was evident well before auditory comprehension for comparable

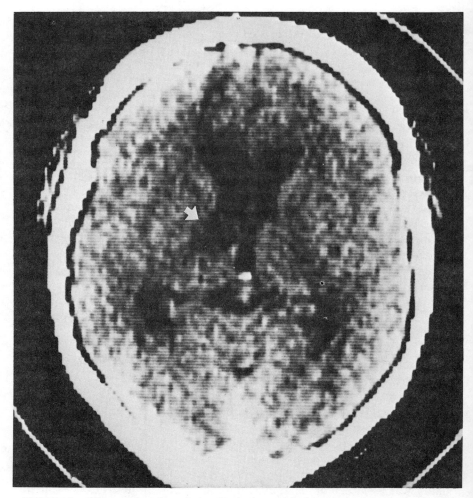

FIGURE 1.   CT scan taken through the anterior horn of the lateral ventricles, indicating the site of the infarction in the left anterior thalamus and part of the adjacent posterior limb of the internal capsule, as marked by the arrow.

material, was functional at a sentence level. Written language was accurate on copying and dictation tasks. Spontaneous written expression was rambling and contained little content. A bilateral upper limb apraxia was evident, though the right side showed greater apraxia than the left side. Despite marked generalized improvement, a paucity of self-initiated activity (including speech, self care, and movement) persisted in general until discharge. At one year, postonset, this patient demonstrated a persistent mild anomic aphasia and decreased verbal memory abilities.

The prognosis for ideational language deficits after thalamic lesions is generally favorable. Most patients recover functional language within several weeks, though formal testing beyond that time may show subtle language deficits (Reynolds, Harris, Ojemann, and Turner, 1978). A few cases with spontaneous thalamic lesions have been reported, however, in which deficits persisted for long periods with little resolution (Ciemins, Case 2, 1970; Reynolds, Harris, Ojemann, and Turner, Case 2, 1978; Brown, 1979).

The generally consistent findings from these thalamic lesion studies include lateralization of language functions and a pattern of language disruption that is different from that seen with cortical lesions. The most prominent, frequently occurring features of the syndrome include fluctuating performance, anomia with perseveration and extraneous intrusions, relatively spared repetition and comprehension, and a general poverty of expression.

Before proceeding, it must be noted that these views are by no means universally accepted (Brown, 1979). There are good reasons for some skepticism. None of the case material derived from the study of thalamic lesions is perfect. The traditional data base of *aphasiology*, the study of language changes after a discrete brain infarct resulting from a constriction of blood supply is almost unknown in thalamic studies. Our case reported above is unusual. Most of the cases of spontaneous thalamic lesions have been due to hemorrhagic bleeding into neural tissue rather than to infarcts. In addition to damage at the site of origin, hemorrhages are likely to have secondary distant effects, in the form of brain distortion, due to the hemorrhagic mass or obstruction of spinal fluid pathways. The question, then, is how much of the syndrome of left thalamic hemorrhage can be ascribed to (1) the local damage, (2) to mass effects on surrounding structures, or (3) to an associated acute hydrocephalus, as has been suggested by several authors (Fisher, 1959; Geschwind, 1967). The syndrome of left thalamic hemorrhage often includes sensory defects which predominate over gross motor disturbances and oculomotor disturbances, particularly with vertical gaze, in addition to the presence of an expressive dysphasia. The study of Reynolds, Harris, Ojemann, and Turner (1978) demonstrated that features of this syndrome could be disassociated by treating the acute hydrocephalus with ventricular drainage. Under those conditions it was the dysphasia that REMAINED unchanged, while other features of the syndrome improved. This suggests that the language changes are more likely the result of a local effect.

Finally, data with regard to metabolic abnormalities in brain areas in aphasic patients have added a new dimension to the study of anatomoclinical correlations in aphasia. Metabolic abnormalities exhibited by position emission computed tomography (PECT) were much more extensive than the CT evidence of structural damage in aphasic patients (Metter, Wasterlain, Kuhl, Hanson, and Phelps, 1981). The local rate of glucose metabolism, an indirect measure of the local neuronal firing rate, was decreased in the thalamus in patients whose primary lesions appeared to be cortical, and decreased in the cortex when the primary lesion appeared to be thalamic. The consistency of these thalamocortical metabolic interactions provides further support that this circuitry has a functional importance in language.

LANGUAGE CHANGES ASSOCIATED WITH STEREOTAXIC THALAMIC LESIONS

Most spontaneous thalamic lesions affect multiple nuclei in the thalamus. For the clinical purposes of reducing tremor and rigidity, stereotaxic lesions have been made in specific thalamic nuclei. The ventrolateral (VL) nucleus, or less often the pulvinar (P), have been targeted for such purposes. Although bilateral lesions are often required, especially in the cases of patients with Parkinson's disease, or multiple sclerosis, lesions are typically made in a two-stage process. The thalamic nuclei contralateral to the most affected side of the body are lesioned at least some months before the nuclei on the ipsilateral side. Additional evidence for participation of the dominant thalamus in language has come from these procedures.

Postoperative aphasia, to varying degrees, has been reported at clinical examination in 34–42% of patients after left stereotaxic thalamic lesions (Krayenbuhl, Siegfried, Kohenof, and Yasargil, 1965; Bell, 1968; Riklan, Levita, Zimmerman, and Cooper, 1969; Selby, 1967). Histologically, the lesions in patients who evidence postoperative dysphasia have been confined to the left ventral lateral nucleus or the left anterosuperior pulvinar nucleus (Samra, Riklan, Levita, Zimmerman, Waltz, Bergmann, and Cooper, 1969; Ojemann, 1976). The deficits do not occur with needling of the brain alone but occur only when a thalamic lesion is made (Riklan and Levita, 1969; Sem-Jacobsen, 1965). Ablation of corresponding nuclei of the right thalamus does not result in postoperative language deficits (Ojemann, 1976).

The postoperative aphasias following stereotaxic thalamotomy have many of the characteristics of those following spontaneous thalamic lesion, including fluency, profuse misnaming with perseveration, and marked fluctuation in performance. In one series of patients tested (Oje-mann, 1975), 62% of patients were found to make anomic errors 48 hours after left VL thalamic lesions. In addition, lesions of the left thalamus but not of the right thalamus have been associated with increased latencies to respond and higher error rates on verbal fluency tasks even after the re-covery of naming and most other language functions (Almgren, Ander-son, and Kullberg, 1969; Riklan, Levita, Zimmerman, and Cooper, 1969; Laitinen and Vilkki, 1977).

The aphasia which follows stereotaxic thalamic lesions is almost al-ways transitory although occasional persistent aphasias have been re-ported (Bell, 1968). However, the later evaluation of patients with surgi-cal left thalamic lesions has often demonstrated persisting verbal memory deficits (Shapiro, Sadowsky, Henderson, and Van Buren, 1973).

Somewhat different, although equally valid criticisms have been lev-eled at the results of stereotaxic lesions in studying thalamic involvement in language. First, although stereotaxic surgical lesions are quite discrete, they are always made in selected patient populations—those with the ap-propriate indication for the operation. One of the most common of such populations are patients with parkinsonism, who sometimes have damage in other areas of the nervous system that may contribute to the language changes seen after the thalamic operation (Van Buren, 1975). Second, it has been argued that the effects on language are related to the diffuse short-term effects of the lesion on surrounding tissue, rather than to the thalamic effect per se. However, the language deficits are often seen in the absence of other symptoms, such as motor deficits, sensory loss, or gaze impairments, which would be apparent if one were observing only mass effects. Finally, in several cases with aphasia, the lesion has been histologically confined to the thalamus, a finding which supports the suffi-ciency of a lesion in this structure with regard to the manifestation of lan-guage disorder (Ojemann, 1976).

THE ROLE OF THE LEFT THALAMUS
IN LANGUAGE: EVIDENCE
FROM ELECTRICAL STIMULATION

The left thalamus' role in language has been further defined by observa-tions of the effects of local electrical stimulation of different thalamic areas on language and related tasks, observations made during the course of stereotaxic thalamotomies. Some of these studies have identified sev-eral subcortical sites where speech articulation is blocked by stimulation. These include the anterior limb of the internal capsule bilaterally (Van

Buren, Li, and Ojemann, 1966) and the portions of the posterior limb of the internal capsule immediately adjacent to the ventral lateral thalamus bilaterally (Guiot, Hertzog, Rondot, and Molina, 1961; Ojemann, 1977). Other studies relate a discrete area of only the left thalamus to the ideational aspects of speech (Ojemann, Fedio, and Van Buren, 1968; Ojemann and Ward, 1971; Ojemann, 1975, 1977). This area includes the medial central portion of the ventral lateral thalamus, extending anteriorly to the lateral thalamic pole and posteriorly into the anterior superior portion of lateral pulvinar. This area of the thalamus is identified by sites where the phenomenon of anomia is evoked during the object-naming task; that is, the patient is unable to name, but demonstrates a retained ability to speak.

Different patterns of naming errors are evoked in different portions of the left thalamus, suggesting that subdivisions of it subserve different functions in the language process. Posteriorly, in the posterior portion of the medial-central left ventral lateral thalamus and the adjacent anterior superior part of the lateral pulvinar, the evoked anomic errors are a mixture of misnamings and omissions, a pattern similar to that obtained from stimulation of cortical areas thought to subserve language functions. More anteriorly, in the medial–central portion of ventral lateral thalamus, are sites where the evoked errors most commonly involve perseveration on a portion of the correct object name, with occasional interspersed misnamings. This pattern of naming errors has not been evoked from the cortex and may indicate a site of interaction between language and motor speech functions as discussed further below. Stimulation more anterior in the lateral thalamus is associated with another pattern of naming errors which is not represented in the cortex. Effects with stimulation there include the production of spontaneous words and phrases (Schaltenbrand, 1965) or the production of a particular repeated word each time naming is required (Ojemann, 1975, 1977). This wrong word is frequently the last object correctly named at subthreshold currents. This phenomenon is thought to reflect the activity of verbal alerting circuits. Intense activation of these circuits by the stimulating current at the time of input of a particular object name makes its subsequent retrieval more likely than any other name.

The study of the evoked changes in short-term verbal memory from these and adjacent areas of the thalamus further defines this thalamic language role (Ojemann, Fedio, and Van Buren, 1968; Ojemann, Blick, and Ward, 1971; Ojemann, 1974, 1975, 1976, 1977, 1979). Although the short-term verbal memory changes are evoked from the same portion of the left thalamus as naming changes, the threshold for memory changes is generally lower. Memory changes can also be evoked from a wider area of the lateral thalamus, beyond the quite discrete area relating to naming. Short-term postdistractional verbal memory has been measured in these studies with a single-item paradigm adopted from Peterson and Peterson (1959). The specific test consists of sixty consecutive trials each having three achromatic slides. The first slide, presented for four seconds, depicts a carrier phrase and an object which the patient reads and names aloud. The slide and object name serve as the input to memory. On the second slide is a two-digit number greater than 30 which the patient reads aloud and then counts backward from by threes. This six-second period of backwards counting acts as a distraction preventing rehearsal of the object name which must be stored during this time in short-term memory. The third slide, on which the word "recall" is printed, acts as a cue for output from verbal memory. The patient says aloud the name of the object pictured earlier in the trial. The patient is trained on this test the night prior to operation. During the operation stimulation is applied through electrodes placed in the VL thalamic target. Stimulation is applied only while showing the input slide on some trials, only while showing the storage slide or output slide on other trials, or while showing both input and output slides on the same trial. Trials without stimulation, used to measure control performance, are interspersed pseudorandomly with stimulation trials.

Stimulation of the left thalamus during performance on this paradigm results in several unusual findings related to memory. When the current is applied at the time of input of information, later retrieval of that information is substantially more accurate than for information with no stimulation during input. Indeed, the recall errors decrease to almost half that seen on control trials. When the same current at the same thalamic sites is applied at the time of retrieval, the latency of retrieval shortens, but the percentage of errors occurring on retrieval of information nearly doubles. Combining stimulation at the time of both input and retrieval produces the algebraic sum of these independent effects and such performance is indistinguishable from control performance. This pattern of memory errors has been modeled as an alteration in an alerting circuit that directs attention to incoming information, while simultaneously blocking retrieval of already internalized material. Thus, this circuit acts like a gate controlling access to or from memory at any point in time. A high level of attention to incoming information increases the likelihood of later retrieval of the information.

This alerting circuit seems to represent a common thalamic mechanism for both language and memory. Indeed, the severity of a postoperative language disturbance after stereotaxic ventrolateral thalamotomy can be largely predicted by a statistic based on the magnitude of the short-term memory changes evoked by stimulation of that nucleus in that patient. The more readily these memory effects are evoked, the less the postoperative language disturbance (Ojemann, 1976). It is this predictive association that suggests that a common mechanism is involved, a mechanism that has been modeled as the specific alerting circuitry. In that model, anomia from thalamic stimulation represents a failure of retrieval of the correct object name from long-term memory, while the appearance of exactly the same word each time naming is called for during stimulation is a combination of the inability to retrieve the correct name and the enhanced ease of retrieval of an item presented during previous stimulation.

The two components of thalamic stimulation on memory can be partially dissociated anatomically. Disturbances in retrieval from short-term memory are more prominent when stimulation is applied at posterior lateral thalamic sites. The effect of input stimulation, enhancement of later retrieval, is more prominent at anterior lateral thalamic sites. This parallels the differences in the types of naming errors observed at posterior naming sites (misnamings and omissions) compared to those at more anterior sites (perseverations and repeated use of the same wrong word). It is not clear whether these attentional effects are a property of lateral, and specifically ventrolateral, thalamic nuclei, or represent changes in *en passage* fibers crossing these areas from the thalamic nuclei usually considered part of the thalamocortical activating system, such as intralaminar or centromedian nuclei.

Thus, one of the left thalamic roles in language is related to alerting mechanisms that have an effect on both naming and verbal memory. The language deficit after left thalamic lesions has been similarly interpreted as a failure of attentional mechanisms (Luria, 1977). The extraneous intrusion type of naming error, so typical of this syndrome, is modeled as a failure of moment-to-moment attention.

The effects of VL thalamic stimulation on short-term memory for nonverbal material (nonsense shapes), as well as for verbal material, were determined in another series of patients using a similar test paradigm (Ojemann, 1977, 1979). The major features of the "specific alerting response," a decrease in errors with stimulation during input and an increase in errors with stimulation during output, were replicated with left stimulation for verbal material and confirmed with right stimulation for nonverbal material. In addition, there was an unexpected effect of left VL thalamic stimulation on the nonverbal test. Right stimulation evoked no changes from control performance on the verbal test, but left VL thalamic stimulation during input to the nonverbal test substantially increased subsequent recognitition errors. The "specific alerting response" evoked with stimulation of the left VL thalamus appears to be highly specific to verbal material, even to the point of blocking attention to relevant nonverbal information when the "alerting" mechanism is activated. This finding suggests dominance of verbal processes within the thalamic-specific alerting system, but may also be a consequence of the test materials. Object pictures have spatial features that may evoke selective attention during right thalamic stimulation, whereas the shapes used in the nonverbal test were specifically selected to have few verbal associations. The apparent "dominance" of a verbal process in comparison to a similar nonverbal process is, of course, not without precedent. With early left hemidecortication, for example, it appears that nonverbal processes suffer more in comparison to verbal processes (Dennis and Whitaker, 1976).

There is some evidence that the effect of stimulation of the VL thalamus during input in enhancing retrieval is apparent not only after a delay of seconds (short-term memory) but is evident at periods up to a week postoperatively, suggesting involvement in the long-term memory process as well. During a period of transient anomia which often occurs after left VL thalamic lesions, the naming of objects which had been associated with stimulation of the left VL thalamus during their presentation at the time of intraoperative testing was significantly more accurate (in six of seven patients tested) than that for objects which had not been presented in conjunction with stimulation (Ojemann, 1979). In addition, better recognition of nonsense CVC syllables five days postoperatively has been identified (in three patients tested) for those syllables presented in conjunction with left VL thalamic stimulation, rather than for those syllables presented without stimulation (Mateer and Ojemann, unpublished data). Stimulation of the right thalamus has had no effect on long term CVC recognition in two patients tested thus far. Similar data implicating an influence of the right thalamic "specific alerting response" in long-term memory for nonverbal material have been collected (Ojemann, 1979; Mateer and Ojemann, unpublished data). In a recognition task using complex multicolored designs, those designs which had been presented in association with right VL thalamic stimulation were recognized postoperatively significantly more often as having been seen before than were shapes presented without stimulation, in three of four patients tested.

The relative activation of this system in the right or left brain may serve to focus attention on verbal or nonverbal features of environmental input, thereby enhancing or diminishing later retrieval or recognition of material-specific input. The simultaneous decrease in ease of retrieval of already internalized material has important implications for the availability of associations to incoming information. Activation of the specific alerting response systems on one side will enhance later retrieval of incoming material-specific information, but block retrieval of the same type of information as a basis for forming new associations. Whether memory systems per se are specifically activated or merely provided with a strong directed signal through increased activation or receptivity in certain cortical areas cannot be determined at this stage.

Another example of the enhancing effects of these alerting mechanisms comes from a study of effects of thalamic stimulation on dichotic listening tasks (Mateer, Fried and Ojemann, unpublished data). Twenty trials of dichotic stimuli (a trial consisting of four monosyllabic words presented to each ear) were presented to eight patients during stereotaxic thalamotomy procedures. On one half of the trials stimulation from an electrode positioned in the VL thalamus was applied. Four patients received left thalamic stimulation and four received right stimulation. The total number of correct responses made during either left or right stimulation was higher than was the total number of correct responses made under nonstimulation conditions. Increased accuracy of performance in this verbal perception-and-report task was particularly dramatic and consistent with left thalamic stimulation. Performance under all four conditions, right and left electrode placement for stimulation and nonstimulation trials, yielded a relative right- over left-ear advantage (REA). However, the REA increased with left thalamic stimulation over nonstimulation conditions and decreased with right thalamic stimulation over nonstimulation conditions. Thus, with stimulation of the thalamus there was a significant increase in correct responses to stimuli presented in the contralateral ear over nonstimulation conditions. Correct responses to ipsilateral stimuli tended to decrease with stimulation, although significance was not reached. These findings further support the specific alerting effect of thalamic activation on cortical receptivity, extending these effects to contralateral channels for input of information in the auditory as well as visual modes. Evidence that the left material-specific alerting effect may be more prominent for auditory than for verbal information has previously been reported (Ojemann, 1975).

Additional evidence of the importance of thalamic attentional mechanisms in language function comes from the study of cortical electrophysiologic changes that are both specific to a language task and language cortex (Fried, Ojemann, and Fetz, 1981). These electrocorticographic changes, slow potential shifts in frontal premotor sites, and focal desynchronization in parietal and temporal sites related to language are the pattern of changes evoked in animals by activation of the thalamocortical activating system (Jasper, 1960; Yingling and Skinner, 1977). This thalamic alerting circuitry may act as a switching mechanism to activate appropriate focal cortical areas involved in language processing.

THE ROLE OF THALAMIC MOTOR
FUNCTIONS IN SPEECH

Another thalamic role in language evident from both stimulation and lesion studies involves the integration of respiratory and other oral motor activities with language. Guiot, Hertzog, Rondot, and Molina (1961) reported decreases in loudness and alterations in the rate of counting with stimulation of either the right or the left thalamus. In those studies where right and left thalamic stimulation have been specifically compared, a variety of motor effects have been asymmetrically evoked. An inhibition of respiration during the expiratory phase has been evoked with a lower threshold of stimulation from the medial left VL thalamus, than from the right VL thalamus (Ojemann and Van Buren, 1967). When one compares the exact sites in the lateral thalamus where this effect is evoked and the sites where naming changes are evoked, it is apparent that they correspond to a similar area of the lateral thalamus. Respiratory changes do not account for the naming changes, particularly perseveration or the production of wrong words, but rather they seem to be evoked from common thalamic sites, suggesting that these sites may act to integrate prolonged expiration with language. Modification of the normal expiratory phase of respiration is considered a fundamental substrate of normal speech production (Darley, Aronsen, and Brown, 1975).

Analysis of speech samples elicited during the naming task in a series of 18 patients revealed that stimulation of the left but not the right VL thalamus was associated with a slowed rate of articulation during correct naming of phrases (Mateer, 1978). The slowed rate was often related to increased slurring and articulatory inaccuracy during left VL thalamic stimulation. Finally, preliminary results from a small series of patients has suggested that the ability to mimic a sequence of oral facial postures can be altered by left but not right thalamic stimulation. Slowing and decreased accuracy of oral facial movement production was found in all of three patients undergoing left VL thalamic stimulation but was not found in either of two patients undergoing right VL thalamic simulation. The overlap within the left thalamus at areas involved in both naming and motor requirements for speech is one of several examples of overlap of motor and language mechanisms of the left hemisphere in the human brain.

In addition to the language changes, a variety of speech-related motor functions have reportedly been altered following thalamotomy. These include decreases in voice loudness, alterations in speaking rate, and articulatory changes in the form of slurring and hesitation (Allen, Turner, and

Gadea-Ciria, 1966; Bell, 1968). There are other motor changes, best described as failures in motor activation, which are not restricted to speech. Hassler (1966) described motor neglect or inattention of the contralateral extremities and facial muscles for one to two weeks following stereotaxic lesions of the posterior VL nucleus. Despite significant relief of tremor, rigidity, and hyperkinetic movement, the patients rarely used the contralateral extremities for automatic and involuntary movements. Limb strength, per se, was not appreciably diminished. We have observed similar features of motor neglect in the contralateral upper limb in a large proportion of cases, particularly in the first postoperative week. The motor neglect appears to extend even to normally bilateral movements. Facial postures can be imitated or produced on command with little or no facial asymmetry while spontaneous facial expression is markedly asymmetrical. Although movements of both eyes may be full on command, spontaneous eye movement in the direction contralateral to the lesion is often not observed.

Unilateral akinesia (particularly for spontaneous movement) has also been reported with lesions of the dorsolateral frontal cortex, supplementary motor cortex, cingulate gyrus, basal ganglia, and mesencephalic reticular formation (Watson, Miller, and Heilman, 1978; Watson and Heilman, 1979; Damasio, Damasio, and Chui, 1980). The VL nucleus of the thalamus appears to be another component in the proposed circuitry for a motor activating mechanism. We have interpreted this transient failure of motor activation or motor neglect following thalamic stereotaxic procedures to be another manifestation of a failure in specific "alerting" or activation of cortical effector regions. This is not a particularly new view of thalamic function. The ventrolateral nucleus of the thalamus has traditionally been considered to be a motor relay nucleus (Ruch, Patton, Woodbury, and Towe, 1965). Its primary afferent inputs are from the cerebellum and from the basal ganglia via the globus pallidus. The primary efferent projections from the VL nucleus are to the ipsilateral motor and premotor cortices. Unexpected are the findings that this motor activation is closely connected to "higher level" cognitive activation or alerting for language and memory tasks.

SUMMARY

Lesions in the dominant thalamus are often associated with (1) a transient aphonia; (2) fluent aphasia, marked by fluctuating levels of responsiveness, anomia with bizarre, often perseverative intrusions, and variable auditory comprehension; (3) verbal memory deficits; and (4) contralateral motor "neglect" or inattention. Stimulation of the dominant thalamus is associated with material-specific asymmetrically evoked alerting effects on verbal and nonverbal recall and with alterations of respiratory and articulatory substrates for speech. How can these findings be synthesized? The aphonia or mutism may reflect an interruption in the corticobulbar pathway adjacent to the thalamus. Alternatively, it may be a profound manifestation (in the early stage after onset of thalamic lesion) of a failure of "cortical activation" or "alerting" mechanisms. The "motor neglect" observed in spontaneous usage of the contralateral limb or reflexive use of the contralateral facial musculature is a more persistent though usually transitory manifestation of a failure in such a motor activating mechanism. The fluctuating nature of the aphasia and the bizarre nature of some of the intrusions and perseverations on naming tasks can also be viewed in a larger context of disturbed "alerting" or activation of the language related cortex. Enhanced material-specific recall and enhanced contralateral ear scores on dichotic tasks with stimulation of the VL thalamus strongly support such a "specific alerting" function. The unusual aspects of aphasia appear to reflect deficient arousal of otherwise intact cortical language mechanisms.

OTHER SUBCORTICAL AREAS

Most of the data relating subcortical structures to language implicate the thalamus, and particularly the lateral thalamus. The original hypothesis of Penfield and Roberts (1959), however, suggested that the left centromedian nucleus was the important thalamic nucleus for language. Although an isolated disturbance of writing has been reported with a lesion there (Sugishita, Ishijima, Hori, Fukushima, and Iwata, 1973), stimulation mapping of sites likely in this nucleus rarely shows language changes. There is also an older literature suggesting that portions of the striatum, particularly the globus pallidus, are involved in language. Ideational language deficits have been documented after left pallidal lesions for dyskinesia (McFie, 1960; Svennilson, Torvik, Lowe, and Leksell, 1960). The specificity of these deficits to the pallidum and the thalamus and not to the intervening internal capsule has also been demonstrated (Herman, Turner, Gillingham, and Gaze, 1966).

More recently, a number of cases of aphasia in association with left-sided capsular/putaminal lesion sites (documented by CT-scan) have been studied (Hier, Davis, Richardson & Mohr, 1977; Naesar, Alexander, Helm-Estabrooks, Levine, Laughlin and Geschwind, 1982). Different patterns of aphasia were observed depending on whether the lesion extended anteriorly or posteriorly into white matter and whether auditory radiations in the temporal isthmus were involved. Although these cases shared certain features of Broca's, Wernicke's, or global aphasia, like the cases with thalamic aphasias we have discussed, the pattern of linguistic disturbance did not completely fit any of the cortical aphasia syndromes. Finally, there is a small body of information suggesting that the caudate may play some role in language-related functions, particularly in verbal memory (Bechtereva, Genkin, Morseeva, and Smirnov, 1967). Confirmation and more precise delineation of the language role of these subcortical structures should prove to be an exciting area in the coming decade.

REFERENCES

Allen, C., Turner, J., and Gadea-Ciria, M. (1966). Investigation into speech disturbances following stereotactic surgery for Parkinsonism. *British Journal of Communication Disorders, 1,* 55–59.

Almgren, P. E., Anderson, A. L., and Kullberg, G. (1969). Differences in verbally expressed cognition following left and right ventrolateral thalamotomy. *Scandinavian Journal of Psychology, 10,* 243–249.

Arseni, C. (1958). Tumors of the basal ganglia. *Archives of Neurology and Psychiatry, 80,* 18–26.

Bechtereva, N. P., Genkin, A., Morseeva, N., and Smirnov, V. (1967). Electrographic evidence of participation of deep structures of the human brain in certain mental processes. *Electroencephalography and Clinical Neurophysiology Supplement, 25,* 153–166.

Bell, D. S. (1968). Speech functions of the thalamus inferred from the effects of thalamotomy. *Brain, 91,* 619–638.

Bertrand, C., Martinez, S., Hardy, J., Molina-Negro, P., and Velasco, F. (1973). Stereotaxic surgery for Parkinsonism. *Progress in Neurological Surgery, 5,* 79–112.

Brown, J. W. (1979). Thalamic mechanisms in language. In M. S. Gazzoniga (Ed.), *Handbook of behavioral neurobiology.* Vol. 2: *Neuropsychology.* New York: Plenum. Pp. 215–238.

Bugiani, O., Conforto, C., and Sacco, G. (1969). Aphasia in thalamic hemorrhage. *Lancet, 1,* 1052.

Cappa, S. F., and Vignolo, L. A. (1979). "Transcortical" features of aphasia following left thalamic hemorrhage. *Cortex, 15,* 121–130.

Cheek, W. R., and Taveras, J. (1966). Thalamic tumors. *Journal of Neurosurgery, 24,* 505–513.

Ciemins, V. A. (1970). Localized thalamic hemorrhage: A case of aphasia. *Neurology, 20,* 776–782.

Damasio, A. R., Damasio, H., and Chui, H. C. (1980). Neglect following damage to frontal lobe or basal ganglia. *Neuropsychologia, 18,* 123–132.

Darley, F. L., Aronsen, A. E., and Brown, J. R. (1975). *Motor speech disorders.* Philadelphia: Saunders.

Dejerine, J., and Roussy, G. (1906). Le syndrome thalamique. *Revue Neurologique* (Paris), *14,* 521–532.

Dennis, M., and Whitaker, H. A. (1976). Language acquisition following hemidecortication: Linguistic superiority of left over right hemisphere. *Brain and Language, 3,* 404–433.

Fazio, C., Sacco, G., and Bugiani, O. (1973). The thalamic hemorrhage. *European Neurology, 9,* 30–43.

Fisher, C. M. (1959). The pathologic and clinical aspects of thalamic hemorrhage. *Transactions of the American Neurological Association, 84,* 56–59.

Fried, I., Ojemann, G. A., and Fetz, E. (1981). Language-related potentials specific to human language cortex. *Science, 212,* 353–356.

Geschwind, N. (1967). Discussions of cerebral connectionism and brain function. In C. H. Milliken and F. L. Darley (Eds.), *Brain mechanisms underlying speech and language.* New York: Grune and Stratton.

Glosser, G., Kaplan, E. and LoVerme, S. (1982). Longitudinal neuropsychological report of aphasia following left-subcortical hemorrhage. *Brain and Language, 15,* 95–116.

Guiot, G., Hertzog, E., Rondot, P., and Molina, P. (1961). Arrest or acceleration of speech evoked by thalamic stimulation in the course of stereotaxic procedures for Parkinsonism. *Brain, 84,* 363–379.

Hassler, R. (1966). Thalamic regulation of muscle tone and the speed of movements. In D. P. Purpura and M. D. Yahr (Eds.), *The thalamus.* New York: Columbia Univ. Press.

Herman, K., Turner, J., Gillingham, F., and Gaze, R. (1966). The effect of destructive lesions and stimulation of the basal ganglia on speech mechanisms. *Confina Neurologica, 27,* 197–207.

Hier, D. B., Davis, K. R., Richardson, E. P. and Mohr, J. P. (1977). Hypertensive putaminal hemorrhage. *Annals of Neurology, 1,* 152–159.

Horenstein, S., Chung, H., and Brenner, S. (1978). Aphasia in two verified cases of left thalamic hemorrhage. *Transactions of the American Neurological Association, 103,* 193–198.

Jasper, H. H. (1960). Unspecific thalamocortical relations. In J. Fields (Ed.), *Handbook of physiology, Section 1, Neurophysiology,* Vol. 2. Baltimore: Waverly Press.

Krayenbuhl, H., Siegfried, J., Kohenof, M., and Yasargil, M. G. (1965). Is there a dominant thalamus? *Confina Neurologica, 26,* 246–249.

Laitinen, L. V., and Vilkki, J. (1977). Observations on physiological and psychological functions of ventral and internal nucleus of human thalamus. *Acta Neurologica Scandinavica, 55,* 198–212.

Luria, A. R. (1977). On quasi-aphasic speech disturbances in lesions of the deep structures of the brain. *Brain and Language, 4,* 432–459.

McFie, J. (1960). Psychological effects of stereotaxic operations for the relief of Parkinsonian symptoms. *Journal of Mental Science, 106,* 512–517.

McKissock, W., and Paine, K. W. E. (1958). Primary tumors of the thalamus. *Brain, 81,* 41–63.

Marie, P. (1906). La troisième circonvolution frontale gauche ne jove aucun rôle spécial dans la fonction du langage. *Semaine Médicale, 26,* 241–247.

Mateer, C. (1978). Asymmetric effects of thalamic stimulation on rate of speech. *Neuropsychologica, 16,* 497–499.

Mateer, C., Fried, I. and Ojemann, G. A. (Unpublished data) Effects of left and right VL thalamic stimulation on the ear advantage for dichotically presented words.

Mateer, C. and Ojemann, G. A. (Unpublished data) Enhancement of long term recognition memory for verbal and nonverbal material by left and right VL thalamic stimulation.

Metter, E. J., Wasterlain, C. G., Kuhl, D. E., Hanson, W. R., and Phelps, M. E. (1981). [18]FDG position emission computed tomography in a study of aphasia. *Annals of Neurology, 10,* 173–183.

Mohr, J. P., Watters, W. C., and Duncan, G. W. (1975). Thalamic hemorrhage and aphasia. *Brain and Language, 2,* 3–17.

Naesar, M. A., Alexander, M. P., Helm-Estabrooks, N., Levine, H. L., Laughlin, S. A., and Geschwind, N. (1982). Aphasia with predominantly subcortical lesion sites: Description of three capsular/putaminal aphasia syndromes. *Archives of Neurology, 39,* 2–14.

Neilson, J. M. (1946). *Agnosia, apraxic, aphasia: Their value in cerebral localization* (2nd ed.). New York: Harper (Hoeber).

Ojemann, G. A. (1974). Mental arithmetic during human thalamic stimulation. *Neuropsychologia, 12,* 1–10.

Ojemann, G. A. (1975). Language and the thalamus: Object naming and recall during and after thalamic stimulation. *Brain and Language, 2,* 101–120.

Ojemann, G. A. (1976). Subcortical language mechanisms. In H. Whitaker (Ed.), *Neurolinguistics,* Vol. 1. New York: Academic Press.

Ojemann, G. A. (1977). Asymmetrical function of the thalamus in man. *Annals of the New York Academy of Science, 299,* 380–396.

Ojemann, G. A. (1978). Organization of short-term verbal memory in language areas of human cortex: Evidence from electrical stimulation. *Brain and Language, 5,* 331–348.

Ojemann, G. A. (1979). Altering memory with human ventrolateral thalamic stimulation. In E. R. Hitchcock, H. T. Ballantine, Jr., and B. A. Meyerson (Eds.), *Modern concepts in psychiatric surgery.* Amsterdam: Elsevier.

Ojemann, G. A., Blick, K., and Ward, A. A. (1971). Improvement and disturbance of short-term memory with human ventrolateral thalamic stimulation. *Brain, 94,* 225–240.

Ojemann, G. A., Fedio, P., and Van Buren, J. M. (1968). Anemia from pulvinar and subcortical parietal stimulation. *Brain, 91,* 99–116.

Ojemann, G. A., and Van Buren, J. M. (1967). Respiratory, heart rate and GSR responses from human diencephalon. *Archives of Neurology, 16,* 74–88.

Ojemann, G. A., and Ward, A. A. (1971). Speech representation in the ventrolateral thalamus. *Brain, 91,* 99–117.

Penfield, W., and Roberts, L. (1959). *Speech and brain mechanisms.* Princeton, N.J.: Princeton Univ. Press.

Peterson, L. R., and Peterson, M. J. (1959). Short-term retention of individual verbal items. *Journal of Experimental Psychology, 58,* 193–198.

Rakic, P. (1972). Embryonic development of the pulvinar–LP complex in man. In I. S. Cooper, P. Riklan, and P. Rakic (Eds.), *The Pulvinar–LP complex.* Springfield, Ill.: Thomas.

Reynolds, A. F., Harris, A. B., Ojemann, G. A., and Turner, P. T. (1978). Aphasia and left thalamic hemorrhage. *Journal of Neurosurgery, 48,* 570–574.

Reynolds, A. F., Turner, P. T., Harris, A. B., Ojemann, G. A., and Davis, L. E. (1979). Left thalamic hemorrhage with dysphasia: A report of five cases. *Brain and Language, 7,* 62–73.

Riklan, M., and Levita, E. (1969). *Subcortical correlates of human behavior.* Baltimore: Williams and Wilkins.

Riklan, M., Levita, E., Zimmerman, J., and Cooper, I. S. (1969). Thalamic correlates of language and speech. *Journal of Neurological Sciences, 8,* 307–328.

Ruch, T. C., Patton, H. D., Woodbury, J. W., and Towe, A. L. (1965). *Neurophysiology.* Philadelphia: Saunders.

Samarel, A., Wright, T. L., Sergay, S., and Tyler, H. R. (1976). Thalamic hemorrhage with speech disorder. *Transactions of the American Neurological Association, 101,* 283–285.

Samra, K., Riklan, M., Levita, E., Zimmerman, J., Waltz, J., Bergmann, L., and Cooper, I. (1969). Language and speech correlates of anatomically verified lesions in thalamic surgery for Parkinsonism. *Journal of Speech and Hearing Research, 12,* 510–540.

Schaltenbrand, G. (1965). The effects of stereotaxic electrical stimulation in the depth of the brain. *Brain, 88,* 835–840.

Selby, G. (1967). Stereotaxic surgery for the relief of Parkinson's disease. II. An analysis of the results of a series of 303 patients (413 operations). *Journal of Neurological Science, 5,* 343–375.

Sem-Jacobsen, C. W. (1965). Depth-electrographic stimulation and treatment of patients with Parkinson's disease including neurosurgical techniques. *Acta Neurologica Scandinavica, Supplement, 13,* 365–377.

Shapiro, D. Y., Sadowsky, D., Henderson, W., and Van Buren, J. (1973). An assessment of cognition function in post-thalamotomy Parkinson patients. *Confina Neurologica, 35,* 144–166.

Sugishita, M., Ishijima, B., Hori, T., Fukushima, T., and Iwata, M. (1973). "Pure" agraphia after left CM-thalamotomy. *Clinical Neurology, 13,* 568–574.

Svennilson, E., Torvik, A., Lowe, R., and Leksell, L. (1960). Treatment of Parkinsonism by stereotaxic thermolesions in the pallidal region. *Acta Psychiatrica et Neurologica Scandinavica, 35,* 358–377.

Van Buren, J. (1975). The question of thalamic participation in speech mechanisms. *Brain and Language, 2,* 31–44.

Van Buren, J., Li, C. L., and Ojemann, G. A. (1966). The frontostriatal arrest response in man. *Electroencephalography and Clinical Neurophysiology, 21,* 114–130.

Watson, R. T., and Heilman, K. M. (1979). Thalamic neglect. *Neurology, 29,* 690–694.

Watson, R. T., Miller, B. D., and Heilman, K. M. (1978). Nonsensory neglect. *Annals of Neurology, 3,* 505–508.

Yingling, C. D., and Skinner, J. E. (1977). Coating of thalamic input to cerebral cortex by nucleus reticularis thalami. *Progress in Clinical Neurophysiology, 1,* 70–96.

CHAPTER 8

The Placement of Experience in the Brain

Robert M. Anderson, Jr., and Demetria C. Q. Leong

INTRODUCTION

Neuroscientists, when discussing the problems important in their field, rarely consider these issues in the context of their most basic assumptions. For example, when they attempt to localize cognitive abilities in the brain, they seldom examine the concept of location and the concomitant notion of space which they are assuming. When they try to place mental activities within the brain, they hardly ever examine the logic of the concepts used to describe the mental activities to determine whether it is even POSSIBLE to locate such activities in the brain. They tacitly assume that the concepts they are using to describe mental activities are sufficiently pliable that they may be used to apply to processes within the brain. They may operationally define their terms so that they have few contact points with commonly used mental terminology or with any other fairly comprehensive view of mental processes. Failure to attend to the logic of mental terminology can result in conceptual confusion because inconsistent properties may be attributed to the same object. Facile re-

liance on operational definitions of mental terms may trivialize any conclusions about location within the brain since the defined term may have
little relation to any relatively complete theory of psychological process.
The localization of mental activities without attention to the concept of
place that is being assumed risks a certain naiveté since one may easily
slip into using an uncritical, NAIVE REALIST concept of space.

It is the purpose of this chapter to spare the neuroscientist some of
these difficulties by providing an analysis of some of the most basic assumptions underlying the localization of mental functions in the brain.
Special attention is given to the problems encountered in localizing experience in the brain. A secondary purpose of this chapter is that of providing some neural candidates for experience. In the first section, problems
standing in the way of localizing experience in the brain, which are due to
the logic of the language used to describe experience, are discussed. Alternative ways of handling these problems are reviewed. The second section contains an analysis of the concept of place. Several kinds of psychological and physical space are distinguished, and the importance of using
the correct kind of space in localizing experience is emphasized. In the
third section the methodological problems which come to the fore in using
research results to support localization hypotheses are examined. As an
illustration, an attempt is made to localize visual experience. Finally, in
the fourth section, having localized visual experience in the brain, neural
candidates for visual experience are scrutinized.

LOGICAL PROBLEMS IN LOCALIZING
EXPERIENCE

"Experience" is a difficult word to define. Its meaning is ambiguous,
and its uses are numerous. It is also a term which is epistemically primitive and fundamental; it is more commonly used to define other terms and
is rarely defined itself. "Experience" may be defined in a way which is
least prejudicial to any philosophical or scientific issues by providing a
few examples. Since the word "experience" is so broadly used (one may
have experiences of ecstasy, of loneliness, and even of correctly performing an analysis of variance for an experiment), we will focus our investigation on sensory experience and, even more particularly, on visual experience. This is convenient since there is a large body of scientific research
on visual experience. Examples of sensory experiences are hearing the
note high C, feeling pressure on one's hand in a handshake, and seeing the
flashes of color (phosphenes) produced when the eyeball is pressed.

Some philosophers, in examining the logic of sensation language, have concluded that sensory experiences cannot be meaningfully localized in space and time (Shaffer, 1961). They contend that the language which we ordinarily use to describe our sensory experiences contains no rules for either asserting or denying that these experiences have a particular location. Although this claim may seem strange to the neuroscientist, it does have some validity. To use a homey example, your friend can sit on a chair in your living room, but she cannot sit on your image or idea of a chair. The reason that she cannot sit on your image of a chair is not that, try as she might, she cannot find the image of the chair; it is that it simply does not make sense for your image of a chair to have location. Her task, therefore, is of the utmost difficulty—it is logically impossible.

This distinction between the physical as spatially localizable and the mental as nonlocalizable is not new to the recent linguistic or ordinary language philosophers (Dennet, 1978). It was reified by Descartes several centuries earlier in his dualist ontology (Copleston, 1958). According to Descartes, the world is composed of two substances: spiritual substance whose principal attribute is thinking and corporeal substance whose principal attribute is extension. This venerable bifurcation of the world into mind versus body, or into nonlocalizable versus localizable aspects presents a serious problem for anyone who wishes to localize experience (or any other mental attribute) in the brain. If it is logically impossible to localize sensory experiences, then one who attempts to do so is taking on a task as unrewarding as that of your friend who tries to sit on your image.

Fortunately there are several ways by which one may attempt to extricate oneself from this logical trap. First of all (Method A), one may take issue with the accuracy of the analysis that has been made of sensory language. J. J. C. Smart (1963), for example, has argued that sensation language is TOPIC NEUTRAL, that its logic says nothing one way or the other about whether sensations are localizable. This gives Smart the freedom to identify sensations and brain processes. Another approach (Method B) is to accept the claim that sensory language entails nonlocalizability but to advocate a revision of sensory language to allow for the possibility of localization. A third and more radical method (Method C) also accepts the claim that the logic of sensation language prohibits the localization of sensory experiences. It views sensation language as beyond hope and advocates its replacement by a new linguistic system fashioned on the basis of scientific discoveries. Richard Rorty (1965), who has championed this method, draws an analogy between sensation language and the explanatory system of demonology prevalent in the Middle Ages. He suggests that just as the conceptual system of demonology disappeared as science advanced, so will our explanatory system based on the sensation concept.

These three methods may be considered whenever an argument is made that a particular mental attribute (M) cannot be physically localizable because such an attribute has properties (P) which are logically incompatible with localization. One's counterargument may run (A) that M does not necessarily have P, or (B) that M should be redefined to not have P, or finally, (C) that M and its entire associated conceptual system should be dropped altogether because the system is inappropriate, and a new conceptual system should be put in its place. Method A is compatible with identity theory—the view that sensory experiences and brain processes are one and the same thing. Method C is inconsistent with identity theory because sensation language is rejected and there is nothing to identify with the brain processes. Method B is a middle-of-the-road technique and is somewhat compatible with identity theory. The essence of the concept is preserved while some of its logically troublesome features are trimmed away. Thus when the newly defined sensation is identified with the brain process, it is not precisely the sensation of our ordinary parlance, but it is close enough.

At this point in the history of philosophy and neuroscience it appears that Method B is the most appropriate to apply to the claim that the logic of sensation language prohibits the localization of sensory experiences. Our present-day sensory language evolved during a period when almost nothing was known about brain functions, so it is not surprising that factual errors were encoded into the grammar of that language. This may be further understood in the light of Herbert Feigl's (1967) concepts of epistemological dualism and ontological monism. According to Feigl, mental processes and brain processes are one and the same thing; this is the ontological monism. But if this is so, then why do there appear to be two different things—the mind and the brain? This is explained by reference to epistemological dualism. There are two ways of knowing the brain: you can know it directly by being identical with your own brain, and you can know it indirectly by observing a person's behavior, by reading about the brain, or by viewing it as a neurosurgeon does. The brain appears differently to us when viewed directly and when viewed indirectly so it appears to us that there are two different kinds of entities when, in fact, there is one. Our misconstrual of the situation becomes formalized in the logic of our sensation language, and we are presented with the logical difficulty just described. Since we now have a great deal of data, however, showing how stimulation of the brain results in sensations, and how lesions of the brain alter sensory capacity, and so on (see section on Methodological Problems of Localization), we can now confidently excise that portion of the logic of sensation language which prohibits localization, while retaining the core of sensory experience terminology captured by examples in the beginning of this section, that part which allows us to refer to internal feeling states.

Another feature alleged of sensations which makes them seem incapable of localization is what has been termed their incorrigibility. It has been claimed that it is logically impossible for one to be mistaken about one's own sensory experiences (Ryle, 1949). How, for example, could one be mistaken about whether or not one was in pain? It seems, by definition, that if one has the sensation of pain, then one must be aware of and know that one has the experience of pain. Of course, a person could be injured so severely that he or she SHOULD be in pain, and yet in the excitement of an emergency not feel any pain until the emergency situation has subsided. In this case we would not say that the person was in pain but that he or she failed to perceive the pain; we would say that he or she was not in pain until he or she felt it after the emergency was over. This same kind of argument has been made for sensations of all varieties.

Besides this logicolinguistic kind of argument, empiricist philosophers have also asserted the incorrigibility of sensations as an indubitable foundation upon which to construct empirical knowledge (Broad, 1960). They reasoned that if we can be absolutely certain about our own experiences, then we can use them as a basis upon which to construct our more inferential and hence more tenuous knowledge of external objects, other minds, and scientific entities such as electrons.

If sensations are incorrigible and cannot be mistaken, then they cannot be identified with brain processes (about which we certainly can be mistaken) and localized in the brain. It turns out, however, that neither the linguistic nor the empiricist arguments hold up under close scrutiny. We can use a combination of Methods A and B. First of all there are numerous examples which suggest that sensory experiences may be more or less accurate (Aune, 1967). Take, for example, these phenomenological discoveries about the visual apparatus. Mariotte discovered and described the blind spot, the break in the retina where the optic nerve enters, in 1668 (Boring, 1950). The horopter, the locus of all points seen as single in binocular vision, was discovered by Aguilonius in 1613. Most of the objects in the visual field are doubled. Purkinje, in 1825, described how colors tend to change in hue continuously from the center to the periphery (Boring, 1942). Millions of people had used their eyes for ages, yet it was not until the seventeenth century that the blind spot was discovered. Although most of the objects in the visual field are doubled, this is rarely noticed. Many of the phenomena described by Purkinje have never been detected by anyone else, and it cannot be certainly held that they depended on the individual peculiarities of this acute observer's visual system. If there can be more or less accurate observers of sensations, then there can be mistakes made with regard to sensations. This argument uses Method A and shows that the logic of sensation language does not require sensations to be incorrigible.

The first author (Anderson, 1972) has constructed a two-stage model of sensory processing which illustrates how mistakes could be made regarding sensations. Early stages of sensory processing in the brain may be scanned by other parts of the brain, thus allowing for the possibility that inaccurate scanning could occur. This application of Method B can show how a future neurophysiology could lead us to call for a redefinition of "sensory experience." By arguing using Methods A and B, we can make a doubly strong case that the logical barrier of incorrigibility does not stand in the way of identifying sensory experiences and brain processes, and thus of localizing them. We argue that either sensations are already defined as corrigible (A) or they should be defined as such (B) and that no matter which is the case, the possibility is left open for localizing them.

One final property which is alleged of sensory experiences and which is incompatible with their localization is that of privacy or subjectivity. No one else can have my sensations in the way that I can have them. They are somehow absolutely private and irreducibly subjective while objects such as brains, which are physically localizable, are public and intersubjective. Method B is appropriate in this instance. It is easy to see why sensation language entails privacy when looked at in terms of Feigl's epistemological dualism. One's sensations seem private because only oneself is identical with one's own brain. No one else has this kind of direct access to the very same entity. It is this asymmetry of access that is encoded in sensation language as irreducible subjectivity. However, the direct–indirect dichotomy reveals how privacy need not be a necessary property of sensations. It is in fact only an illusion due to perspectival differences (Gunderson, 1970). Thus we see how by drawing on our knowledge of rudimentary facts about the processing of information in the brain and our own perceptual experience, we can use Methods A and B to examine and revise the terminology of sensory experience and allow for the possibility of localizing experience in the brain.

THE NATURE OF PLACE

Another set of assumptions which is essential to examine when attempting to localize experience or any mental process is that which surrounds the concept of space. It is tempting to use a NAIVE REALIST concept when localizing mental processes. According to naive realism, the world is pretty much what we perceive it to be. The American flag, for example, has the colors red, white, and blue in it that we perceive it to have. It may seem obvious that the American flag is red, white, and blue. This turns out, however, to be false when we examine how we perceive an American flag. Consider the case of a flag waving in the sunlight at midday with you, the observer, admiring it. Photons, travelling from the sun, strike the flag. When they strike certain stripes of fabric, most are absorbed, and only those with a wavelength of approximately 600 mμ are reflected. This is due to the molecular structure of the strip of fabric. Some of the reflected photons travel to the retina of your eye where they are absorbed by cones containing the appropriate photopigment. Absorption by the cones changes the neural impulses travelling to the occipital lobe. Somewhere in the occipital lobe a neural process occurs (see section on Methodological Problems of Localization), and there is an experience of red, resulting from stimulation by a stripe on the American flag. Note, however, that at no time during this description of the perceptual process, including the description of the flag itself, was it necessary to mention the color "red." Only at the end of the causal chain when one reaches the visual area of the brain is it necessary to mention "red."

From this kind of example philosophers have concluded that naive realism is false and that the world is not quite as it appears to us. The American flag is not red; it is a molecular structure. The red lies in our experience. In place of naive realism a representational theory of perception called structural realism has been proposed (Maxwell, 1970, 1976).

This theory is a descendent from the causal theories of perception advanced by Galileo, Locke, and Bertrand Russell (1954, 1967). There are two distinctions important for structural realism: the distinction between direct and indirect perception which we have alluded to in the context of epistemological dualism, and the distinction between intrinsic and structural properties. Our own experience is direct while our perception of the external, physical world is indirect. We perceive the physical world via representations or models of it in our immediate experience. The similarity of the representation to the object is structural. This is where the distinction between the intrinsic and structural becomes important. Intrinsic or first order properties pick out classes. Structural properties, second order and up, pick out classes of classes, and so on. They are definable in terms of logical constants and intrinsic properties. A perception of "red" is an intrinsic property; "two" is a structural property.

In our mental life we are acquainted with both intrinsic and structural properties. Since we only know of the external world by inferences from changes (samenesses and differences) produced in our perceptions by the external world, we can only know its structural properties. Our knowledge of the external world is like the hypothetico-deductive knowledge the physicist has of the world of unobservable elementary particles. Just as the physicist's knowledge is in the form of a theory containing correlation laws which connect the behavior of macroscopic bodies with that of submicroscopic entities, our knowledge of the external world can be put in the form of a theory with correlation laws which connect the variations of our perceptions with the behavior of external objects. The correlation of the physicist's theory provides it with a naive-realistic confirmation base; those of the theory of the external world provide it with a confirmation base in experience. Both theories can be logically reconstructed by translating them into Ramsey sentences, which allow a clear demarcation to be drawn between the observational content of a theory (predicates referring to macroscopic objects or our impressions), on the one hand, and logical structures representing theoretical entities (logical constants and uninterpreted predicates), on the other. Again, it is only the structure of the external world that matters in explaining and predicting our percepts; its particular intrinsic properties do not matter.

One consequence of structural realism is a solution to the mind–body problem. As we have seen, some of the laws in our theory of the external world are correlation laws, laws such as: "All instances of perceptions of red are neural discharges of such and such a type." Such laws assert that certain kinds of perceptions stand to certain kinds of neural activity as content to structure. For a structure to exist there must be intrinsic properties of which it is a structure. The correlation laws attach intrinsic properties to their structures.

A problem that opponents of mind–body identity have always thought a particular difficulty for the theory can be shown to be innocuous from the point of view of structural realism. When a neurosurgeon operates on a patient's occipital lobe while the patient is seeing a green patch, why does the surgeon not see the green patch in the patient's brain? He sees only grey matter, white matter, and red blood. If, however, the identity theory is true, and sensations of green are identical with brain processes, then the neurosurgeon should see green when he looks into the brain.

The resolution of this difficulty is accomplished in two stages, corresponding to the two distinctions important for structural realism. In the first place, the neurosurgeon need not necessarily see the green in his patient's brain since he does not see the brain directly. What the neurosurgeon sees (directly), however, does bear a structural similarity to his patient's brain (here is where the intrinsic–structural property distinction becomes relevant). If, for example, the patient were presented with a circle, it would (in suitable circumstances) produce something structurally similar in the brain of the patient which could in turn produce a structure in the brain of the surgeon. Colors, however, being intrinsic and not structural, are a different matter. Quality is not preserved over causal chains; only sameness or difference in quality is preserved. Thus when the patient sees green, there is no reason to suppose that the neurosurgeon's representation of the patient's brain should contain green among its elements. If, however, the patient sees a patch of green contiguous with a patch of red, the neurosurgeon (under proper conditions) should also be able to see a corresponding difference in quality.

Structural realism emphasizes the importance of distinguishing between physical space, psychological space, and the space of naive realism which has sometimes been termed common-sense space (Russell, 1954). Psychological space is the space of our direct experience. It may be divided into sensory space and perceptual space. Sensory space involves simple extension in a sensory experience; two parts of a blue patch in the visual field are seen to be a distance apart. When one observes the array of furniture around a room, one is experiencing perceptual space. Psychological space is located inside the brain in physical space. Physical space can be characterized structurally, with the exception of our experiences, which must be characterized as intrinsic properties of the brain. It is instructive to try and localize psychological space more precisely in the brain. It might be thought to be in the primary projection areas where an initial mapping of perceived space occurs. Or it might be believed to be in the parietal area where lesions result in disorders of spatial perception (Lynch, 1980). On the other hand, the prefrontal area appears to be importantly involved in the object concept which is related to the concept of space (Anderson, Hunt, Vander Stoep, and Pribram, 1976), and the hippocampus has been claimed to form spatial maps (O'Keefe and Nadel, 1978). It is impossible to localize psychological space more precisely in the brain because space is such an all-pervasive concept—to be in the universe is to be in space. It is not surprising therefore that all of the brain does spatial processing.

The concept of physical space assumed in localizing experience in the brain is radically different depending on whether the physics assumed is Newtonian, relativistic, or quantum theoretical. If space is Newtonian or relativistic then a fairly straightforward mapping may account for much of perceptual processing, and objects, including brains, are precisely differentiable in space. If, however, space is quantumlike, as maintained by David Bohm (1973), then perception may involve Fourier transformations and space may require a more holistic representation, such as those provided by the hologram (Pribram, 1977; Anderson, 1977, 1978a).

In localizing experience or any other mental phenomena, it is important to be aware of exactly which concept of space one is using. It is even more important to be sure that one is not using the naive realist, common-sense concept of space which inappropriately mixes psychological and physical space and places directly perceived intrinsic properties in the external, physical world. With this warning in mind we may now proceed to considering some methodological problems associated with localizing experience.

METHODOLOGICAL PROBLEMS OF LOCALIZATION

In 1808 Gall and Spurzheim postulated that specific regions of the cerebral cortex subserved specific mental functions. Although their concept of phrenology was generally rejected by scientists, the notion that different faculties of the mind were located in different areas of the brain became popularized, spawning the advent of experimental research efforts on the localization of specific functions in the brain. The discovery of Broca's (1861) "speech area" located in the posterior region of the frontal lobes and Fritsch and Hitzig's (1870) mapping of the motor cortex as specific for motor function when stimulated, argued for the notion of localization. Other experimenters of the nineteenth century, notably Flourens (1824) and Goltz (1884) claimed on the basis of their data that instead of localization of function, the cortex appeared to be a homogeneous entity in which the same degree of recovery of function was obtained regardless of which parts of the cortex were damaged. Lashley (1929), following an extensive series of lesion studies in his laboratory, also concluded that the extent of the lesions determined the level of impairment of function ("law of mass action") as complex functions are distributed over various cortical areas ("law of equipotentiality").

As more and more data have accumulated, strict localization of function resulting in exclusive and unique representations of the cortex have not been unambiguously specified. Because strict localization is now untenable, the brain should no longer be viewed as a collection of separate centers or organs in which each center mediates a specific function. Luria (1966) and others (Gregory, 1961; Thatcher and John, 1977; Uttal, 1978; Weiskrantz, 1968) suggest that if an adequate theory of localization is to be developed, our concepts of "centers" and "functions" must be altered. Furthermore, as discussed in the first and second sections, our notions of localization and place must also be clarified. However, conceptualizations of "centers," "function," "localization," and "place" differ. For example, Gregory (1961) would argue that to say that some feature is localized in a region of the brain means that a NECESSARY CONDITION for this behavior is specified in a certain portion of the brain. This means that we must know how the system works before we can specify the necessary condition(s). According to Luria's (1966) conception of localization of function, "function" is the product of an underlying nervous system whose integration and analysis of afferent input via excitations and inhibitions are united in a common task so that an equilibrium of the organism and the environment is achieved. Hécaen and Albert (1978) maintain that cortical regions function not in isolation, but that they may contain many functional potentialities even though there may be a predominance of one particular behavioral "skill."

Modern research techniques have contributed much to these more recent concepts of cerebral localization of function. Neuroscientists, armed with selective staining techniques, retrograde and anterograde labeling, ablation and stimulation techniques, evoked-potential and single-cell recording techniques, and split-brain preparations, to name a few, have provided a wealth of information about the macroanatomy, microanatomy, and physiology of the nervous system. However, these techniques have certain limitations for the localization issue. One of the clearest expositions of such limitations has recently been made by Uttal (1978). For example, cortical lesions made by aspiration of cortex, or selective transections of fiber tracts, may result in the loss of certain functions, but does this prove that the lesioned area was responsible for those functions? Uttal (1978) effectively argues that the only conclusion that can be made is that the lesioned area was probably involved in processing some of those functions, but not that those functions *per se* were localized in that area. Furthermore, if a lesion FAILS to affect a certain task or behavior, it cannot be concluded that the lesioned area is unimportant in normal subjects (Chow, 1967).

Another difficulty for the localization issue is the failure of some scientists to differentiate between brain function and psychological function (Glassman, 1978). Even when such attempts are made, the psychological processes and constructs are often ill-defined (Uttal, 1978). Furthermore, even if one could assume that psychological functions are well defined, the problem of isolating the specific neural units is still a major one because of the lack of precision in identification means for determining the appropriate size and level of the neural units for that given psychological function (Uttal, 1978). Even the use of microelectrodes for localization of neural units is insufficient because not only do regions vary among individuals, but also within an individual's nervous system. Microregions, like macroregions, are not homogeneous in cytoarchitecture, and at both micro and macro levels, the interconnectivity of neurons may function as feedback and feedforward loops to simultaneously affect multiple neurons in varying loci. Such interconnectivity may result in such a complicated response pattern that our current methods of analysis may not suffice to decipher it (Uttal, 1978, illustrates this well). Also related to the high levels of interconnectivity are the possibilities that a specific neuron may subserve more than one function, and that many neurons may share the same function; redundancy in the nervous system may therefore result in masking a deficit in several ways, only to lead to the wrong experimental conclusion (Uttal, 1978).

In recent years much has been learned about the neuronal response to injury (Woolsey, 1978). Not only is there degeneration of axons severed in a lesion (and in many cases soma degeneration as well), but also in some cases transneuronal degeneration (where degeneration "crosses" the synapse) occurs (Cowan, 1970). Some researchers now report that severed neurons may reestablish former anatomical connections (Liu and Chambers, 1958; Raisman, 1969; Lynch, Matthews, Mosko, Parks, and Cotman, 1972) via regenerative sprouting or collateral sprouting (Stenevi, Björklund, and Moore, 1973). Other investigators have reported "recovery of function" (Goldman, 1974, 1976; Schneider, 1970, 1973), although in most cases the age at which the animals sustained their lesions appears to be the relevant factor for the recovery of function. Yet, others have shown that progressive serial lesions in contrast to one-stage bilateral lesions may produce a sparing effect (Rosen, Stein, and Butters, 1971) in some cases (see Finger, Walbran, and Stein, 1973, for a review of serial lesions) but greater deficits and poorer recovery of function in other cases (Patrissi and Stein, 1975). LeVere and Weiss (1973) maintain that the sparing effect may be limited to neocortical structures and more recently LeVere (1975) postulates that the recovery of function may be produced by spared tissue and not related to reorganization or regrowth. Uttal (1978) points out that LeVere's analysis implicitly suggests that a particular behavior may be larger and less localized than originally conceptualized.

Although there are many problems and objections that a contemporary localizationist must overcome, we have suggested earlier in the first and second sections that localization of experience in the brain is logically possible. We now wish to consider the correspondence of brain states with visual experience. We begin our inquiry by asking what is necessary for visual experience, but our ultimate goal is to establish what is necessary AND sufficient for visual experience to occur. If we accomplish this goal it may be possible to speculate which physiological tissues correspond to and are alleged by the identity theorist to be identical with experience.

For people to have visual experiences they must have properly functioning hearts, for if the heart stops, the body dies and experience ceases. However, heart failure does not cause visual experience to cease instantaneously since several seconds must pass before blood ceases to flow and tissues are depleted of oxygen (Smith and Kampine, 1980); furthermore, complete removal and substitution of a "new" heart does not result in alterations of visual experience. Although the heart is part of the life-support system providing oxygen and other nutrients for the human body, we are not tempted to become Aristotelians and identify the heart as the seat of visual experience. Thus, while the heart may be necessary for visual experience to occur, it is not sufficient in and of itself to produce visual experience. The causal distance between heart and visual experience is relatively large (i.e., there is a relatively long chain of causation between the heart and visual experience), and to satisfy the identity theory there should be no causal distance.

.A part of the body which is minimally causally distant to visual experience is, of course, the nervous system. But can we narrow down the causal distance within the nervous system? To our knowledge, sectioning of the spinal cord below the level of the cranial nerves does not produce loss of visual experience. But lesions in certain areas of the brain and certain cranial nerves DO affect the processing of visual input and will be discussed in the following pages, as we attempt to reduce causal distance and determine the necessary and sufficient neural tissues for visual experience.

The centrencephalic system, consisting of the diencephalon, mesencephalon, and rhombencephalon, might be that part of the brain most closely associated with consciousness and hence sensory experience. According to neurosurgeon Wilder Penfield (1952) the centrencephalon might serve to integrate the highest levels of brain function. This part of the brain is known to control sensory input including visual afferents (Bartlett and Doty, 1974), and lesions within the mesencephalon produce a permanent loss of consciousness (Ingvar and Sourander, 1970). There is, however, evidence to the contrary. The mesencephalon in humans has been selectively inhibited in its activity by the introduction of amobarbital, and consciousness was not affected (Alema, Perria, Rosadini, and Zattoni, 1966). The injection of amobarbital into the carotid artery of patients having only one brain hemisphere, however, produced an immediate loss of consciousness when it affected the cortex but not the mesencephalon (Obrador, 1964). But according to Doty (1975) the most compelling evidence that the forebrain contains the "locus of consciousness' is Sperry's patients with transections of the corpus callosum and anterior commissures.

The mesencephalon, although it does not directly participate in the brain processes that correspond to sensory experience, does probably function as a consciousness-support system in a manner analogous to the heart's acting as a life-support system. It may serve as a support system by neurally bombarding the cortex.

Having eliminated the mesencephalon as a candidate for the "locus of consciousness" and therefore as sufficient for visual experience, we can now close on our quarry and begin to examine one of the more specialized areas of the brain, that is, the visual system. The major pathway of the visual system consists of the eye (with its lens and retina, including rods, cones, bipolar, and ganglion cells), the optic nerve (which carries neural impulses from the retina to the lateral geniculate nucleus of the thalamus), the thalamus, the optic radiations, and the striate cortex. However, projections from the ganglion cells of the retina also terminate in other regions of the brain, and the roles these may play in vision are not to be overlooked. These projections include the suprachiasmatic nuclei of the hypothalamus, the accessory optic nuclei, the pretectum, the superior colliculus, and the ventral and dorsal lateral geniculate nuclei (Rodieck, 1979). It is the latter nucleus which projects via the optic radiations to the striate cortex of the occipital lobe (Brodmann's area 17) and whose pathway is most thoroughly studied. The extrastriate visual cortex (areas 18 and 19) and the inferotemporal cortex (area 37) are also involved in visual processes.

The problem is to determine which of these parts of the visual system is necessary AND sufficient for the production of visual experience, given functioning support systems and some kind of neural stimulation. Thus, if any of the components of the visual system is destroyed and yet visual experience can still occur, that particular structure could not be necessary for visual experience. Both the eye and the optic nerve fall in this category. People who are blind, by reason of complete transections of the optic nerves or irreversible damage to the entire retina, have spontaneous hallucinations and can be electrically stimulated in the visual area of the cortex to see phosphenes (Brindley, Donaldson, Falconer, and Rushton, 1972; Dobelle, Mladejorsky, and Girvin, 1973). If the eyes and optic nerves are intact but all other projection areas from the retina are destroyed, it is plausible to hypothesize that stimulation of the eye or nerve alone would not produce visual experience. It is true that recently Weiskrantz, Warrington, Sanders, and Marshall (1974) have shown that subjects with damage to the striate cortex are able to point accurately to spots of light which lie within their scotomata, but in this case, the subjects insist that they see (i.e., visually experience) nothing. Weiskrantz has hypothesized, and his hypothesis seems warranted on the basis of monkey (Weiskrantz and Cowey, 1970) and lower mammal research (Schneider, 1969), that this ability is due to the processing of visual information in the subjects' superior colliculi. Perinin and Jeannerod (1975) also found that subjects with postgeniculate lesions could point with fairly good accuracy to a light source whereas subjects with pregeniculate lesions could not. The tenuousness of Weiskrantz's hypothesis is illustrated by the fact that a plausible alternative explanation of these results has been made by Hécaen and Albert (1978). They suggest that the striate cortex may be necessary for higher perceptual processing, and with this region damaged, the subjects can possibly retain capacities for visual perception but not be aware of what they see. In general, however, data on whether stimulation of the eye or nerve alone will produce visual experience are hard to come by. This problem, which is particularly acute in any effort to localize experience, exists because relevant experimentation on human subjects is ethically intolerable and experimentation on animals is generally of doubtful significance. It is difficult if not impossible, for example, to distinguish behaviorally whether a monkey is hallucinating a fly it is trying to grasp or whether it is simply making a grasping motion. In general it is difficult to determine behaviorally whether an animal is having a visual experience or merely making a simple motor response. For this reason much of our attempt to localize experience must be speculative.

It is tempting to dismiss the lateral geniculate nucleus of the thalamus as a mere information-processing way station between the eye and the striate cortex where nothing corresponding to visual experience exists. However, little direct evidence exists on this point. Some tenuous indirect evidence can be extracted from the study by Weiskrantz, Warrington, Sanders, and Marshall (1974) previously mentioned. According to that study visual information was correctly processed but there was no concomitant visual experience. In all likelihood these patients had intact superior colloculi. An outstanding anatomical feature of the superior colliculus is that it is a nucleus or globular structure of nerve cell bodies. On the basis of these facts we may make an extremely weak generalization to there being no sensory experience and perhaps no consciousness at all directly associated with globular nuclei. Therefore, it is possible that the lateral geniculate nucleus, being a globular arrangement of cell bodies, is not sufficient for visual experience.

A comprehensive analysis and summary of alternative cortical structures, such as the inferotemporal and extrastriate regions and other association cortex regions, has been done by Hécaen and Albert (1978) and should be consulted for descriptions of various patients' clinical pictures and anatomical correlates. But because lesions in any two human subjects are rarely bilateral and even more rarely identical in location and depth of lesion, no conclusions can be drawn at this time as to whether these cortical structures are necessary for visual experience. Whether the inferotemporal cortex is sufficient for the production of visual experience is also unknown. Stimulation of the inferotemporal cortex does produce visual experience, but the striate area has always been intact so it is not known whether the stimulation indirectly activates the visual cortex (Penfield, 1966).

Although the evidence is extremely weak, we will assume for the sake of argument that the striate cortex (along with, to some degree, the borderline area 18) appears to be part of the brain that is minimally causally distant from visual experience. We can then ask whether the brain processes that are to be correlated with visual experience occur only in the cortical sheet of grey matter, or whether they extend into the cortico-cortical white matter fibers. Since the visual field is experienced as a unity and not as two halves corresponding to the two hemispheres of the brain, and since this unity is accomplished by the white matter tracts of the corpus callosum, we must answer that the processes corresponding to visual experience do extend into white matter (Anderson and Gonsalves, 1981). This is further evidenced by the fact that the fovea of the retina and the center of the visual field correlate with the far edges of the striate cortex, yet the center of the visual field is experienced as a unity (Teuber, Battersby, and Bender, 1960).

What is it about the cellular architecture of the primary visual projection area (area 17) that makes it suitable for visual experience? Is it that its cells are functionally and anatomically organized in columns (Chow and Leiman, 1970)? (In the striate cortex neurons with similar orientation response properties are found in columns perpendicular to the cortical sheet.) This is probably not the answer since the superior colliculus is also arranged in a columnar manner (Sprague, Berlucci, and Rizzolatti, 1973). Perhaps it is due to the fact that the striate cortex is a sheet of neurons rather than a globular structure. The cerebellum, however, is also a cortical sheet, and yet consciousness is not associated with its activity. Our "cortical sheet hypothesis" also does not appear to be necessary and sufficient for visual experience. Thus, although we have found some likely candidates for the tissues wherein the processes corresponding to visual experience occur, we are not able, at this time, to determine just what it is about these parts of the cortex that gives them this unique property.

THE NATURE OF QUALIA

As difficult as it may be to localize sensory experiences in the brain, it is at least an equally formidable task to give a neurophysiological account of the qualitative aspects of experience. There are two related problems which are particularly important to deal with in attempting to accomplish this latter task. These are (1) to give an account of qualitative difference — the problem of modality and (2) to give an account of qualitative sameness — the grain problem. Qualitative difference across modalities presents a problem because there does not seem, on first glance, to be a difference in neural processes comparable to the vast difference experienced between hearing, seeing, and touching. This issue has been argued at length in a recent issue of *The Behavioral and Brain Sciences* (Puccetti and Dykes, 1978). The first author (Anderson, 1978b), and many others, argued that there was no reason to believe that qualitative difference could not be explained by reference to the relative place and kind of processing being performed in the nervous system. In making this argument, we were merely repeating what Hayek (1952, 1969) had said years earlier. Our argument in this case was a functionalist style of argument (Fodor, 1981; Dennet, 1978). According to the functionalist, mental properties simply are sets of relations among things.

The problem of quality-sameness has received less attention lately. This problem stems from the fact that we can have visual experiences which are qualitatively uniform. For example, if you lie on the beach on a sunny day without sun glasses, you will see a uniform red expanse as the sunlight is filtered through the blood vessels in your eyelids. The problem then lies in accounting for this uniform spread of quality in terms of a set of numerous, discrete neural components (Sellars, 1963). To the authors' knowledge, no satisfactory account has yet been given, although many attempts have been made. One approach is to take experience as primary and to construct the physical from it. This is the method advocated by Russell (1954) and Maxwell (1970). Russell maintains that, in the brain, experiences which are compresent, that is, experiences which overlap each other in a certain way, can define elementary particles such as electrons and, hence, also define neurons. By taking experiences as primary, Russell is able to theoretically construct electrons and other elementary particles out of experiences. Since neurons, in turn, are composed of elementary particles, he is able to construct neurons out of experiences. The problem with this approach is its premise that experience is primary. Thus far no satisfactory demonstration has been made of how an electron in the brain could be composed of experience.

A functionalist approach to the problem is of no avail since, as Keith Gunderson (1971) has pointed out, qualia are nonprogrammable properties. It is part of their nature that they cannot be captured solely in terms of an organism's input and output and the relations holding between the organism's internal states. If, for example, one person's perceived spectrum of colors were inverted with respect to another's, there would be no way to detect this in the individual's behavior (Shoemaker, 1975). You might experience my green and call it "red" and experience my red and call it "green," and we could never find this out from our behavior. This claim that qualia are nonprogrammable is similar to the claim of structural realism that intrinsic qualities cannot be propagated over causal chains.

Gunderson (1974) attempts to resolve the grain problem by examining how one would establish the identity of what is seen when an object appears to have two different textures when viewed from two separate perspectives. A poster blow-up of the cartoon character Dick Tracy, for example, when viewed from eight feet away is seen as the smooth-grained expanse which we call Dick Tracy's face. At eight inches, however, hundreds of different little dots are seen. What one sees up close is different from what one sees from a greater distance. They are perspectively nonidentical. If, however, we buy the poster, we buy both the dots and the face at the same time. The dots and the face are nonperspectively identical. One of the important reasons we can accept identity in this case is that we can walk up on the face and see the dots emerge. Gunderson points out that in the case of our neural states and mental states, we cannot perform such a feat. We cannot walk up to or back from our own mental states; we are locus-bound to them. Understanding this limitation, according to Gunderson, should resolve our perplexity and the grain problem. Unfortunately, however, this argument completely misses the point of the grain problem. Its force lies in the fact that we do not really have a perspective on our own experiences. We do not view them from a distance but are ONE AND THE SAME with them. Thus we cannot resolve the problem by reference to spatially separate perspectives from which we view them.

One of the most promising approaches to the grain problem has been proposed by Stephen Pepper (1966). Pepper proposed that a threshold of neural activity must be reached in order to have sensory awareness. If such units overlap, then the continuity of qualia may be explained. He described the threshold as a threshold of fusion of awareness. Such an approach is plausible since neural complexity appears to be required for consciousness. The more complex a creature's nervous system, the higher its level of awareness. This solution has been worked out by the first author in detail (Anderson, 1976). Unfortunately it is highly speculative and probably almost impossible to test.

With this last suggestion, our attempt to localize experience must now come to a close. Although we sought the necessary and sufficient structure in the brain for experience, we had to be satisfied with but a possible candidate which was only minimally causally distant.

As with most tasks of profound import, in trying to localize sensory experience we have been lucky merely to get a better understanding of the difficulty of the task, without even coming close to accomplishing our mission. We have examined some of the many pitfalls—both logical and methodological—that lie in wait for anyone who attempts to localize experience in the brain. We can only hope that we have eased the way for others who wish to try to capture the mind within the brain.

REFERENCES

Alema, G., Perria, L., Rosadini, G., Rossi, G. F., and Zattoni, J. (1966). Functional inactivation of the human brain stem related to the level of consciousness. *Journal of Neurosurgery, 24,* 629–639.

Anderson, R. M. (1972). *An essay in neuroepistemology.* Ann Arbor, Michigan: University Microfilms.

Anderson, R. M. (1976). *The illusions of experience.* In R. S. Cohen, C. A. Hooker, A. C. Michalos, and J. W. van Eura (Eds.), *PSA 1974,* Dordrecht, Holland: D. Reidel.

Anderson, R. M. (1977). A holographic model of transpersonal consciousness. *Journal of Transpersonal Psychology, 9,* 119–128.

Anderson, R. M. (1978a). Quandaries of mind, brain, and cosmos. *International Philosophical Quarterly, 18,* 215–222.

Anderson, R. M. (1978b). Relativistic color coding as a model for quality differences. *The Behavioral and Brain Sciences, 3,* 45–46.

Anderson, R. M., and Gonsalves, J. F. (1981). Sensory suppression and the unity of consciousness. *The Behavioral and Brain Sciences, 4,* 99–100.

Anderson, R. M., Hunt, S. C., Vander Stoep, A., and Pribram, K. H. (1976). Object permanency and delayed response as spatial context in monkeys with frontal lesions. *Neuropsychologia, 14,* 481–490.

Aune, B. (1967). *Knowledge, mind, and nature.* New York: Random House.

Bartlett, J. R., and Doty, R. W. (1974). Influence of mesencephalic stimulation on unit activity in striate cortex of squirrel monkeys. *Journal of Neurophysiology, 37,* 642–652.

Bohm, D. (1973). Quantum theory as an indication of a new order in physics. Part B. Implicate and explicate order in physical law. *Foundations of Physics, 3,* 139–168.

Boring, E. G. (1942). *Sensation and perception in the history of experimental psychology.* New York: Appleton.

Boring, E. G. (1950). *A history of experimental psychology.* New York: Appleton.

Brindley, G. S., Donaldson, P. E. K., Falconer, M. A., and Rushton, D. N. (1972). The extent of the region of the occipital cortex that when stimulated gives phosphenes fixed in the visual field. *Journal of Physiology, 225,* 57P–58P.

Broad, C. D. (1960). *The mind and its place in nature.* Patterson, N.J.: Littlefield, Adams, and Co.

Broca, P. (1861). Remarques sur le siège de la faculté du langage articulé suivies d' une observation d' aphémie. *Bulletins de la Société Anatomique de Paris, 36,* 330–357.

Chow, K. L. (1967). Effects of ablation. In G. C. Quarton, T. Melnechuk, and F. O. Schmitt (Eds.), *The neurosciences: A study program.* New York: Rockefeller Univ. Press.

Chow, K. L., and Leiman, A. L. (1970). Aspects of the structural and functional organization of the cortex. *Neuroscience Research Program Bulletin, 8,* 153–200.

Copleston, F. C. (1958). *A history of philosophy.* Westminster, Maryland: Newman Press.

Cowan, W. M. (1970). Anterograde and retrograde transneuronal degeneration in the central

212   ROBERT M. ANDERSON, JR., AND DEMETRIA C. Q. LEONG

and peripheral nervous system. In W. J. H. Nauta and S. O. B. Ebbesson (Eds.), *Contemporary research methods in neuroanatomy*. New York: Springer-Verlag.

Dennett, D. C. (1978). Current issues in the philosophy of mind. *American Philosophical Quarterly, 15,* 249–261.

Dobelle, W. L., Mladejorsky, M. G., and Girvin, J. P. (1973). Artificial vision for the blind: Electrical stimulation of visual cortex offers hope for a functional prosthesis. *Science, 183,* 440–449.

Doty, R. W., Sr. (1975). Consciousness from neurons. *Acta Neurobiologiae Experimentalis, 35,* 791–804.

Feigl, H. (1967). *The "mental" and the "physical"*. Minneapolis: Univ. of Minnesota Press.

Finger, S., Walbran, B., and Stein, D. G. (1973). Brain damage and behavioral recovery: Serial lesion phenomena. *Brain Research, 63,* 1–18.

Flourens, M. J. P. (1824). *Recherches expérimentales sur les propriétés et les fonctions du système nerveux dans les animaux vertébrés*. Paris: Crerot.

Fodor, J. A. (1981). The mind-body problem. *Scientific American, 244,* 124–132.

Fritsch, G., and Hitzig, E. (1870). Über die electrische Erregbarkeit des Grosshirns. *Archiv für Anatomie, Physiologie, und wissenchaftliche Medizin, 37,* 300–332.

Gall, F. J., and Spurzheim, L. C. (1808). Recherches sur le système nerveux en général, et sur celui du cerveau en particulier. *Académie des Sciences*. Paris: Memoirs.

Glassman, R. B. (1978). The logic of the lesion experiment and its role in the neural sciences. In S. Finger (Ed.), *Recovery from brain damage: Research and theory*. New York: Plenum.

Goldman, P. S. (1974). An alternative to developmental plasticity: Heterology of CNS structures in infants and adults. In D. G. Stein, J. J. Rosen, and N. Butters (Eds.), *Plasticity and recovery of function in the central nervous system*. New York: Academic Press.

Goldman, P. S. (1976). The role of experience in recovery of function following orbital prefrontal lesions in infant monkeys. *Neuropsychologia, 14,* 401–412.

Goltz, F. (1884). Uber die Verrichtungen des Grosshirns. *Pflüger's Archiv Gesamte Physiologie, 26.*

Gregory, R. L. (1961). The brain as an engineering problem. In W. H. Thorpe and O. L. Zangwill (Eds.), *Current problems in animal behavior*. Cambridge: Cambridge Univ. Press.

Gunderson, K. (1970). Asymmetries and mind–body perplexities. In M. Radner and S. Winokur (Eds.), *Minnesota studies in the philosophy of science*. Vol 4. Minneapolis: Univ. of Minnesota Press.

Gunderson, K. (1971). *Mentality and machines*. Garden City, New York: Anchor.

Gunderson, K. (1974). The texture of mentality. In R. Bambrough, (Ed.), *Wisdom: Twelve essays*. Totowa, N.J.: Littlefield.

Hayek, F. A. (1952). *The sensory order*. Chicago: Univ. of Chicago Press.

Hayek, F. A. (1969). The primacy of the abstract. In A. Koestler and J. R. Smythies (Eds.), *Beyond reductionism*. New York: Macmillan.

Hécaen, H., and Albert, M. L. (1978). *Human neurophysiology*. New York: Wiley.

Ingvar, D. H., and Sourander, D. (1970). Destruction of the reticular core of the brainstem. *Archives of Neurology, 23,* 1–8.

Lashley, K. S. (1929). Learning. I. Nervous mechanisms in learning. In C. Murchism (Ed.), *The foundations of experimental psychology*. Worcester, Mass.: Clark Univ. Press.

LeVere, T. E. (1975). Neural stability, sparing, and behavioral recovery following brain damage. *Psychological Review, 82,* 344–358.

LeVere, T. E., and Weiss, J. (1973). Failure of seriation dorsal hippocampal lesions to spare spatial reversal behavior in rats. *Journal of Comparative and Physiological Psychology, 82,* 205–210.

Liu, C.-N., and Chambers, W. W. (1958). Intraspinal sprouting of dorsal root axons. *Archives of Neurology and Psychiatry, 79,* 46–61.

Luria, A. R. (1966). *Higher cortical functions in man.* New York: Basic Books.

Lynch, G., Matthews, D. A., Mosko, S., Parks, T., and Cotman, C. (1972). Induced acetylcholinesterase-rich layer in rat dentate gyrus following entorhinal lesions. *Brain Research, 42,* 311–318.

Lynch, J. C. (1980). The functional organization of posterior parietal association cortex. *The Behavioral and Brain Sciences, 3,* 485–534.

Maxwell, G. (1970). Theories, perception, and structural realism. In R. G. Colodny (Ed.), *The nature and function of scientific theories.* Pittsburg: Univ. of Pittsburgh Press.

Maxwell, G. (1976). Scientific results and the mind–brain issue: some afterthoughts. In G. Globus, G. Maxwell, and I. Savodnik (Eds.), *Consciousness and the brain.* New York: Plenum.

Obrador, S. (1964). Nervous integration after hemispherectomy in man. In G. Schaltenbrand and C. N. Woolsey (Eds.), *Cerebral localization and organization.* Madison: Univ. of Wisconsin Press.

O'Keefe, J., and Nadel, L. (1978). *The hippocampus as a cognitive map.* Oxford: Oxford Univ. Press.

Patrissi, G., and Stein, D. G. (1975). Temporal factors in recovery of function after brain damage. *Experimental Neurology, 47,* 470–480.

Penfield, W. (1952). Epileptic automatism and the centrencephalic integrating system. *Association for Research in Nervous and Mental Disease, 30,* 513–528.

Penfield, W. (1966). Speech, perception and the uncommitted cortex. In J. C. Eccles (Ed.), *Brain and conscious experience.* New York: Springer-Verlag.

Pepper, S. C. (1966). *Concept and quality.* La Salle, Ill.: Open Court.

Perinin, M. T., and Jeannerod, M. (1975). Residual vision in cortically blind hemifields. *Neuropsychologia, 13,* 1–7.

Pribram, K. H. (1977). Holonomy and structure in the organization of perception. In U. M. Nicholas (Ed.), *Images, perception and knowledge.* Dordrecht, Holland: D. Reidel.

Puccetti, R., and Dykes, R. W. (1978). Sensory cortex and the mind–brain problem. *The Behavioral and Brain Sciences, 3,* 337–344.

Raisman, G. (1969). Neuronal plasticity in the septal nuclei of the adult rat. *Brain Research, 14,* 25–48.

Rodieck, R. W. (1979). Visual pathways. *Annual Review of Neuroscience, 3,* 193–225.

Rorty, R. (1965). Mind–body identity, privacy, and categories. *Review of Metaphysics, 19,* 24–54.

Rosen, J. J., Stein, D. G., and Butters, N. (1971). Recovery of function after serial ablation of prefrontal cortex in the rhesus monkey. *Science, 173,* 353–356.

Russell, B. (1954). *The analysis of matter.* New York: Dover.

Russell, B. (1967). *Human knowledge: Its scope and limits.* New York: Simon and Schuster.

Ryle, G. (1949). *The concept of mind.* New York: Barnes and Noble.

Schneider, G. E. (1969). Two visual systems. *Science, 163,* 895–902.

Schneider, G. E. (1970). Mechanisms of functional recovery following lesions of visual cortex or superior colliculus in neonate and adult hamsters. *Brain, Behavior, and Evolution, 3,* 295–323.

Schneider, G. E. (1973). Early lesions of the superior colliculus: Factors affecting the formation of abnormal retinal projections. *Brain, Behavior, and Evolution, 8,* 73–109.

Sellars, W. (1963). *Science, perception, and reality.* New York: Humanities.

Shaffer, J. (1961). Could mental states be brain processes? *Journal of Philosophy, 58,* 813–822.

Shoemaker, S. (1975). Functionalism and qualia. *Philosophical Studies, 27,* 291–315.

Smart, J. J. C. (1963). *Philosophy and scientific realism.* New York: Humanities.

Smith, J. J., and Kampine, J. P. (1980). *Circulatory physiology: The essentials.* Baltimore: Williams and Wilkins.

Sprague, J. M., Berlucci, G., and Rizzolatti, G. (1973). The role of the superior colliculus and pretectum in vision and visually guided behavior. In R. Jung (Ed.), *Handbook of sensory physiology,* Vol. 7. New York: Springer-Verlag.

Stenevi, U., Björklund, A., and Moore, R. Y. (1973). Morphological plasticity of central adrenergic neurons. *Brain, Behavior, and Evolution, 8,* 110–134.

Teuber, H.-L., Battersby, W. S., and Bender, M. B. (1960). *Visual field defects after penetrating missile wounds of the brain.* Cambridge: Harvard Univ. Press.

Thatcher, R. W. and John, E. R. (1977). *Foundations of cognitive processes.* Hillsdale, N.J.: Laurence Erlbaum Associates.

Uttal, W. R. (1978). *The psychobiology of mind.* Hillsdale, N.J.: Lawrence Erlbaum Associates.

Weiskrantz, L. (1968). Treatments, inferences and brain function. In L. Weiskrantz (Ed.), *Analysis of behavioral change.* New York: Harper.

Weiskrantz, L., and Cowey, A. (1970). Filling in the scotoma: A study of residual vision after striate cortex lesions in monkeys. In E. Stellar and J. M. Sprague (Eds.), *Progress in physiological psychology,* Vol. 3. New York: Academic Press.

Weiskrantz, L., Warrington, E. K., Sanders, M. D., and Marshall, J. (1974). Visual capacity in the hemianopic field following a restricted occipital ablation. *Brain, 97,* 709–728.

Woolsey, T. A. (1978). Lesion experiments: Some anatomical considerations. In S. Finger (Ed.), *Recovery from brain damage: Research and theory.* New York: Plenum.

ACKNOWLEDGMENTS

In writing this chapter we have benefited from discussions with Anthony Marsella, Grover Maxwell, Karl Pribram, and Bill Uttal, who are not in any way culpable for its contents. This chapter was written while the first author was being supported as a clinical psychology intern by the Department of Health of the State of Hawaii. We should like to thank Diana Stephens, librarian of the Hawaii State Hospital Medical Library, for a computerized literature search.

PART III

On the Requirements of a Developmental Theory of Lateralization

CONFUSING MATURATION WITH DEVELOPMENT

When developmental trends occur in neuropsychological patterns, it is tempting to treat them as maturational, that is, as being due to factors associated with biological growth. Often, of course, this is an appropriate interpretation. For example, if compensation for brain damage is indeed more likely in the immature organism (although cf. Isaacson, 1975; St. James-Roberts, 1981), the mechanism may be entirely maturational, such as the relatively rapid death of neurons and growth of dendrites during the first year after birth (Huttenlocher, 1979). To a large extent, this is how the developmental lateralization hypothesis is viewed—that there is a biologically determined critical period for the establishment of language representation in the left hemisphere (Lenneberg, 1967). However, test scores that reflect neuropsychological organization, whether from clinical cases or from normal subjects, can show developmental changes for many reasons, not all related to neuropsychological organization of the brain. Compared to adults, children have organizational structures that differ neurophysiologically, psychologically and, therefore neuropsychologically.

NEUROPHYSIOLOGICAL CHANGES

Obviously there are hosts of neurophysiological changes during childhood. Some have direct bearing on neurolinguistic measures. For example, the postbirth myelination of the cortex and rapid growth of dendrites and axons have been held accountable for the change in control from subcortical to cortical structures. Thus, primitive reflexes that disappear shortly after birth are thought to be subcortically mediated (e.g., Brackbill, 1971). When a movement asymmetry is found to be related to speech stimulation in infants, the neurolinguistic significance may be a subcortical asymmetry rather than a cortical one (Segalowitz and Chapman, 1980). Similarly, the link between manual behavior and speaking in children and adults (Hiscock and Kinsbourne, 1980) may reflect different processes when compared with a similar link in infants. A seeming continuity may thus be a discontinuity.

Similarly, the even later myelination of the corpus callosum has demonstrable effect on lateralized input in young children (Galin, Johnstone, Nakell, and Herron, 1979) and may account for hemisphere asymmetry effects that are stronger in infancy than in adulthood (e.g., Molfese, Freeman, and Palermo, 1975).

PSYCHOLOGICAL CHANGES

The vast qualitative changes during childhood in needs, competencies, and attitudes form the basis for developmental psychology. The psychodynamic changes during infancy certainly have some influence on communicative and psycholinguistic needs and strategies. More importantly however, cognitive changes may directly affect performance on neurolinguistic tasks. For example, a gradual increase with age of an attentional bias to the right side of the body (Porac, Coren, and Duncan, 1980) enhances the chances of finding a greater right-ear advantage in a dichotic listening task in older children than in younger (Bryden and Allard, 1981). Procedural changes can compensate for such biases and can thus produce different developmental patterns (Bryden and Allard, 1981; see also Bryson, Mononen, and Yu, 1980). Other developments that are reflected in changes of cognitive strategy in a more global framework, such as Piaget's, should certainly affect how children of different ages address the neurolinguistic tasks given them. For example, Kraft (Mimeo; Kraft,

Mitchell, Languis, and Wheatley, 1980) proposes that there is a specific relationship between asymmetric brain activities and cognitive development within a Piagetian framework. None of the developmental changes in cognition should affect the brain organization for linguistic skills in one sense: if it were possible to ensure identical cognitive strategies from children of different ages, then a true developmental pattern could be extracted. But, if we admit that such changes influence the way our neurolinguistic tasks are performed, then the age variable is hopelessly confounded with such strategy variables. Another instance involves the place of verbal stimuli in the child's cognitive framework, whether as isolated speech sounds, as linguistic entities (word images), or as items with semantic force (Porter and Berlin, 1975; Belmore, 1980; see also Valsiner's discussion of Hrizman's work (this volume)).

The shape of a developmental neuropsychological theory should include, then, both neurological and psychological factors. The two sets of factors, however, sometimes compete. For example, the generally accepted notion that there is a critical period for second language acquisition, and that after puberty such acquisition is qualitatively more difficult, is used to support the notion that either brain lateralization for language is complete by that age or that some fundamental aspect of brain plasticity is lost (Lenneberg, 1967). Alternatively, it can be argued that the onset of formal operations at around puberty dramatically changes the learning style of the child. The less intuitive approach developing at that time works against language learning, not to mention the dramatic increase in self-consciousness (Krashen, 1973). It is conceivable that the two sets of factors may be related in some way, but even if not, any explanation of this critical period will have to include both sets.

CHANGES IN CEREBRAL ASYMMETRIES WITH AGE

Let us accept, then, the notion that any good summary of developmental changes in neurolinguistic organization will rest on developmental changes in psycholinguistic processing. That is, rather than brain organization for linguistic activities changing over time, it is the psycholinguistic strategies and skills of the child that change, and that any concomitant neurolinguistic developments are probably due to these changes. This principle probably works from age 2 or 3 onward, that is, from the beginning of overt language acquisition. Before this time, there are such maturational changes in the cortex that we should not extrapolate backward from the experimental data with children to neonates. This position was well articulated and supported in the review by Witelson (1977), where she summarized the developmental literature on lateralization up until 1976. Since then, there have been concurring reports that support the hypothesis of no growth in cerebral asymmetries with age. For example, consistent right-ear advantages are reported in dichotic listening studies using verbal stimuli (digits, words, or consonant–vowel syllables) with preschoolers (Piazza, 1977; Bryson, Mononen, and Yu, 1980; Hiscock and Kinsbourne, 1980) and with school children from kindergarten through grade six and college students (Bakker, Hoefkens, and Vander-Vlugt, 1979; Borowy and Goebel, 1976; Bryden & Allard, 1981; Hiscock and Kinsbourne, 1980; Hynd and Obrzut, 1977; Schulman-Galambos, 1977). In all these studies, no interaction between degree of ear asymmetry and age was found. Other paradigms and nonverbal stimuli have received less attention. When they have, though, the same lack of developmental effect is reported: Piazza (1977) found a left-ear effect in a dichotic listening task with nonverbal, environmental sounds in 3- to 5-year-olds, and found parallel results in the same sample with the unimanual finger-tapping time-sharing technique with verbal and with nonverbal tasks; similarly Hiscock and Kinsbourne (1978, 1980) found no age-asymmetry interaction in children 3 to 12 years of age in a verbal unimanual time-sharing paradigm. In a dichhaptic task with nonsense shapes, no interaction with age was found in the range of grade one to adults (Flanery and Balling, 1979); in a series of tasks involving listening to speech, music, and watching spatial stimuli, children aged 6 months to 9 years produced asymmetric alpha EEG patterns that paralleled those in adults (Nava and Butler, 1977).

A small number of researchers have addressed themselves to the issue of possible changes in hemispheric dominance in later adulthood. Brown and Jaffe (1975) propose that there is a gradual increase in left-hemisphere dominance with aging. Consistent with this hypothesis are reports by Clark and Knowles (1973) and Johnson, Cole, Bowers, Foiles, Nikaido, Patrick, and Woliver (1979) who find a decrease in left-ear report scores on dichotic listening tasks for both verbal and melodic stimuli. In a better-controlled study, however, Borod & Goodglass (1980) find no age change in ear asymmetry for linguistic or melodic processing over the age range of 24–79 years. The issue of hemisphericity changes during adulthood merits further examination before final conclusions can be drawn.

CONCLUSIONS

During the period of childhood, there is no compelling evidence to suggest a change in neurolinguistic organization—when one controls for task demands, no asymmetry-by-age interaction is found. The question of how brain organization reflects the mental development of the child is a complex one, for that mental development must be taken into account in any suitable answer.

REFERENCES

Bakker, D. J., Hoefkens, M., and VanderVlugt, H. (1979). Hemispheric specialization in children as reflected in the longitudinal development of ear asymmetry. *Cortex, 15,* 619–625.

Borod, J. C., and Goodglass, H. (1980). Lateralization of linguistic and melodic processing with age. *Neuropsychologia, 18,* 79–83.

Belmore, S. M. (1980). Depth of processing and ear differences in memory for sentences. *Neuropsychologia, 18,* 657–663.

Borowy, T., and Goebel, R. (1976). Cerebral lateralization of speech: The effects of age, sex, race, and socioeconomic class. *Neuropsychologia, 14,* 363–370.

Brackbill, Y. (1971). The role of the cortex in orienting: Orienting reflex in an anencephalic human infant. *Developmental Psychology, 5,* 195–201.

Brown, J. W., and Jaffe, J. (1975). Hypothesis on cerebral dominance. *Neuropsychologia, 13,* 107–110.

Bryden, M. P., and Allard, F. A. (1981). Do auditory perceptual asymmetries develop? *Cortex, 17,* 313–318.

Bryson, S., Mononen, L. J., and Yu, L. (1980). Procedural constraints on the measurement of laterality in young children. *Neuropsychologia, 18,* 243–246.

Clark, L. E., and Knowles, J. B. (1973). Age differences in dichotic listening performance. *Journal of Gerontology, 28,* 173–178.

Flanery, R. C., and Balling, J. D. (1979). Developmental changes in hemispheric specialization for tactile spatial ability. *Developmental Psychology, 15,* 364–372.

Galin, D., Johnstone, J., Nakell, L., and Herron, J. (1979). Development of the capacity for tactile information transfer between hemispheres in normal children. *Science, 204,* 1330–1332.

Hiscock, M., and Kinsbourne, M. (1978). Ontogeny of cerebral dominance: Evidence from time-sharing asymmetry in children. *Developmental Psychology, 14,* 321–329.

Hiscock, M., and Kinsbourne, M. (1980). Asymmetries of selective listening and attention switching in children. *Developmental Psychology, 16,* 70–82.

Huttenlocher, P. R. (1979). Synaptic density in human frontal cortex—developmental changes and effects of aging. *Brain Research, 163,* 195–205.

Hynd, G. W., and Obrzut, J. E. (1977). Effects of grade level and sex on the magnitude of the dichotic ear advantage. *Neuropsychologia, 15,* 689–692.

Isaacson, R. L. (1975). The myth of recovery from early brain damage. In N. E. Ellis (Ed.), *Aberrant development in infancy.* New York: Wiley.

Johnson, R. C., Cole, R. E., Bowers, J. K., Foiles, S. V., Nikaido, A. M., Patrick, J. W. and Woliver, R. E. (1979). Hemispheric efficiency in middle and later adulthood. *Cortex, 15,* 109–119.

Kraft, R. H. *Asymmetric brain specialization: Proposed relationship between its development and cognitive development.* Mimeographed manuscript.

Kraft, R. H., Mitchell, O. R., Languis, M. L., and Wheatley, G. H. (1980). Hemispheric asymmetries during six- to eight-year-olds' performance of Piagetian conservation and reading tasks. *Neuropsychologia, 18,* 637–643.

Krashen, S. D. (1973). Lateralization, language learning, and the critical period: Some new evidence. *Language Learning, 23,* 63–74.

Lenneberg, E. H. (1967). *Biological foundations of language.* New York: Wiley.

Molfese, D., Freeman, R. B., and Palermo, D. S. (1975). The ontogeny of brain lateralization for speech and nonspeech stimuli. *Brain and Language, 2,* 356–368.

Nava, P. L., and Butler, S. R. (1977). Development of cerebral dominance motivated by asymmetries in the alpha rhythm. *Electroencephalography Clinical Neurophysiology, 43,* 582.

Piazza, D. M. (1977). Cerebral lateralization in young children as measured by dichotic listening and finger tapping tasks. *Neuropsychologia, 15,* 417–425.

Porac, C., Coren, S., and Duncan, P. (1980). Life-span age trends in laterality. *Journal of Gerontology, 35*(5), 715–721.

Porter, R. J., and Berlin, C. I. (1975). On interpreting developmental changes in the dichotic right-ear advantage. *Brain and Language, 2,* 186–200.

St. James-Roberts, I. (1981). A reinterpretation of hemispherectomy data without functional plasticity of the brain. *Brain and Language, 13,* 31–53.

Schulman-Galambos, C. (1977). Dichotic listening performance in elementary and college students. *Neuropsychologia, 15,* 577–584.

Segalowitz, S. J., and Chapman, J. S. (1980). Cerebral asymmetry for speech in neonates: A behavioral measure. *Brain and Language, 9,* 281–288.

Witelson, S. F. (1977). Early hemisphere specialization and interhemispheric plasticity: An empirical and theoretical review. In S. J. Segalowitz and F. A. Gruber (Eds.), *Language development and neurological theory.* New York: Academic Press.

CHAPTER 9

Cerebral Asymmetries for Speech in Infancy

Sidney J. Segalowitz

INTRODUCTION

The developmental question concerning cerebral dominance for speech used to be, When is lateralization for language complete? The notion that cerebral dominance for language should develop over time rested on two premises, one logical, the other empirical. The logical one goes as follows: Babies are born without language, therefore they cannot be lateralized for it and, at the earliest, can only become lateralized for language once they have some. Thus, various important milestones in the development of cerebral dominance have been suggested: puberty, since after this date a second language becomes difficult to master; or at 2 years of age, when language production begins; or perhaps at 5 years of age, when the primary structures of the native language are reputed to be well internalized. Neurological data have been variously used to support such milestones (Lenneberg, 1967; Krashen, 1973). The empirical argument for the development of speech dominance concerns recovery from brain damage: it appears that restoration of language skills after brain damage depends to

LANGUAGE FUNCTIONS
AND BRAIN ORGANIZATION

some extent on the age at which the injury was sustained; also, language loss through right-hemisphere as well as left-hemisphere damage has been reported to be more likely in younger children (Basser, 1962). The implication here is that both hemispheres initially support language functions and gradually the left hemisphere alone becomes capable of this. Thus, only in the early years would right-sided damage disrupt speech, and only in the early years could the other hemisphere be free to take on the relearning of language.

Both these arguments have now been shown to be either logically unnecessary or empirically untrue. The empirical argument fails for two reasons: (1) the relative incidence of right-hemisphere damage producing language loss is not higher in young children than in adults when one examines the literature carefully (Dennis and Whitaker, 1977), and (2) the evidence for recovery of function need not be taken as evidence for lack of specialization: specific brain tissue can be dominant for specific cognitive functions within the context of whole-brain functioning, while still being capable of supporting other functions if the need arises through a drastic change in the balance of the entire brain's activity. The restorative ability may, therefore, be a function of factors other than specialization, such as the rate of natural cell death and dendritic growth (Huttenlocher, 1979).

The logical argument has also been shown to be insufficient. First of all, every cognitive function has other cognitive and behavioral precursors. No behavior appears in the repertoire of the child without a developmental history. It is entirely possible that the precursors of language, whatever they may be, could have an asymmetrical neurological substrate not unlike that for language. Second, there is now considerable evidence that very young babies do have some perceptual knowledge of speech sounds, very possibly from birth. It is possible that certain aspects of this knowledge are asymmetrically served in the neonatal brain, just as some sensory and motor functions are asymmetrically represented. The knowledge of speech demonstrated by young infants include phoneme discriminations, such as /b/ from /p/, /d/, /g/ and /m/ (Miller and Eimas, 1979); /d/ from /t/ (Trehub and Rabinovitch, 1972); /r/ and /l/ (Eimas, 1975); and others. Now, some of these discriminations (/b, p, d, t, g, k/) involve "encoding," that is, some complex knowledge about how the sounds are produced. The complexity refers to how the acoustic properties of these sounds vary with the context in which they appear (Lieberman, Cooper, Shankweiler, and Studdert-Kennedy, 1967). It is also just these sounds that produce the most robust asymmetries in dichotic listening tasks (Studdert-Kennedy and Shankweiler, 1970). It may be the case, then, that some of this early knowledge of speech sounds is lateralized in infants as well as in adults, producing a lateralized predisposition for all speech, and therefore for language. There have been a number of reports of lateralization for speech perception in infants, although it is not clear yet which acoustic factors determine the asymmetry. In the remainder of this chapter, we will review the evidence for lateralization of speech perception in infancy and explore the implications of these data for a developmental model of cerebral dominance for language.

EVIDENCE FOR SPEECH LATERALIZATION IN INFANCY

Studies of speech lateralization in children have been hampered by the problem of measures: for obvious reasons, the main techniques used with adults, dichotic listening and visual half-field reports, are inappropriate for use with infants. When such techniques are used with young children, there are no developmental changes in asymmetry (see the introductory material to Part III in this volume). Yet, extrapolation to infancy is not possible on the basis of the child data. New measures are required, and some have been found to produce asymmetries for speech in the expected direction. Although not all of the dozen or so studies available have reported speech lateralization, it is interesting to note that where speech does produce a lateral asymmetry, it is always the left hemisphere that shows the advantage. Note also that since the studies involve infants, it is impossible to tell whether or not the subjects were all right-handed, thus increasing the likelihood of nonhomogeneous results (see Chapter 15 in this volume).

Most of the reports of infant lateralization to date have dealt with speech versus music. In some cases, the response measure is such that running speech could be used; often, single syllables or words are used. For example, Gardiner and Walter (1977) and Segalowitz and Chapman (1980) found asymmetric responses to running speech versus music using very different measures, the former a slow-wave EEG response, the latter a limb tremor measurement. See Table 1 for details of all the studies discussed here. More common, however, has been the comparison between the processing of speech bursts versus nonspeech sounds. Molfese, Freeman, and Palermo (1975) and Molfese (1977) report left–right asymmetries in neuroelectric event-related potentials (ERPs) to speech syllables compared with piano cords and pure tone stimuli. Molfese and Molfese (1979) similarly found evidence for a specifically left-hemisphere differen-

Table 1

SUMMARY OF STUDIES ON CEREBRAL ASYMMETRIES FOR SPEECH IN INFANTS ORDERED BY AGE OF SUBJECTS

Authors	Ages tested	Stimuli used	Measures used	Results	Inferences
Molfese and Molfese (1980)	Preterm babies: mean gestational age = 35.9 weeks $n = 11$ males	/bæ/ and /gæ/ in phonetic and nonphonetic (first and third formant transition downgliding instead of upgliding) form; sign wave (bandwidth = 1 Hz) analogs of these	Principal components analysis (PCA) of ERPs over T_3 and T_4 referred to linked earlobes	One PCA factor showed that the LH differentiated phonetic and nonphonetic forms of the stimuli while the RH did not; another factor showed this again for the stimuli with speech formant characteristics.	LH for certain speech–nonspeech differentiations
Segalowitz and Chapman (1980)	Preterm babies: mean gestational age = 36 weeks at testing $n = 80$ males, 73 females	Speech: nursery rhymes Nonspeech: orchestral	Reduction to limb tremors after exposure to stimuli for varying lengths of time	Greater relative reduction in right limb tremors after listening to speech compared to control (no stimulation) and music	LH for speech input compared to music and control
Molfese and Molfese (1979)	mean = 21 hours $n = 8$ males, 8 females	Speech: /bæ/, /gæ/ Nonspeech: analogous to speech CVs with 1-Hz bandwidth for both formants	PCA of ERPs over T_3 and T_4 referred to linked earlobes	One factor in the PCA differentiated formant structure (/b/ versus /g/) in both hemispheres. Another factor showed only the LH capable of differentiating CVs with normal bandwidth (i.e., real speech).	Evidence for LH making greater distinction between syllables
Molfese (1977)	Exp. 1: same as Molfese, Freeman, and Palermo (1975) Exp. 2: same as Molfese, Nunez, Siebert, and Ramanaiah (1976) Exp. 3: 24 hr $n = 4$ males, 4 females 6 adults (3 males, 3 females)	Speech: /ba₄₀/, /ba₂₀/, /pa₄₀/, /pa₄₀/, synthesized with 0, 20, 40, and 60 msec voice onset time (VOT) respectively, and (da₀)	Exp. 2 & 3: amplitude change in ERPs over T_3 and T_4 referred to linked earlobes	Exp. 2: Wide bandwidth (speech) produced greater LH amplitude change; narrow bandwidth (nonspeech) produced > RH. Exp. 3: infants: LH and RH dishabituate to VOT changes across voicing boundary (/ba₂₀/–/ba₄₀/). No dishabituation for consonant change or VOT within boundary change. adults: LH dishabituation for all phonemic changes	Exp. 2: LH for wide bandwidth RH for narrow bandwidth Exp. 3: no asymmetry in AEP dishabituation for VOT or consonant change
Molfese, Nunez, Siebert, and Ramanaiah (1976)	48 hr $n = 7$ males, 7 females	Speech: /gæ/, æ/ Nonspeech: pure tones corresponding to central frequencies of formants for the speech syllables, and a 500-Hz tone	PCA of ERPs over T_3 and T_4 referred to linked earlobes	One factor in the PCA differentiated hemispheres on all stimuli; no interaction between hemispheres and either bandwidth or formant transition variables.	No evidence of asymmetry for the acoustic cues of bandwidth or formant transitions.
Hammer (1977) also in Turkewitz (1977)	>24 hr n unspecified	Speech White noise	Lateral eye movements made in response to simultaneous input to both ears of speech or white noise	Significantly more right movements to speech and more left movements to noise	LH activation with speech; RH activation with white noise input

Study	Subjects	Stimuli	Measure	Findings	Conclusion
Vargha-Khadem and Corballis (1979)	Exp. 1: 4.0–13.2 weeks, mean = 7.7 weeks; Exp. 2: 4.0–14.5 weeks, mean = 9.4 weeks; n = 12 males, 12 females in each experiment	Speech: /ma/, /ba/, /da/, /ga/	As in Entus (1977)	No significant asymmetry in dishabituation	No asymmetry found
Entus (1977)	Exp. 1: mean = 75 days; Exp. 2: mean = 70 days; n = 24 males, 24 females in each experiment	Speech: dichotic CV tapes (Exp. 1); Nonspeech: dichotic presentations of 440 Hz on various musical instruments (Exp. 2)	Infants habituate to repetitive dichotic stimuli, and control the rate of presentation by nonnutritive high amplitude sucking (HAS). Stimulus is changed in one ear only and dishabituation is measured (by recovery rate of HAS).	Greater dishabituation to right-ear change with speech. Greater dishabituation to left-ear change with music	LH for speech. RH for music
Merryweather (1978)	12 weeks; n unspecified	Speech: CV syllables; Nonspeech: piano chords	EEG alpha (8–13 Hz) asymmetry	No difference	No asymmetry with alpha
Glanville, Best, and Levenson (1977)	Mean = 3.4 months; n = 5 males, 7 females	Speech: CV syllables; Nonspeech: 440-Hz tones played by various instruments	Heart rate recovery response to novel stimuli presented to one ear or the other	Greater response to right-ear change in speech stimuli. Greater response to left-ear change in musical stimuli	LH for speech. RH for music
Molfese, Freeman, and Palermo (1975)	1 week to 10 months, mean = 5.8 months; n = 4 males, 6 females	Speech: /ba/, /dæ/, /bɔi/, /dɔg/; Nonspeech: piano chords	ERPs over T_3 and T_4 referred to linked earlobes	Greater amplitude change over left for speech. Greater amplitude change over right for piano chords	LH for speech. RH for music
Gardiner and Walter (1977)	6 months; n = 3 males, 1 female	Conversational speech; unspecified music from tape or radio	EEG power in 3–5-Hz band over P_3 and P_4, and over Wernicke's area on left and right, referred to linked earlobes	Greater relative power on right for speech and left for music. N.B. increased power means increased synchronizing implying reduced cognitive processing	LH for speech. RH for music
Barnet, Vicentini, and Campos (1974)	5–12 months, mean = 8 months; Exp. 1: n = 45; Exp. 2: n = 19	Exp. 1: clicks; Exp. 2: name of child tested	ERPs over C_3 and C_4; degree of asymmetry in P_2N_2 amplitude differences as a percentage of average amplitude	Exp. 1: nonsignificant RH bias for clicks.[a] Exp. 2: > amplitude in LH for names ($p < .001$) interaction: $p < .001$	LH for verbal input compared to clicks

[a] T-tests were calculated with arcsin transformation from original data of Barnet, Vicentini, and Campos (1974).

tiation of syllables with normal bandwidth (i.e., real speech) compared to pure tone stimuli. Later, they also found that preterm infants (average gestational age less than 36 weeks) showed a left-hemisphere ERP component that differentiates phonetic and nonphonetic forms of /b/ and /g/ (Molfese and Molfese, 1980).

Not all electrophysiological studies with infants have found asymmetries. For example, Merryweather (1978) found no asymmetries in alpha wave recording when comparing speech and piano chords. However, one could argue that in young infants, slower frequencies are more appropriate reflectors of degree of cerebral processing (John, 1977; Gardiner and Walter, 1977). Molfese, Nunez, Siebert, and Ramanaiah (1976) report no ERP asymmetries in neonates for two specific acoustic characteristics of speech: bandwidth and formant transition. These factors, then, are not straightforwardly responsible for the other ERP asymmetries found in neonates.

Behavioral measures of speech lateralization in infants sometimes make use of variants on the dichotic listening paradigm. Glanville, Best, and Levenson (1977) found heart-rate dishabituation greater when speech was played to the right ear of infants, and conversely greater dishabituation when musical tones were played to the left ear. Similarly, Entus (1977) reports dishabituation, reflected in increased sucking rate after a change in stimulus when the sucking behavior of infants controls the rate of presentation of speech or musical stimuli: greater dishabituation occurs when the change in speech stimulus occurs in the right ear and when the change in musical stimulus occurs in the left ear. Vargha-Khaden and Corballis (1979) failed to replicate Entus's results. Their only change in procedure was the use of a mechanical arm to insert the pacifier in the infant's mouth instead of the experimenter placing it. This change was made in order to avoid inadvertent bias communicated through the placement of the pacifier. (Since it is entirely possible for the human interaction to remain with the person involved being blind to the experimental condition, replication of Entus's original paradigm as is would be interesting.)

Other reports of asymmetries in speech representation in infants include Hammer (1977) and Barnet, Vincentini, and Campos (1974). Hammer found a greater incidence in eye movements to the right when listening to speech (through both ears) and more left movements when listening to white noise. Similarly, Barnet, Vicentini, and Campos found a significant interaction between stimuli—speech (name of child being tested) versus a click—and the ERP amplitude over the left and right hemispheres: the name produced a greater left-sided ERP, the clicks a greater right-sided ERP.

WHAT IS LATERALIZED IN INFANCY?

That speech stimuli can produce asymmetries in infants much as in adults (also with failures to replicate as in adults) suggests to us that the brain codes something about speech asymmetrically. However, from the evidence gained so far, we can barely speculate on the nature of the mechanism producing the asymmetry. Speech and musical tones (or clicks) differ in a variety of ways: speech stimuli show nonharmonic formant structures, formant transitions, wide bandwidth, nonsimultaneity of onset and offset of parts of the signal, and so on, when compared to the nonspeech stimuli used. It is by no means clear exactly what it is in the signal that cues an asymmetry in processing in adults, much less in infants. With infants, however, at least we do not have semantic and possibly syntactic factors to deal with. Molfese and his colleagues have tried to look at specific factors in the speech signal—bandwidth and formant transition—to no avail. One possibility untested with infants is the harmonic structure, found to be highly correlated with ear advantage in dichotic listening with adults (Sidtis, 1980).

IMPLICATIONS FOR A DEVELOPMENTAL MODEL OF LATERALIZATION

As Witelson (1977) argued, if brain lateralization exists in early childhood or even at birth, then a developmental model must be one concerned with qualitative changes in the child's cognitive and perceptual abilities with growth. Some speech lateralization effects would be clearly unobtainable with infants simply because they have not yet learned requisite aspects of language. For example, Van Lancker and Fromkin (1973) report that phonemic tone in Thai produces a dichotic right-ear advantage only in Thai speakers. If knowledge (in this case, of Thai) is sometimes necessary for an asymmetry, presumably those aspects of thinking that are tied to initially lateralized functions will also appear lateralized. Thus, lateralization for linguistic functions may reflect only that the speech signal is favored, for some unknown reason, by the left hemisphere. Linguistic systems that utilize other modalities may not be as tied to the left hemisphere. Indeed, this is what is found in deaf individuals who use ASL (see Ross, this volume). Similarly, the direction of a visual half-field asymmetry for reading Japanese words depends, at least in part, on whether the word is an ideogram or is sound-based.

As is argued in the Introduction to this section of the book, the question of whether lateralization for speech increases with age is misplaced. Rather, we should enquire about the patterns of qualitative change in asymmetries with age. Some cognitive skills may appear to increase in the asymmetry of representation, but this may be only because they come to rely on initially lateralized functions. Conversely, some functions may decrease in asymmetry with age (as Molfese, Freeman, and Palermo, 1975, found) because with expanding cognitive horizons, the child comes to be able to execute that function without as much reliance on the initially lateralized function. The question addressing the developmental neurolinguistic issue is, then, For what aspects of language does the infant have an asymmetric cerebral representation?

REFERENCES

Barnet, A. B., Vincentini, M., and Campos, S. M. (1974). EEG sensory evoked responses (ERs) in early infancy malnutrition. Paper presented at Society for Neuroscience, St. Louis, Missouri.

Basser, L. S. (1962). Hemiplegia of early onset and the faculty of speech with special reference to the effects of hemispherectomy. *Brain, 85,* 427–460.

Dennis, M., and Whitaker, H. A. (1977). Hemispheric equipotentiality and language acquisition. In S. J. Segalowitz and F. A. Gruber (Eds.), *Language development and neurological theory.* New York: Academic Press.

Eimas, P. D. (1975). Auditory and phonetic coding of the cues for speech, discrimination of the [r-l] distinction by young infants. *Perception & Psychophysics, 18,* 341–347.

Entus, A. K. (1977). Hemispheric asymmetry in processing of dichotically presented speech and nonspeech stimuli by infants. In S. J. Segalowitz and F. A. Gruber (Eds.), *Language development and neurological theory.* New York: Academic Press.

Gardiner, M. F., and Walter, D. O. (1977). Evidence of hemispheric specialization from infant EEG. In S. Harnad, R. W. Doty, L. Goldstein, J. Jaynes, and G. Krauthamer (Eds.), *Lateralization in the nervous system.* New York: Academic Press.

Glanville, B. B., Best, C. T., and Levenson, R. (1977). A cardiac measure of cerebral asymmetries in infant auditory perception. *Developmental Psychology, 13,* 54–59.

Hammer, M. (1977). *Lateral differences in the newborn infant's response to speech and noise stimuli.* Unpublished Ph.D. dissertation, New York University. *Dissertation Abstracts International, 38*(2), 1439-B.

Huttenlocher, P. R. (1979). Synaptic density in human frontal cortex—developmental changes and effects of aging. *Brain Research, 163,* 195–205.

John, E. R. (1977). *Neurometrics: Clinical applications of quantitative electrophysiology.* Hillsdale, N.J.: Lawrence Erlbaum.

Krashen, S. D. (1973). Lateralization, language learning, and the critical period: Some new evidence. *Language Learning, 23,* 63–74.

Lenneberg, E. H. (1967). *Biological foundations of language.* New York: Wiley.

Lieberman, A. M., Cooper, F. S., Shankweiler, D. P., and Studdert-Kennedy, M. (1967). Perception of the speech code. *Psychological Review, 74,* 431–461.

Merryweather, J. F. (1978). *The development of hemispheric specialization in infants and young children.* Unpublished Ph.D. dissertation, University of Denver. *Dissertation Abstracts International, 39*(2), 1016-B.

Miller, J. L., and Eimas, P. D. (1979). Organization in infant speech perception. *Canadian Journal of Psychology, 33,* 353–367.

Molfese, D. L. (1977). Infant cerebral asymmetry. In S. J. Segalowitz and F. A. Gruber (Eds.), *Language development and neurological theory.* New York: Academic Press.

Molfese, D., Freeman, R. B., and Palermo, D. S. (1975). The ontogeny of brain lateralization for speech and nonspeech stimuli. *Brain and Language, 2,* 356–368.

Molfese, D. L., and Molfese, V. J. (1979). Hemisphere and stimulus differences as reflected in the cortical responses of newborn infants to speech stimuli. *Developmental Psychology, 15,* 505–511.

Molfese, D. L., and Molfese, V. J. (1980). Cortical responses of preterm infants to phonetic and nonphonetic speech stimuli. *Developmental Psychology, 16,* 574–581.

Molfese, D. L., Nunez, V., Siebert, S. M., and Ramanaiah, N. V. (1976). Cerebral asymmetry: Changes in factors affecting its development. *Annals of the New York Academy of Sciences, 280,* 821–833.

Segalowitz, S. J., and Chapman, J. S. (1980). Cerebral asymmetry for speech in neonates: A behavioral measure. *Brain and Language, 9,* 281–288.

Sidtis, J. J. (1980). On the nature of the cortical function underlying right hemisphere auditory perception. *Neuropsychologia, 18,* 321–330.

Studdert-Kennedy, M., and Shankweiler, D. (1970). Hemispheric specialization for speech perception. *Journal of the Acoustical Society of America, 48,* 579–594.

Trehub, S. E., and Rabinovitch, M. S. (1972). Auditory–linguistic sensitivity in early infancy. *Developmental Psychology, 6,* 74–77.

Van Lancker, D., and Fromkin, V. A. (1973). Hemispheric specialization for pitch and "tone": Evidence from Thai. *Journal of Phonetics, 1,* 101–109.

Varga-Khadem, F., and Corballis, M. C. (1979). Cerebral asymmetry in infants. *Brain and Language, 8,* 1–9.

Witelson, S. F. (1977). Early hemisphere specialization and interhemispheric plasticity: An empirical and theoretical review. In S. J. Segalowtiz and F. A. Gruber (Eds.), *Language development and neurological theory.* New York: Academic Press.

CHAPTER 10

Hemispheric Specialization and Integration in Child Development

Jaan Valsiner

INTRODUCTION

In this chapter, I argue that developmental issues in brain organization for speech have reached an impasse because of a number of methodological and philosophical difficulties. It is the contention of the present author that no progress in the given area of research can be achieved, unless these problems are explicitly outlined and attempts made to solve these problems. The problems outlined subsequently are not specific to the field of research on hemispheric functions in the working brain, but instead, they constitute some basic epistemological questions important for all psychology in its quest for the status of "science". The present chapter, however, deals only with those issues that are important for the understanding of the development of brain functions.

231

LANGUAGE FUNCTIONS
AND BRAIN ORGANIZATION

THE INPUT—OUTPUT PROBLEM AND THE
TWO-COMPARTMENT BLACK BOX

The theoretical paradigms used in the research on the ontogenesis of hemispheric functional asymmetry are usually derived from research on adults. In a sense, these studies have dealt with the brain as a two-compartment "black box," demonstrating the presence of one or another cognitive or linguistic functions in one or the other of these compartments. The experimental support comes from behavioral paradigms, such as the dichotic listening, visual half-field, and finger-tapping time-sharing tasks.

The basic methodological–conceptual problem with this experimental approach is that from the RELATIVE efficiency of one or the other hemisphere in some task (e.g., in dichotic listening tests, or the relative dominance of one of the ears over the other), the presence of ABSOLUTE psychological mechanisms in the hemispheres is inferred. Thus, from the experimental data on the right-ear advantage (REA) in dichotic tasks, it is frequently inferred that language functions are solely organized in the left hemisphere of the brain. (This conceptual fallacy is, naturally, found in other fields of the behavioral sciences as well.) In experiments on hemispheric lateralization, the input (stimulus) lateralization is usually controlled: the dichotic tasks are used in order to stimulate the right hemisphere (RH) and the left hemisphere (LH) simultaneously, and visual stimulation of the left and right visual hemifields guarantees the input to the LH and RH due to the decussation of optic fibers in the optic chiasm. However, we cannot be sure of what happens to the lateralized input stimulation once it is being processed by the brain. Blocking of the interhemispheric transfer (by commissurotomy) or switching off one of the hemispheres temporarily (by unilateral electric shock) can reveal the compensatory possibilities of the brain to face such distortions. Whether the central information-processing mechanisms in the brain possess clearcut lateral specifications in the case of normal subjects, is in question. Attempts to modify the functioning of the intact brain by introducing concurrent tasks that are expected to interfere with the main information-processing task (e.g., time-sharing, sensory overload of the input of one of the hemispheres) need not necessarily interfere with the work of the particular area of the brain which is occupied by processing the "main" information. It may rather modify the activity of the whole brain in some way so as to influence the information processing, keeping in mind that both the processing as well as experimentally created "distractors" may function in both hemispheres of the intact brain at the same time. Relative differences in the effect of a distractor to one hemisphere versus the other are not a sufficient reason to infer that the processing of a particular type of stimulus material takes place in the hemisphere where the distractor was more effective. Inference from differences between hemispheres in their distractibility tend to be based upon an implicit assumption that the hemispheres function as separate systems, independent of each other. The frequent oversimplification of the hemispheric asymmetries in the form of arguments "left hemisphere controls functions X and right hemisphere deals with Y" is an example. The hemispheres, much the same way as any subsystems of a biological system, can be independent of each other only in the minds of the investigators, if the latter adhere to an atomistic *Weltbild*. The relative asymmetry in the reaction of the two-hemisphere system to lateralized distractors may be the result of the interaction of the hemispheres, rather than a demonstration of the differential vulnerability of separately functioning hemispheres. How information is processed both inter- and intrahemispherically is a problem beyond the scope of stimulus–response-type experiments. This conceptual difficulty is worth emphasizing, since inferences from input lateralization (stimulating different hemispheres) to information-processing lateralization as an explanation of stimulation effects have been implicitly present in the researchers' minds. And, finally, there is the problem of output (response) lateralization: to what extent are the responses required from subjects due to the asymmetry of motor control of the particular motor functions?

However difficult these conceptual issues seem to be, two points are worth emphasizing: (1) the brain is a far more complex system than a simple, two-compartmental (LH and RH) "black box" and (2) there is no reason to generalize from input lateralization to lateralization of information processing. More complex a picture of the functioning of the brain could be obtained if the development of structural–functional characteristics of the brain is studied through a combination of neurophysiological and psychological methods (see the following).

DEVELOPMENT AS A NONLINEAR SEQUENCE

The research on the ontogeny of hemispheric functional asymmetries has operated on a very simplistic model of child development. The two major opposing hypotheses—initial equipotentiality of RH and LH in newborns and infants which develops into differentiated LH–RH asymmetry, and (2) the idea of hemispheric asymmetries being present in children from the birth onwards—tend to look upon child development as a process of linear growth, rather than as a complex process of development of an organism. Even though this is the simplest way to look at it, child development is not necessarily a linear increase of psychological functions over age, to come to resemble those of the adults as the end result. Rather, it is a complex process of changes where the development of some new psychological functions coincides with the disappearance of some others. The initial range of the infant's and child's possibilities of development is invariably wider than the end result of the development (cf. Bower, 1974; Trehub, 1976; Beltyukov, 1977; Beltyukov and Salahova, 1973, 1975). The development of differences in the functioning of the left and right hemispheres in children is too easily oversimplified. Given the apparent complexity of the brain mechanisms, the question so often asked—Does hemispheric lateralization develop?—may turn out to be an irrelevant oversimplification.

Such oversimplifications are certainly present in research fields that have extended the study of certain phenomena (in this case—functional specificities of the two hemispheres) from the analysis of the status quo of the grown-up organisms, to the ontogenetic process of development. The simplest (and the least adequate) theoretical model for the development is a quantitative linear growth function. This model of development may be perhaps an adequate way of thinking about some time-related changes in some simple phenomena in isolated closed systems of the physical world. However, biological organisms and social systems are better characterized as open systems, for which linear growth models are inadequate. The processes of development provide a problem for scientific explanation that requires utilization of models that are adequate for the description of open systems (cf. von Bertalanffy, 1952). Recently, a number of attempts in that direction have been undertaken (Prigogine, 1976, Kugler, Turvey and Shaw, 1982, London and Thorngate, 1981).

Another crucial problem in the study of the development of interrelationships of the two hemispheres in the working brain is the question of what constitute the "date" in empirical research. The overwhelming number of studies of the ontogeny of hemispheric lateralization are based on cross-sectional studies of groups of children of different ages. Inference about development in "the child" (an abstract, average child, who is supposedly a model for each and every concrete child of the given age) is made from such samples. On epistemological grounds, inference from group data to an understanding of the development of individuals (even in the form of an "abstract individual") is unwarranted—and, it is not surprising that the numerous studies of the ontogeny of hemispheric lateralization have produced very little that goes beyond the "does hemispheric lateralization develop, or not?" question. It has been stressed in the past (Lewin, 1933, p. 591) that the reduction of the question of development of individual children to the study of "average" children of different age groups can, in principle, provide no solutions for the problems of understanding development. In any developing child, it is the particular brain of that child that functions in the child's life and development, and not "an average brain" of a group of children; nor "average brains" of children's groups of different ages. Therefore, the *longitudinal analysis of individual cases* in their environmental contexts of development is the only adequate research strategy to employ in the study of the development of the functional independence (and, connected to it, interdependence) of the two hemispheres. This research approach has been notably absent from the developmental studies of hemispheric lateralization (and from child psychology at large). In neuropsychology, however, the "syndrome analysis" of individual cases has been proposed, and used, by A. R. Luria (cf. Luria and Artemyeva, 1970). The usage of syndrome analysis in developmental neuropsychology may provide an adequate picture of the developmental process itself, rather than reduce that process to statistical abstractions, in which psychology has generally become increasingly involved (cf. G. W. Allport, 1940).

THE STUDY OF THE DEVELOPING BRAIN: THE SOVIET ELECTROPHYSIOLOGICAL APPROACH

The controversy around the hypothesis of development of hemispheric lateralization for language (Lenneberg, 1967) merits looking at different functional systems of the brain and how these appear during ontogenesis, rather than trying to explain (via the simple two-compartment black box model of the brain) the developmental decrease of compensatory potential of the brain when unilateral brain damage is incurred. The possibility is popular among some Russian neuropsychologists (Simernitskaya, 1978a; Simernitskaya, Alle, and Havin, 1977; Simernitskaya, Rostotskaya, and Alle, 1978) that qualitative changes in the brain of the developing child, rather than the equipotentiality of the hemispheres in early childhood, account for the lack of aphasia in the case of left-hemisphere lesions. Compensatory mechanisms of the brain are not necessarily built upon one or another type of functional specialization. They may rather be a result of a greater plasticity of the brain in forming subcortical–cortical connections in the process of development. If this is the case, it is necessary to consider different theoretical options in building our interpretations of how brain mechanisms develop.

Considering all the limitations of the existing behavioral methods of measuring left–right brain asymmetries, electrophysiological responses measured from the scalp may provide the investigators with complementary opportunities to understand what is going on in the hemispheres during information processing. Electrophysiological measures may allow us to ask questions about the information processing per se, instead of trying to construct models from input–output measures obtained with behavioral methods. Generalizations from electrophysiological results have been hampered in the past·because of their great sensitivity to procedural factors and slight changes in the behavior of the subjects (e.g., body movement "noise" in experimental conditions), and the use of different procedures in obtaining the electrophysiological measures (cf. Varner, Peters, and Ellingson, 1978) from newborns and infants.

It is a tradition in Russian neuropsychology and neurophysiology to look at the functioning brain as a complex system of interdependent subsystems. This kind of general, holistic systems-theory approach has been accepted both in psychology (Vygotsky, 1960; Luria, 1963, 1969, 1970, 1973) as well as in neurophysiology (Anokhin, 1968; Livanov, 1972; Kogan, 1964, 1970, 1973; Batuyev, 1975; Bekhtereva, Bundzen, and Gogolitsyn, 1977). Speech, according to this systems-theory view, is organized in the brain by a complex functional system uniting different levels and different areas within a level of brain processes. Neurophysiologically, different functional systems are based upon complex aggregations of neurons (Kogan, 1964, 1970, 1973). The activity of different parts of the brain in the system carrying out a certain function can be measured via an analysis of spatial synchronization of brain biopotentials in these different brain centers (Livanov, 1972). The developmental aspect of the functional systems is looked upon as the development of brain systems (especially thalamocortical systems) that help to integrate different analyzers to increase the adaptability of an organism to the environment (Batuyev, 1975).

Among Russian neurophysiologists, Hrizman (1973, 1974a,b, 1975, 1978) has been the most active investigator of the developmental processes in the child's brain. She has studied the EEG of children of different ages (between 1 month and 7–8 years) using 12 symmetric electrode placements on the scalp (taken to measure electric potentials in Broadman fields 10 [frontal lobe], 44 and 45 [Broca's speech center], 4 and 6 [precentral, motor projection zone], 39 [parietal], 17 and 18 [occipital], and 41 [temporal, acoustic projection zone]. Complex pictures of brain activity, using correlation matrices of activity in different centers, were obtained by Hrizman in different stimulation conditions: no stimulation, simple sensory input, processing of familiar (child's name) and unfamiliar words, and piano playing. Four- to five-year-old children were also given commands ("look," "listen to . . . ," "put . . . together," "say . . . ," "attention").

In her thorough study of children with normal hearing, Hrizman (1978) has revealed a complex picture of brain functioning in children of different ages (1 month–7 years). Her data are also pertinent to the problems of hemispheric function differences between the left and right sides of the brain. Hrizman (1978, p. 51) found that the ontogenetic development of interrelationships between different brain centers (especially between the frontal lobes and other centers) is a nonlinear process with varying speeds of development of these interrelationships at different ages. The percentage of significant correlation coefficients between the frontal areas of the child's brain and other areas was found to vary in an interesting manner (in situations with no stimulation): 28.7% in 1–2-month-olds, decreasing to 19.4% among 3–4 month olds. This is followed by another increase at the age of 7–8 months (30.4%), followed by another decrease (to 15.7%) for babies 9–12 months old. Thus, the role of the frontal cortex seems to have two periods of high involvement in brain functioning during the first year of life (at 2 and at 7–8 months of age), revealed by Hrizman's EEG study. Naturally, the relative role of the frontal cortex grows immensely after the first year of the infant's life has passed—the development of the frontal area's role in general brain activity increases drastically during the second to third years of life and again during the sixth to seventh years.

Hrizman's (1978) data on the development of brain functions in childhood have revealed a quite complex picture of the ontogenetic brain development.

These data point to different important aspects in the development of hemispheric specialization. First, interconnections among centers of one hemisphere may be different from their counterparts in the other hemisphere both quantitatively (the change from baseline to the stimulus condition) as well as qualitatively (in the direction of change: increase or decrease—cf. Table 1, 1–1½-year-olds). Different corresponding centers of the hemispheres may display right or left asymmetries independently of one another, and may be dependent upon stimulus characteristics (not only verbal versus nonverbal acoustic stimuli, but also more subtle, individual characteristics, such as familiarity and meanings of the words; cf. Bekhtereva, Bundzen, and Gogolitsyn, 1977). Second, any hemispheric asymmetry may be dependent upon the baseline of these potentials in the no-stimulation situation. In a study of EEG theta-rhythms in children 5–7 years old (Tarakanov, 1979, 1980), there was a general right-hemisphere theta-rhythm prevalence in baseline (no task) conditions. However, when the subjects had to solve a verbal task, the theta-waves' activity in frontal and temporal lobes become higher in the LH. In an attempt to specify further what tasks cause this RH to LH theta-rhythm dominance transfer, Tarakanov (1980) found that this shift occurred only on verbal tasks with minimum emotional concomitants (verbal–cognitive tasks). Nonverbal–

Table 1

DEVELOPMENT OF RELATIONSHIPS OF LH/RH BRAIN CENTERS IN CHILDREN[a]

Age	Left Hemisphere	Right Hemisphere
3–5 mo	**AUDITORY STIMULUS**	
	FAMILIAR WORD—Child's Name	
	Temporal area: the greatest increase in the number of correlations with other areas (compared to baseline)	*Temporal area:* analogous to LH temperature increase, but less in magnitude
	UNFAMILIAR WORD	
	In both LH and RH: Lowered correlations of frontal, parietal, and temporal zones	
		Trend expressed more in RH
5–6 mo	**ANY AUDITORY STIMULATION (WORDS, PIANO CHORDS)**	
	General: doubles intercorrelations between brain centers (in comparison to no-stimulus baseline)	
	WORDS	
	Temporal area: insignificant increase	*Temporal area:* twofold increase in synchronization
	PIANO CHORDS	
	In the whole LH: Increase in the between-areas' connections (esp.: frontal–temporal, temporal–occipital)	No change in comparison to no-stimulus baseline
	FAMILIAR WORD—Child's Name	
	Temporal and parietal areas: similar increase in the simultaneity of potential in LH and RH	
	from 28.6% baseline to 80%	from 14.2% to 71.1%
	UNFAMILIAR WORD—"matryoshka"	
	Produced asymmetry between hemispheres, no simultaneity of reaction, similar increase to the between temporal and parietal familiar word (to 75%) areas—temporal areas *precede* parietal in 60% of cases	
6–12 mo	*FAMILIAR WORDS AS STIMULI*	
	Twofold increase in temporal-parietal correlations over RH (both compared with the baseline)	
	Both hemispheres: number of connections of the *inferior parietal* zones with other areas is increased. No asymmetry	
	Temporal regions: Increase in the number of connections with other areas (from 18.2% to 26.6%)	No increase

Table 1 (*continued*)

Age	Left Hemisphere	Right Hemisphere
1–1½ yrs	*Qualitative changes in the brain activation patterns from earlier ages:* in addition to *increase* in the number of interconnections, their *decrease* may occur.	

FAMILIAR WORDS

Increase in the spatial synchronization of brain responses, in both hemispheres (especially in the frontal–parietal connections of LH)

UNFAMILIAR WORDS

Decrease the spatial synchronization of brain responses, especially in the LH
Decreases number of connections—
   correlations between areas—in LH
   2.5 times

| 4–5 yrs | *General:* number of significant correlations between brain areas continues to decrease by that age. Asymmetry in the number of significant between-areas' correlations between LH and RH | |

BASELINE

2.5 times more relationships in LH
   than in RH

FAMILIAR WORD

Previously mentioned asymmetry disappears—number of relationships in LH and RH becomes equal

^a Data from Hrizman, 1978.

cognitive tasks (construction tasks) increased the number of theta-waves in both hemispheres (compared to the baseline, no-activity period). Emotional arousal accompanying success at finding a solution to the problem increased theta-activity in the RH more than in the LH.

An overview of these electrophysiological data brings us to a methodologically relevant conclusion: that hemisphere dominance and asymmetries in the functions of the hemispheres are a highly FLEXIBLE phenomenon dependent upon the environmentally given task and the present emotional–motivational state of the subject. This double relationship may be the reason why behavioral indicators of hemisphere dominance have proved to be unreliable within the subjects: whereas the test situations may have been the same, the state of an individual subject may have been highly variable. Group averages which are insensitive to individual fluctuations of that kind, unless the distribution of these in the group is changed, could remain constant over time.

Thus, it may be heuristically worthwhile to study the conditions under which hemispheric dominance for certain tasks can be shifted from one side to the other, and vice versa. Shifts in the lateralization could be well connected to the ease of assimilation of the stimulus. This assimilation process (establishing a connection between the new stimulus and former experience) can be based on individual interest of the subject as well as learning. Language acquisition by a child is a combination of these both. Russian neurophysiologists, being influenced by the Pavlovian traditions, have tended to look upon the language development in childhood as the establishment of conditioned reflexes (Koltsova, 1958, 1967, 1973). Zaitseva (1975, 1976) has found that after the development of the "signal function" of words in two-year-old children, a word with "signal function" increases the number of functional correlations between frontal and parietal EEG activity. She found that these correlations in the LH increased 5.5 times over the baseline control condition. When the child was dealing with the same word, but when it did not possess the "signal function," the increase was only threefold. The learning of new skills and sophistication in children may well change, with age, the pattern of hemispheric lateralization in a nonpermanent fashion.

Animal experiments have shown (Kogan, Kurayev, and Reps, 1980) that functional asymmetry in the brains of some animals (cats) can be produced by conditioning these animals to solve new tasks. This sensitivity to learning effects and susceptibility to changes due to the internal state changes in the subject make the functional specialization of hemispheres a very complex and dynamic phenomenon in child development. The complexity of this phenomenon cannot easily be understood by asking simple questions such as Do hemispheric asymmetries develop? or Do the RH and LH have similar or different speed in their development? The functional development of the brain, as well as behavior, may vary in growth tempo from one age range to another. Entus (1977, p. 71) has argued that the usage of concepts of prospective significance, prospective potency, and determination will be helpful in dealing with the development of hemispheric asymmetry. The prospective significance of each hemisphere would be to mediate specific functions under the particular circumstances of state and task. In the event of some blockage of one hemisphere (e.g., lesion), the intact hemisphere may have the prospective potency to take over the functions of the blocked organ. In fact, these kinds of compensatory potentials must be an essential part of any highly developed organ. In humans, in fact, determination (the age after which prospective potency is no longer in effect) may develop gradually and be a result of complex brain development, rather than of a simple loss of compensatory abilities of one or the other hemisphere.

CONCLUSIONS

The present chapter was devoted to the analysis of the conceptual frameworks and experimental methods of research on hemispheric lateralization, and emphasized in particular the problems involved in finding out what kind of asymmetric or symmetric functions develop in the child's brain.

Behavioral methods that operate on varying input stimulation (dichotic listening tasks and tachistoscopic methods) or output behaviors (time-sharing procedures) have provided interesting data on hemispheric lateralization at the level of left and right sides of the brain. However, the results based on these methods have revealed that there seems to be no GENERAL change in the degree of right–left side differences during childhood. However, this negative conclusion emanating from the data—that hemispheric (or, better, brain) asymmetries do not develop—may be unwarranted, since the measures which have been useful in studying adults need not be suitable for more elaborate questions concerning development of the phenomena which we are used to look upon in adults' brains. Child development, as well as the ontogenesis of a great majority of organisms, is not a preformationistic case of linear growth from a small organism (with a small brain, with two small hemispheres) into an adult organism where the brain and its hemispheres are simply much bigger. On the contrary, ontogenetic development seems to be better understood as nonlinear reorganization of different subsystems of the developing brain, and traditional behavioral lateralization measures are evidently too general to be a basis for further specification of these subsystems in development.

The electrophysiological measure of the development of brain mechanisms in children seems to reveal a complex picture where asymmetries between different zones of the two hemispheres may occur at different age periods. These differences seem to be closely connected to speech functions of the child: the familiarity–unfamiliarity of a certain stimulus word is found to be an essential variable in these asymmetries. The general conceptual framework of our understanding of the development of hemispheric asymmetries is oversimplistic. What can be said about the development of brain lateralization is that it is definitely not the development of simple left–right asymmetries. This is what developmental applications of behavioral measures of lateralization have shown. However, we are sufficiently far away from a positive understanding of the develop-

ment of lateralization, and—as we tried to show in the present chapter—
it is not only data but also conceptual understanding that we are lacking.

REFERENCES

Allport, G. W. (1940). The psychologist's frame of reference. *Psychological Bulletin, 37*,(1), 1–28.

Anokhin, P. K. (1968). *Biology and neurophysiology of conditioned reflexes*. Moscow: "Nauka" (In Russian.)

Batuyev, A. (1975). *Evolution of frontal lobes and integrative function of the brain*. Leningrad: "Nauka" (In Russian.)

Bekhtereva, N., Bundzen, P., and Gogolitsyn, Y. (1977). *Brain codes of psychic activity*. Leningrad: "Nauka" (In Russian.)

Beltyukov, V. I. (1977). The role of the auditory analyzer in the process of assimilation by children of the sound aspect of speech. *Problems of Psychology, 22*(2), 105–113. (In Russian.)

Beltyukov, V. I., and Salahova, A. D. (1973). Babble of the hearing child. *Problems of Psychology, 18*(2), 105–116. (In Russian.)

Beltyukov, V. I., and Salahova, A. D. (1975). On the assimilation of the sound system of language by the child. *Problems of Psychology, 20*(5), 71–80. (In Russian.)

Bertalanffy, L. von. (1952). Theoretical models in biology and psychology. In: D. Krech and G. S. Klein (Eds.), *Theoretical models and personality theory*. Durham, N.C.: Duke University Press.

Bower, T. G. R. (1974). *Development in infancy*. San Francisco: Freeman.

Entus, A. K. (1977). Hemispheric asymmetry in processing of dichotically presented speech and nonspeech stimuli by infants. In S. J. Segalowitz and F. A. Gruber (Eds.), *Language development and neurological theory*. New York: Academic Press.

Hrizman, T. P. (1973). *Child's movements and electric activity of the brain*. Moscow: "Pedagogika" (In Russian.)

Hrizman, T. P. (1974). On the connection of inferio-parietal associative areas of the brain to speech perception in children. *Journal of Higher Nervous Activity. 23*(4), 758. (In Russian.)

Hrizman, T. P. (1974b). The reflexion of semantic basis of words in the organization of spatio-temporal relationship of biorhythms in children's brains. In *Informational meaning of brain's electric potentials*. Leningrad: "Nauka" (In Russian.)

Hrizman, T. P. (1975). Organization of spatio-temporal relationships in child's brain while stimulated by verbal commands. *Journal of Higher Nervous Activity, 25*(4), 690. (In Russian.)

Hrizman, T. P. (1978). *The development of the functions of the child's brain*. Leningrad: "Nauka"

Kogan, A. B. (1964). Probabilistic–statistical principles of neuronal organization of the brain's functional systems. *Proceedings of the Academy of Sciences* (USSR), 5, (In Russian.)

Kogan, A. B. (1970). On the principles of the organization of brain systems on the basis of neuronal elements. In *Cybernetic aspects of the study of brain functions*. Moscow: "Nauka" (In Russian.)

Kogan, A. B. (1973). On the principles of neuronal organization of control mechanisms of functional systems. In *The principles of system organization of functions*. Moscow: "Nauka" (In Russian.)

Kogan, A. B., Kurayev, G. A., and Reps, F. E. (1980). The role of functional asymmetry of brain hemispheres in the organization of instrumental alimentary conditioned reflex in cats. *Journal of Higher Nervous Activity, 30*(1), 37–42. (In Russian).

Koltsova, M. M. (1958). *On the formation of higher nervous activity in children.* Leningrad: "Nauka" (In Russian.)

Koltsova, M. M. (1967). *Generalization as a function of the brain.* Leningrad: "Nauka" (In Russian.)

Koltsova, M. M. (1973). *Movements and the development of brain functions in children.* Moscow. (In Russian.)

Kugler, P. N., Turvey, M. T., and Shaw, R. (1982). Is the "cognitive penetrability" criterion invalidated by contemporary physics? *The Behavioral and Brain Sciences, 5*(2), 303–306.

Lenneberg, E. H. (1967). *Biological foundations of language.* New York: Wiley.

Lewin, K. (1933). Environmental forces. In C. Murchison (Ed.), *A handbook of child psychology.* 2nd ed. Worcester: Clark University Press.

Livanov, M. (1972). *Spatial organization of brain processes.* Moscow: (In Russian.)

London, I. D., Thorngate, W. (1981). Divergent amplification and social behavior: some methodological considerations. *Psychological Reports, 48,* 203–228.

Luria, A. R. (1963). Human brain and psychic processes. Vol. 1. Moscow: Academy of Pedagogical Sciences Press, (In Russian.)

Luria, A. R. (1969). *Higher cortical functions* (2nd ed.). Moscow: Moscow State University Press, (In Russian.)

Luria, A. R. (1970). *Human brain and psychic processes.* Vol. III Moscow: Pedagogika, (In Russian.)

Luria, A. R. (1973). *Foundations of neuropsychology.* Moscow: Moscow State University Press, (In Russian.)

Luria, A. R. and Artemyeva, Y. U. (1970). On two ways of reaching validity of the psychological investigation (the validity of the fact and syndrome analysis). *Problems of Psychology, 16*(3), 105–112. (In Russian.)

Prigogine, I. (1976). Order through fluctuation: self-organization and social system. In C. H. Waddington (Ed.), *Evolution and consciousness: Human systems in transition.* Boston: Addison-Wesley.

Simernitskaya, E. (1978a). *Cerebral dominance.* Moscow: Moscow State University Press (In Russian.)

Simernitskaya, E. (1978b). Neuropsychological analysis of brain organization of psychic processes in children. *Problems of Psychology, 23*(1), 110–113. (In Russian.)

Simernitskaya, E., Alle, A., and Havin, A. (1977). A study of hemisphere dominance in children with dichotic listening test. *Vestnik of Moscow University: Psychology, 1*(2), 76–81. (In Russian.)

Simernitskaya, E., Rostotskaya, V., and Alle, A. (1978). On the destruction of dichotic perception of speech in children with local brain lesions. *Vestnik of Moscow University: Psychology, 2*(3), 69–74. (In Russian.)

Tarakanov, P. (1979). Interhemispheric EEG asymmetry in 5–7-year-old children. *Journal of High Nervous Activity, 29*(2), 227–231. (In Russian.)

Tarakanov, P. (1980). Emotional effects on interhemispheric asymmetry of brain theta-activity in 5–7-year-old children. *Journal of Higher Nervous Activity, 30*(4), 844–847. (In Russian.)

Trehub, S. E. (1976). The discrimination of foreign speech contrasts by infants and adults. *Child Development, 47,* 466–472.

Varner, J. L., Peters, J. F., and Ellingson, R. (1978). Interhemispheric synchrony in the EEGs of full-term newborns. *EEG and Clinical Neurophysiology, 45,* 641–647.

Vygotsky, L. S. (1960). *The development of higher psychic functions*. Moscow: Publishing House of the Academy of Pedagogical Science, (In Russian.)

Zaitseva, L. M. (1975). About neurophysiological mechanisms of words' signal function formation in early childhood. *Journal of Higher Nervous Activity, 24*(4), 681. (In Russian.)

Zaitseva, L. M. (1976). Spatial relationships of electric activity during reproducing word signals in children. *Journal of Higher Nervous Activity, 25*(5), 1032–1040. (In Russian.)

CHAPTER *11*

Relationships among Brain Organization, Maturation Rate, and the Development of Verbal and Nonverbal Ability

C. Netley and Joanne Rovet

INTRODUCTION

In this chapter, we attempt to integrate a number of different empirical findings and theoretical propositions bearing on individual differences in intellectual functioning. Topics included are sex differences in ability for verbal and nonverbal tasks, genetic processes governed by the sex chromosomes, and functional hemispheric organization. In each case, relationships among these areas are examined in the context of developmental changes from early in life to maturity. It is probably true that much of what is proposed and concluded may turn out to be simply wrong. However, it is our position that a satisfactory synthesis of these disparate areas will be made at some point and even premature attempts in this direction have heuristic value.

LANGUAGE FUNCTIONS
AND BRAIN ORGANIZATION

SEX-RELATED BIOLOGICAL FACTORS
AND INTELLECTUAL ABILITY

There is considerable evidence that the two sexes differ in their intellectual characteristics with mature males being relatively more able spatially and mature females, more able verbally (Maccoby and Jacklin, 1974). Although there is a sizable body of research which implicates the importance of sociocultural factors in determining these sex differences (Nash, 1979), a variety of observations suggest that biological factors are also involved (McGlone, 1980). Although our emphasis is on an examination of biological influences, it is not our intention to suggest that these are of exclusive importance. We recognize that both are probably influential, and it is simply our purpose to examine the validity of competing theories that make reference to biological determinants.

The plausibility of biological factors in relation to the origin of sex differences in ability is suggested by a variety of observations. It is, of course, true that normal males and females have differences in sex chromosome complement which, hypothetically at least, could be responsible for their differing intellectual characteristics. Males with 46,XY constitutions are without the second X of 46,XX females, a fact which has prompted some investigators to postulate a recessive X-linked mode of inheritance for superior spatial ability (e.g., Stafford, 1961). Under this theory, the spatial advantage of males would arise because for them, only one X chromosome with a gene for spatial superiority need be present in contrast to females who would require two, a less probable occurrence. This view has received support from reports indicating that some correlations between parents and children were higher for differing sex combinations (e.g., mother–son) than same sex combinations (e.g., father–son) (Bock and Kolakowski, 1973). A recent review of this literature has, however, indicated that these patterns of familial correlations may not be reliable findings and this theory seems open to question (Vandenburg and Kuse, 1979).

Additional evidence supporting the importance of biological factors in intellectual differences between the sexes has been obtained from studies of individuals with a variety of pathological or abnormal conditions. Some have had sex hormone abnormalities, while others have been males and females with genetic anomalies of the sex chromosomes. Investigations of such individuals are really experiments of nature and, at times, there are problems in relation to such matters as "tightness" of experimental design. However, frequently they do permit an examination of issues which, otherwise, would not be possible when research is restricted to normal subjects. For example, examination of individuals with abnormal numbers of X chromosomes have provided data that create problems for X-linked recessive theories concerning the inheritance of spatial ability. Turner syndrome females who have a 45,X constitution and who, therefore, lack one of the normally occurring two X chromosomes would, under this theory, be expected to resemble normal males, whereas repeated psychometric studies have indicated that they have specific deficits in spatial or nonverbal ability (Garron, 1977; Money, 1973; Netley, 1977).

We have recently studied the specific deficits of 45,X females (Rovet and Netley, 1982). The results of this study are interesting since they suggest that Turner females have primary deficits on spatial tasks but do not differ from controls in the kinds of cognitive processes they employ to solve them. We examined the performance of Turner females on a mental rotation task of the kind used by Shepard in various studies (Shepard, 1978). The results of this investigation (shown in Figure 1) indicated that Turner girls did not differ from their controls in those aspects of the task which required little or no subjective manipulation of a mental image. Their performance was, however, considerably poorer than controls when such processes were called for. Significantly, however, both Turner subjects and controls had identical performance characteristics in that lin-

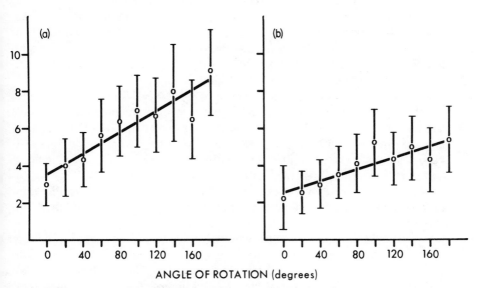

FIGURE 1. The relationship between reaction time and angular discrepancy in a mental rotation task for (a) Turner syndrome subjects and (b) female controls (copyright © American Psychological Association).

ear relationships between the angular discrepancies of the figures and response latencies were found for both. Thus, it is plausible to conclude that if we accept the arguments and findings of Shepard, both Turner and normal females use a mental rotation strategy in dealing with these problems.

Some recent studies of individuals with a supernumerary X chromosome have provided additional evidence in support of the notion that abnormalities in numbers of X chromosomes are associated with atypical intellectual development. It should be pointed out that both males and females may be born with an extra X chromosome. In the case of males, the chromosome complement is 47,XXY and for the females, 47,XXX. As children, neither group shows clinically evident disorders which would, in the normal course of events, result in those investigations which would identify their chromosome abnormality. As a result, the best data in this area are provided by studies of subjects who have been identified in large scale neonatal screening surveys where the presence of clinical disorder, either of a medical or behavioral kind, has not influenced subject selection. There have been comparatively few such surveys carried out, but those that have been done in the United States, Europe, and Canada, have consistently pointed to the presence of a specific deficit in verbal ability in children with a supernumerary X chromosome (Robinson, Lubs, and Bergsma, 1979). Our psychometric data, which are representative of those obtained by others, are presented in the following figures. Wechsler results for 33 47,XXY boys tested between 8 and 9 years of age as well as the results obtained from sibling controls, are presented in Figure 2. The principal difference between the groups was that the 47,XXYs performed at a lower level than controls on all verbal subtests excepting Arithmetic. No differences between them were found on the performance subtests.

A similar picture emerged in the case of 13 47,XXX females. Again, the Wechsler test was administered when they were between 8 and 9 years of age and these data were compared to sibling controls. The results of this test are presented in Figure 3. Significant differences were found for most verbal subtests (Information, Arithmetic, Vocabulary, Comprehension) and none of the performance subtests.

We have also conducted an investigation using a sentence verification task procedure which indicates the presence of specific verbal deficits in supernumerary X-chromosome children as well. Two primary questions were examined in this study: (1) whether extra-X children were impaired on a verbal task which required little in the way of verbal expression (Walzer, Wolff, Bowman, Silbert, Bashir, Gerald, and Richmond (1978) had suggested that receptive language skills were relatively intact in 47,XXY boys) and (2) whether the chronometric model of verbal comprehension developed by Carpenter and Just (1975) provided any insight into the nature of the verbal disorders of extra-X children.

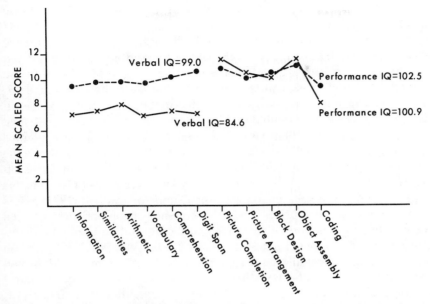

FIGURE 2.   Wechsler Intelligence Scale for Children: revised test results for 47,XXY children (×) and related controls (●).

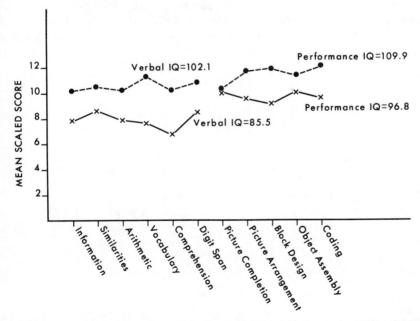

FIGURE 3.   Wechsler Intelligence Scale for Children: revised test results for 47,XXX children (×) and related controls (●).

Subjects were presented with a series of slides which depicted two people engaged in some activity, such as chasing, kicking, and so on, and with one person as the active agent. Each slide contained a sentence which was read to the child when the slide was exposed. Sentences could be affirmative or negative and also either true or false in relation to the picture. The child's task was to specify whether the sentence was true or false. Performance was assessed in terms of accuracy and response latency.

It is well established that the various conditions of this task are unequal in difficulty when this is assessed by the response latency measure. True Affirmatives (TA) are executed most quickly followed in turn by False Affirmatives (FA), False Negatives (FN), and True Negatives (TN). It is this kind of result that is accounted for by the Carpenter and Just (1975) model which, in essence, postulates a cyclical sequence of comparisons between the constituent elements of sentence and picture until all are congruent. The results of an assessment of 22 47,XXY boys and 8 47,XXX girls, all aged 9 years, indicated that the chromosomally abnormal children had great difficulties with this task. In fact, so few were able to exceed chance levels of performance even in the simplest TA condition that no analyses of reaction times were possible. The percent correct data are presented in Figures 4 and 5.

In order to ensure that these results were not simply due to limitations in the abilities of the children to understand the task, a second set of analyses, limited to those subjects who performed beyond chance in the sim-

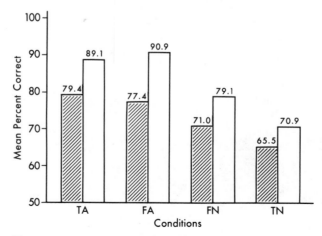

FIGURE 4. Mean percent correct scores in the four conditions of a sentence-verification task for 47,XXY children (shaded bar: $n = 22$) and controls (white bar: $n = 22$).

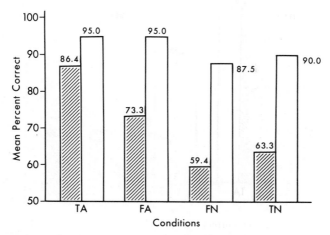

FIGURE 5. Mean percent correct scores in the four conditions of a sentence-verification task for 47,XXX children (shaded bar; $n = 8$) and controls (white bar; $n = 8$).

plest TA condition, was also done. The results, shown in Figure 6, indicate that the 47,XXY boys performed at lower levels than controls in the FA and FN conditions. An identical analysis of 47,XXX girls produced similar results, that is, lower levels of performance in the FA, FN, and TN conditions (see Figure 7).

Because of the high error rates, it was not possible to analyze the re-

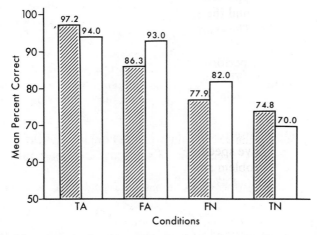

FIGURE 6. Mean percent correct scores in the sentence-verification task for those 47,XXY children (shaded bar; $n = 13$) and control children (white bar; $n = 20$) who exceeded chance in the True Affirmative (TA) condition.

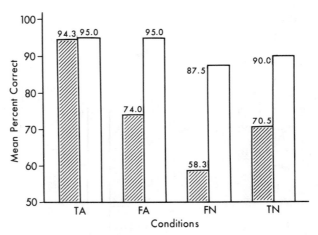

FIGURE 7. Mean percent correct scores in the sentence-verification task for those 47,XXX children (shaded bar; $n = 6$) and control children (white bar; $n = 8$) who exceeded chance in the True Affirmative (TA) condition.

sults further by methods used by Carpenter and Just (1975) to verify their sentence verification model, since these depend on reaction times for correct responses. However, it was true that the gradation of accuracy scores of the extra-X children (and the controls) in the four conditions of the task corresponded to the gradation that would be predicted by their model. This suggests that they used essentially the same verbal strategies. This conclusion is supported by the results of correlations between the Wechsler test scores and the sentence verification accuracy measures of the extra-X subjects. These are shown in Tables 1 and 2, and indicate generally that verbal IQs are more closely associated with sentence verification test scores than performance IQs. While not conclusive, this does imply that individual differences in verbal ability are strong determinants of the performance levels of the extra-X children in this experimental task.

Taken together, these results support the conclusions that:
1. 45,X females have specific deficits in spatial and perhaps other areas of nonverbal problem solving.
2. 47,XXY males and 47,XXX females have specific deficits in verbal ability.
3. The strategies employed by these subjects appear similar to those of chromosomally normal and intellectually unimpaired controls.
4. The patterns of intellectual deficit observed in children are not explained by X-linked models for the inheritance of spatial ability.

Table 1

CORRELATIONS BETWEEN WISC-R IQS AND SENTENCE-VERIFICATION TEST RESULTS FOR XXY SUBJECTS[a]

	Whole group (N = 22)		Non-chance TA subgroup (N = 13)	
	Verbal IQ	Performance IQ	Verbal IQ	Performance IQ
TA	.424**	.561***		
FA	.334	.193	.633**	−.292
FN	.376*	.228	.581**	−.252
TN	.378*	.024	.136	−.402

[a] Intercorrelation between WISC-R IQs and sentence-verification percent correct scores for 47,XXY subjects (children exceeding a chance level of performance in the TA condition shown separately).
   * $p < .10$
  ** $p < .05$
 *** $p < .01$

Since there is evidence that sex differences in ability patterns are more easily verified and possibly larger after puberty than before (Maccoby and Jacklin, 1974), it is reasonable to ask whether sex hormone activity, which also becomes distinctly different in males and females at puberty, may contribute to normal sex differences in cognition (Petersen, 1979). Direct evidence is limited, although some studies have indicated that intrasex variations in intellectual efficiency may be related to day-to-day

Table 2

CORRELATIONS BETWEEN WISC-R IQS AND SENTENCE-VERIFICATION TEST RESULTS FOR XXX SUBJECTS[a]

	Whole group (N = 8)		Non-chance TA subgroup (N = 6)	
	Verbal IQ	Performance IQ	Verbal IQ	Performance IQ
TA	.224	.324		
FA	.172	−.037	.354	−.028
FN	.303	.108	.884**	.340
TN	.428	.065	.603	−.180

[a] Intercorrelation between WISC-R IQs and sentence-verification percent correct scores for 47,XXX subjects (children exceeding a chance level of performance in the TA condition shown separately).
  ** $p < .05$

fluctuations in levels of sex hormones after puberty (Klaiber, Broverman, Vogel, and Kobayashi, 1974). However, the literature relating verbal and nonverbal abilities to postpubescent sex hormones is not sufficiently consistent to conclude that normal sex differences in ability are direct consequences of gender-specific sex hormone activity. One immediate problem is that after birth, the sexes are identical or nearly identical in sex hormone levels (Forest, Cathiard, and Bertrand, 1973), yet may already differ in their cognitive characteristics (Molfese and Hess, 1978; Witelson, 1976). Furthermore, those studies which have related intellectual functioning to sex hormone levels (either measured or inferred from such indices as somatotypes or secondary sex characteristics) have produced results which are difficult to reconcile with normal sex differences in ability. For example, Broverman has found in various studies that subjects with higher levels of testosterone activity were less proficient at tasks of a spatial nature than they were at such tasks as automatized naming (Broverman and Klaiber, 1969; Broverman, Klaiber, Kobayashi, and Vogel, 1968). Petersen's results with male subjects were similar in that individuals less stereotypically male were more able spatially than less androgynous males. With female subjects, better spatial ability was found in less feminine or more androgynous subjects (Petersen, 1976). It is difficult to see how these findings account for the normal spatial advantage of males or the verbal superiority of females.

A different perspective on the relationship between sex hormone activity and individual differences in ability is provided by the work of Waber (1976, 1977). She questioned whether normal sex differences in ability were artifactual and really due to differing rates of maturation for males and females. Typically, males reach sexual maturity later than females and it was Waber's hypothesis that variations in pubertal onset would be associated with ability patterns in both male and female subjects. Her investigations of early- and late-maturing subjects of both sexes supported this proposition, at least partially. Her findings indicated that, regardless of sex, individuals who were delayed in pubertal onset were more able spatially than verbally. Early maturers tended to show the opposite pattern of intellectual strengths. It was her conclusion that maturation rate primarily affected the development of spatial ability since early and late maturers differed only in this area. It should be noted, however, that this hypothesis does not account for results obtained with children with abnormal numbers of X chromosomes, since their deficits are evident before pubescence and in the absence of sex hormone abnormalities (Stewart, Netley, Bailey, Haka-Ikse, Platt, Holland, and Cripps, 1979).

Do the observations of Waber (1976, 1977), Petersen (1976) and Broverman, Klaiber, Kobayashi, and Vogel (1968) reflect any common mechanism underlying the development of intellectual abilities? Quite simply, we do not know, but it is true that with one assumption, they can be explained collectively by a single principle. The required assumption is that delayed pubertal onset in males and females, low testosterone levels in males and androgynous somatotypes in both sexes are the result of a slowness in the general process of maturation which is reflected in diminished sexual differentiation. If so, then it is conceivable that variations in spatial ability relative to verbal ability are the result of individual differences in rates of maturation. That is to say, it can be hypothesized that delays in development lead to higher spatial than verbal ability and rapid development to the reverse.

Before leaving the issue of sex hormone activity and normal differences in ability between the sexes, it should be noted that some theories have laid particular stress on events occurring prenatally. Male and female neonates have had different histories of exposure to sex hormones during intrauterine life. In fact, it is these differences which are responsible for the neonates' clearly distinguishable male and female phenotypes (Mittwoch, 1973). It has been suggested that the male and female sex hormones have specific effects on brain development with the result that, at birth, the two sexes have functionally different brains leading ultimately to their observed differences in ability (Reinisch, Gandelman, and Spiegel, 1979). This theory, however, cannot explain observations obtained from children with sex chromosome abnormalities. Individuals with 47,XXY complements have similar verbal deficits to those with 47,XXX genotypes, yet each group has been exposed to sufficient differences in sex hormones during prenatal life to develop clearly distinguishable and normal male and female phenotypes (Stewart, Netley, Bailey, Haka-Ikse, Platt, Holland, and Cripps, 1979).

GROWTH, THE X CHROMOSOME, AND CEREBRAL ORGANIZATION

In this section we are concerned with examining a set of hypotheses which have been developed to explain the cognitive deficits of children with abnormal numbers of X chromosomes. The principal issues are to account for the nonverbal or spatial deficits of 45,X Turner syndrome females and the verbal impairments of 47,XXY males and 47,XXX females. The materials reviewed thus far have indicated that theories based either on a simple X-linked pattern of inheritance for spatial ability or on sex hormone activity occurring prenatally or at puberty are inadequate in this context. However, the literature did suggest that maturation rate was a useful concept, and provided a basis for interpreting a variety of findings obtained with chromosomally normal subjects.

Is there reason to believe that growth rate is of any value in understanding the cognitive deficits of individuals with abnormalities of sex chromosome complement? It is known that growth is disturbed in individuals with abnormal numbers of X chromosomes. In the case of extra-X subjects, bone age is delayed (Stewart, Netley, Bailey, Haka-Ikse, Platt, Holland, and Cripps, 1979), whereas in the case of 45,X females, the index of growth hormone activity provided by the sulphation factor is elevated (Almgvist, Linsten, and Lindvall, 1963). Indeed, it has been reported that rates of mitotic cell division are influenced by numbers of X chromosomes, being high in 45,X conditions and low in supernumerary X states (Barlow, 1973; Mittwoch, 1973). Interestingly, variations in numbers of Y chromosomes are not associated with either specific intellectual deficits or disturbances of mitotic cell division (Mittwoch, 1973; Witkin, Mednick, Schulsinger, Bakkestrom, Christiansen, Goodenough, Hirschborn, Lundsteen, Owen, Philip, Rubin, and Stocking, 1976). These observations suggest that the arrested growth in extra-X subjects and accelerated growth in 45,X individuals might be responsible for their differing patterns of intellectual deficit. Although this hypothesis resembles that offered by Waber (1976, 1977), it differs from hers in that the influence of growth rate is not restricted to puberty but is operative from the earliest times of life. This extension of principle is made necessary by the fact that the cognitive impairments of these children can be observed well before puberty (Robinson, Lubs, and Bergsma, 1979).

Our evidence for this hypothesis is derived from several different investigations conducted on children identified at birth as having abnormal numbers of X chromosomes (Bell and Corey, 1974). Since only a few 45,X Turners were found in this series, the bulk of our evidence is derived from extra-X male and female children. Our first study was prompted by the observation that variations in bone age maturity in 45,X, 47,XXY, and 47,XXX subjects correspond to group differences in the balance of verbal IQ to performance IQ (Stewart, Netley, Bailey, Haka-Ikse, Platt, Holland, and Cripps, 1979). That is, extra-X subjects with verbal deficits have more pronounced delays in bone age than 45,X females who are spatially impaired. Our first direct examination of this issue was, however, more stringent. In this, we assessed whether degree of bone age retardation was associated with the severity of verbal impairment WITHIN groups of 47,XXY and 47,XXX subjects. For the boys, this was done by dividing them into two groups on the basis of their degree of bone age retardation assessed between 3 and 6 years. "Severe" was defined as 1.5 (SD) standard deviations or more below age expectancy levels. Of a sample of 20 boys, 5 were found to have this degree of slow growth. Fifteen fell into a more moderately delayed group. An examination of the Wechsler Test results indicated that 4 of the 5 severely delayed subjects had a significantly lower verbal IQ than performance IQ. This occurred in only 4 of the remaining 15 less severely growth-retarded subjects. This difference approached statistical significance ($p = .054$, Fisher Exact Test). A similar analysis of the relation between bone age delays and Wechsler IQ discrepancies within the 47,XXX group did reach significance ($p = .047$, Fisher Test). Thus, bone age retardation assessed during the preschool years was significantly associated with the degree of verbal disability at 8 to 9 years of age.

In a second study, we examined the relationship between prenatal growth and intellectual ability in extra-X subjects. In this investigation, we made use of that index of prenatal growth rate which is provided by total finger ridge counts (TFRC). Dermal ridges differentiate by mid-fetal life and do not change after this time (Holt, 1968). It is generally accepted that they provide an index of growth rate during this period, with high numbers of ridges reflecting fast growth, and low counts, slow growth (Barlow, 1973). As would be expected, ridge counts are higher than normal in 45,X females and lower than normal in 47,XXX females and 47,XXY males (Valentine, 1969). Our study was limited to supernumerary X subjects, 24 of whom had ridge count analyses done independently by anthropologists at the University of Toronto (Hzreczko and Sigmon, 1980). This group was divided into two subgroups: 12 with the lowest total dermal finger counts and the 12 with higher counts. Our dependent measures were Wechsler verbal and performance IQ discrepancies and scores on the sentence verification task described previously. These two groups were found to differ in the sentence verification task in the most difficult TN condition. These findings are shown in Figure 8. They also differed in their Wechsler test results. The low ridge count group had significantly lower verbal IQ scores (relative to performance IQ) than the high count group (19.4 versus 10.6 IQ points). Thus, it does appear that prenatal growth rate has an influence on the degree of verbal impairment shown by these extra-X children.

If growth rate has an influence on the development of abilities, what is the mechanism? The answer is suggested by various findings which indicate that normal males and females differ in functional cerebral organization. Some of this evidence has been derived from studies of normals using such procedures as dichotic listening, half-field tachistoscopic, and dichhaptic recognition tasks. These techniques share the common characteristic of presenting a subject with different stimulus information to left-

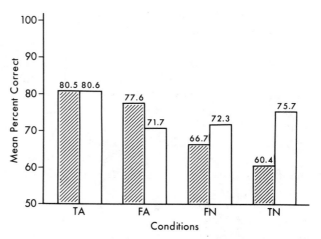

FIGURE 8. Mean percent correct scores in the four conditions of the sentence-verification task for low (shaded bar) and high (white bar) finger ridge count extra-X subjects.

sided and right-sided sensory channels, thereby permitting a comparison of the processing efficiency of the two cerebral hemispheres. The usual result when verbal material is used is that information presented to right-sided sensory channels is better processed than material presented to the left, since the former is more directly available to the verbally proficient left cerebral hemisphere. Conversely, left-sided presentation of nonverbal material has advantages over right-sided presentation, since it is directed primarily to the spatially proficient right hemisphere (Bryden, 1978; Witelson, 1977).

Several studies have indicated that normal males and females perform differently on these tasks. McGlone and Davidson (1973), for example, found that females tended to have a greater right-field advantage for tachistoscopically presented dots than males. Kimura (1969) obtained a similar result, while Witelson (1976) reported that males show a left-hand advantage on a nonverbal dichhaptic recognition task at a younger age than females. These and other results obtained with dichotic listening procedures (although the evidence here has been somewhat mixed [Witelson, 1977]) have frequently been interpreted to indicate that males have more functionally specialized cerebral hemispheres than do females (Bryden, 1979; McGlone, 1980; Waber, 1976, 1977).

Additional evidence in support of this conclusion is found in the literature on the differential effects of brain lesions on males and females. For example, Lansdell (1968) reported that right temporal lobectomies had a greater disruptive effect on spatial functions for males than females. In addition, he found that only in the case of male patients was it possible to demonstrate a correlation between the degree of spatial impairment and the amount of neural tissue removed. McGlone and Kertesz (1973) reported that the spatial impairments of females were related to the degree of their verbal deficits. They suggested that females employ left-hemisphere-based verbal strategies for spatial tasks, in contrast to males, who use right-hemisphere-based nonverbal processes. McGlone (1977, 1978) provided additional evidence on this issue in her studies of unilaterally brain-damaged adult males and females. She found that males had a greater incidence of verbal deficits following left-sided lesions than females. She suggested that males had a greater degree of hemispheric specialization than females. In a subsequent comprehensive review article, McGlone (1980) persuasively managed (in the face of the considerable critical comment generated by *The Behavioral and Brain Sciences* format) to conclude again that brain specialization tends to be different in the two sexes.

If male and female brains do differ in functional specialization, it is, of course, quite possible that this is due to some biological process. Some authors have suggested that the difference is the consequence of processes which begin early in life and which are mediated by the impact of rates of development on neural maturation. Waber (1979a), for example, has proposed that several maturational factors are responsible for the different levels of verbal and spatial ability in males and females. Her position is that in the case of certain language and neuromotor functions, prepubescent females are superior to males because they mature faster than boys of the same ages. These differences, in her view, are not due to the impact of growth on hemispheric specialization, since she judges the evidence too inconsistent to permit such a conclusion. Higher levels of spatial functions, however, do depend on hemispheric specialization (the more the better) and, further, result from slowness in the maturational processes associated with puberty. These conclusions, of course, are tenuously based, since they rest on between-group comparisons of males and females, as well as on early-maturing and late-maturing subjects.

Corballis and Morgan (1978) took a different view of the significance of maturational factors. They proposed a left–right maturational gradient, such that left-sided functions and their underlying neural substrates develop more quickly than right-sided functions. This left-sided advantage provides the left hemisphere with a greater readiness to acquire complex skills and also inhibits right-sided acquisition of these same functions. The result is that, for most individuals, such complex abilities as language are established predominantly in the left hemisphere and only minimally in the right. Only in the case of left hemisphere injury, where the inhibiting process is interfered with, does the right hemisphere have an opportunity to acquire and mediate language. Under ordinary circumstances, right-hemisphere functions emerge essentially by default, that is, they consist of those cognitive processes which the left hemisphere (committed to more complex cognitive activities) is unable to execute.

Taylor (1969) offered a theory similar to that of Corballis and Morgan (1978), but with a reversal of hemispheric maturational advantage. He argued, on the basis of hemispheric differences in vulnerability to epileptic seizures, that the left hemisphere was more inactive than the right during the first few months of life. He added that this was most pronounced in the case of males. His conclusions, therefore, were that left-hemisphere development is delayed relative to the right and most particularly so in the case of males.

Clearly, although all these theorists are making use of similar concepts, they are quite disparate in their conclusions and predictions. All appear to imply that the maturational rate has an impact on the development of hemispheric organization and ability levels. They differ quite distinctly, however, in the nature and, indeed, in the direction of the relationships involved.

Our theory concerning the relationships between growth rate, hemispheric specialization, and the development of intellectual functions makes use of some concepts advanced by Levy (Levy, 1969; Levy and Reid, 1976). In the first of these publications, she concludes, on the basis of intellectual differences between left-handers and right-handers, that greater degrees of bilateral language interfere with the development of spatial ability. In her later paper this proposition is extended to include females whose spatial ability, just as that of left-handers, suffers as a result of the interfering effects of verbal development. The essence of Levy's model (Levy, 1969) is that hemispheric specialization for verbal and nonverbal abilities occurs as a result of a competitive interactive process between these two cognitive processes for available neural tissue. If verbal ability is relatively advanced throughout development, then spatial ability is arrested and more tenuously represented in neural terms. In the case of slow verbal development, the process is reversed, with the result that spatial ability would emerge under conditions of minimal interference and be better represented in neural structures.

It is conceivable that variations in these developmental processes are mediated through the impact of growth rate on those left hemisphere sites identified in adults (Geschwind and Levitsky, 1968) and newborns (Wada, Clarke, and Hamm, 1975; Witelson and Paillie, 1973) and which plausibly may be regarded as providing the structural bases for language acquisition (Hécaen, 1977). If this is correct, then the development of hemispheric specialization can be viewed as being due to the relative physical and functional maturity of left-hemisphere-based language centers. The growth of the right hemisphere's spatial functions would, under this model, depend on the degree to which they are interfered with by the left hemisphere's verbal processes.

There have been few empirical investigations of the relationship of maturation rate to hemispheric specialization. Waber (1976, 1977) studied the issue in her series of early- and late-pubertal-onset children by means of the dichotic listening procedure. She reported that those with delayed puberty were more asymmetrical in their reports of left- and right-channel verbal material than were the earlier-maturing children, which is consistent with the idea expressed above. Studies of the individuals with abnormal numbers of X chromosomes are particularly valuable in this context. Their deviations in growth are apparent from early in life (possibly even from conception) and have been related to atypical mitotic cell division rates. Thus, these subjects permit an examination of the question of whether early-onset growth disturbances in prepubertal children affect functional brain organization. Several studies of 45,X Turner syndrome females have indicated that these subjects respond atypically to dichotically presented verbal information. Netley (1977), Netley and Rovet (1982) and Waber (1979b) have found that 45,X Turner subjects showed an attenuation of the usual right-ear advantage, a finding which is consistent with the notion that, in the absence of a normally occurring X chromosome, hemispheric specialization is diminished.[1]

As mentioned in a previous section, extra-X subjects have diminished rates of prenatal and postnatal maturation, as indicated by bone age measures and total finger ridge counts. It was also reported that the degree of growth delay in these prepubescent subjects was associated with the severity of their verbal impairments. Under the model advanced earlier, it would be predicted that the performance of these subjects would differ from normal controls on indices of hemispheric specializations. Since their maturation rate is slower than normal, their left hemisphere sites concerned with language acquisition would be relatively underdeveloped. Under the extension of the theory of Levy (1969) elaborated earlier, this would minimize the interference of verbal processes on the development of spatial functions, which as a result would become strongly established

---

[1] Further, in two of these investigations (Netley, 1977; Netley and Rovet, 1982), it was found that presence of asymmetries in dichotic reports were associated with lesser degrees of spatial impairment. Taken together, these investigations suggest that 45,X Turner syndrome subjects (who, it will be recalled, have higher-than-normal rates of mitotic cell division) have spatial deficits because their right-hemisphere spatial functions have not developed to normal levels.

262   C. NETLEY AND JOANNE ROVET

Table 3[a]

	XXY (N = 34)	XY	P
Dichotic digits	.179	.015[b]	<.05
Dichhaptic	.043	−.057[b]	<.10
T-scope: Dots	.160	.027[c]	<.05
T-scope: Letters	−.01	−.036[c]	NS

[a] Mean asymmetry scores (as indicated by phi) of 47,XXY and control children found in dichotic listening, dichhaptic, and lateralized tachistoscopic tests. See text page 262 for explanation of computation of phi.
[b] Matched VIQ and PIQ.
[c] Unmatched.

in the right hemisphere. Thus, the final result would be an exaggeration of the functional hemispheric specialization of individuals with a supernumerary X chromosome.

We have investigated this issue in a series of studies of 34 47,XXY boys, all prepubescent, who were compared to age-matched and in some cases IQ-matched controls. The experimental procedures were a dichotic listening task using digit series (Netley, 1977), a tachistoscopic half-field recognition task involving either dots or letters (Kimura and McGlone, 1979), and a dichhaptic recognition task (Witelson, 1976). The asymmetry of the response pattern of each subject was computed to reflect the expected result in each task. Thus, a positive phi indicated a right-ear advantage in the dichotic task and a right-field advantage in the tachistoscopic letter recognition task. Conversely, a positive phi result for the

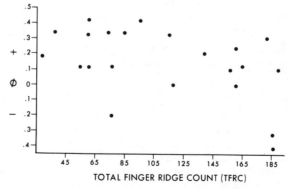

FIGURE 9. Relationship between total finger ridge count and the asymmetry of performance in a tachistoscopic dot enumeration task for 47,XXY boys ($r = -.428; p < .05$).

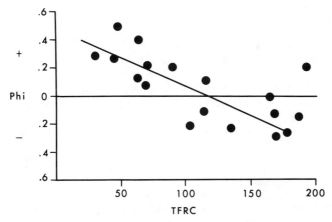

FIGURE 10. Relationship between total finger ridge count and the asymmetry of perform-ance in a dichhaptic recognition task for 47,XXY boys ($r = -.699$; $p < .01$).

dichhaptic and the tachistoscopic dot tests indicated a left-sided response bias. The results are shown in Table 3, and generally conform to the ex-pectation that extra-X boys have larger-than-normal lateral asymmetries.

We also examined the relationship between prenatal growth rate and hemispheric function by means of correlational analyses. Our interest here was twofold. First, we wished to determine whether the ridge count measure was related to hemispheric specialization in our subjects since, if it were, this would suggest that the development of functional neural orga-nization begins very early in life. Our second interest was to establish whether individual differences in hemispheric specialization and growth rate could be demonstrated within a sample of growth-delayed children. In the cases of the dichhaptic and half-field tachistoscopic dot recognition tasks, this proved to be so (see Figures 9 and 10).

CONCLUSION

This chapter began with the statement that biological factors cannot be regarded as exclusive determinants of the differences in abilities of nor-mal males and females. At this point, it is necessary to add that the partic-ular biological disturbances, and their behavioral correlates seen in chil-dren with abnormal numbers of X chromosomes, may not apply to findings obtained with chromosomally normal individuals. However, it does appear reasonable to advance the hypothesis that variations in growth rate, beginning in prenatal life, may, through their impact on neural maturation and hemispheric organization, be responsible for some part of the cognitive distinctiveness of the two sexes.

REFERENCES

Almgvist, S., Linsten, J., and Lindvall, N. (1963). Linear growth, sulfation factor activity and chromosomal constitution in 22 subjects with Turner's syndrome. *Acta Endocrinologica, 42,* 168–186.
Barlow, P. (1973). The influence of inactive chromosomes on human development. *Humangenetik, 17,* 105–136.
Bell, A., and Corey, P. (1974). A sex chromatin and Y body survey of Toronto newborns. *Canadian Journal of Genetics and Cytology, 16,* 239–250.
Bock, R., and Kolakowski, D. (1973). Further evidence of sex-linked major gene influence on human spatial visualizing ability. *American Journal of Human Genetics, 25,* 1–14.
Broverman, D., and Klaiber, E. (1969). Negative relationships between abilities. *Psychometrika, 34,* 5–20.
Broverman, D., Klaiber, E., Kobayashi, Y., and Vogel, W. (1968). Roles of activation and inhibition in sex differences in cognitive abilities. *Psychological Review, 75,* 23–150.
Bryden, M. P. (1978). Strategy effects in the assessment of hemispheric asymmetry. In G. Underwood (Ed.), *Strategies of information processing.* New York: Academic Press.
Bryden, M. P. (1979). Evidence for sex-related differences in cognitive functioning. In M. Wittig and A. Petersen (Eds.), *Sex-related differences in cognitive functioning.* New York: Academic Press.
Carpenter, P., and Just, M. (1975). Sentence comprehension: A psycholinguistic processing model of verification. *Psychological Review, 82,* 45–73.
Corballis, M. C., and Morgan, M. J. (1978). On the biological basis of human laterality: I. Evidence for a maturational left–right gradient. *The Behavioral and Brain Sciences, 2,* 261–336.
Forest, M. G., Cathiard, A., and Bertrand, J. (1973). Total and unbound testosterone levels in the newborn and in normal and hypogonaded children: Use of a sensitive radioimmunoassay for testosterone. *Journal of Clinical Endocrinology and Metabolism, 36,* 1132–1142.
Garron, D. (1977). Sex-linked recessive inheritance of spatial and numerical abilities and Turner's syndrome. *Psychological Review, 77,* 147–152.
Geschwind, N., and Levitsky, W. (1968). Human brain: Left–right asymmetries in temporal speech region. *Science, 161,* 186–187.
Hécaen, H. (1977). Language representation and brain development. In S. Berenberg (Ed.), *Brain Fetal and Infant.* The Hague: Nijhoff.
Holt, C. (1968). *The genetics of dermal ridges.* Springfield, Ill.: Thomas.
Hzreczko, T., and Sigmon, B. (1980). The dermatoglyphics of a Toronto sample of children with XXY, XXYY, and XXX aneuploidies. *American Journal of Physical Anthropology, 52,* 33–42.
Kimura, D. (1969). Spatial localization in left and right visual fields. *Canadian Journal of Physiology, 23,* 445–448.
Kimura, D., and McGlone, J. (1979). *Neuropsychology test procedures* (used at University Hospital, London, Ontario). Unpublished booklet available from authors.
Klaiber, E., Broverman, D., Vogel, W., and Kobayashi, Y. (1974). Rhythms in plasma MAO activity, EEG, and behavior during the menstrual cycle. In M. Ferin, F. Halberg, R. Richart, and R. Vande Wiele (Eds.), *Biorhythms and human reproduction.* New York: Wiley.

Lansdell, H. (1968). Effect of extent of temporal lobe ablations on two lateralized deficits. *Physiology and Behavior, 3,* 271–273.

Levy, J. (1969). Possible basis for the evolution of lateral specialization of the human brain. *Nature, 224,* 614–615.

Levy, J., and Reid, M. (1976). Variations in writing posture and cerebral organization. *Science, 200,* 1291–1292.

Maccoby, E., and Jacklin, C. (1974). *The psychology of sex differences.* Stanford: Stanford Univ. Press.

McGlone, J. (1977). Sex differences in the cerebral organization of verbal functions in patients with unilateral brain lesions. *Brain, 100,* 775–793.

McGlone, J. (1978). Sex differences in functional brain asymmetry. *Cortex, 14,* 122–128.

McGlone, J. (1980). Sex differences in human brain asymmetry: A critical survey. *The Behavioral and Brain Sciences, 3,* 215–263.

McGlone, J., and Davidson, W. (1973). The relation between cerebral speech laterality and spatial ability with special reference to sex and hand preference. *Neuropsychologia, 11,* 105–113.

McGlone, J., and Kertesz, A. (1973). Sex differences in cerebral processing of visuo spatial tasks. *Cortex, 9,* 313–320.

Mittwoch, U. (1973). *Genetics of sex differentiation.* New York: Academic Press.

Molfese, D., and Hess, T. M. (1978). Hemispheric specialization for VOT perception in the preschool child. *Journal of Experimental Psychology, 26,* 71–84.

Money, J. (1973). Turner's syndrome and parietal lobe functions. *Cortex, 9,* 385–393.

Nash, S. (1979). Sex role as a mediator of intellectual functioning. In M. Wittig and A. Petersen (Eds.) *Sex-related differences in cognitive functioning.* New York: Academic Press.

Netley, C. (1977). Dichotic listening of callosal agenesis and Turner's Syndrome patients. In S. J. Segalowitz and F. A. Gruber (Eds.), *Language development and neurological theory* New York: Academic Press.

Netley, ⌐., and Rovet, J. (1982). Atypical hemispheric lateralization in Turner's syndrome (in press).

Petersen, A. (1976). Physical androgyny and cognitive functioning in adolescence. *Developmental Psychology, 12,* 524–533.

Petersen, A. (1979). Hormones and cognitive functioning in normal development. In M. Wittig and A. Petersen (Eds.), *Sex-related differences in cognitive functioning.* New York: Academic Press.

Reinsich, J., Gandelman, R., and Spiegel, F. (1979). Prenatal influences on cognitive abilities: Data from experimental animals and human genetic and endocrine syndromes. In M. Wittig and A. Petersen (Eds.), *Sex-related differences in cognitive functioning.* New York: Academic Press.

Robinson, A., Lubs, H., and Bergsma, D. (1979). Sex chromosome aneuploidy: Prospective studies on children. *Birth Defects: Original Article Series, 15,* No. 1.

Rovet, J., and Netley, C. (1982). Processing deficits in Turner syndrome. *Developmental Psychology, 18,* 77–94.

Shepard, R. (1978). The mental image. *American Psychologist, 33,* 125–137.

Stafford, R. (1961). Sex differences in spatial visualization as evidence of sex-linked inheritance. *Perceptual and Motor Skills, 13,* 428.

Stewart, D., Netley, C., Bailey, J., Haka-Ikse, K., Platt, J., Holland, W., and Cripps, M. (1979). Growth and development of children with X and Y chromosome aneuploidy: A prospective study. *Birth Defects: Original Article Series, 15,* 75–114.

Taylor, D. (1969). Differential rates of cerebral maturation between sexes and between hemispheres. *Lancet,* July 1969, 2, 140–142.

Valentine, G. H. (1969). *The chromosome disorders: An introduction for clinicians.* 2nd ed. Philadelphia: Lippincott.

Vandenburg, S. G., and Kuse, A. R. (1979). Spatial ability: A critical review of the sex-linked major gene hypothesis. In M. Wittig and A. Petersen (Eds.), *Sex-related differences in cognitive functioning.* New York: Academic Press.

Waber, D. (1976). Sex differences in cognition: A function of maturation rate. *Science, 192a,* 572–574.

Waber, D. (1977). Sex differences and rate of physical growth. *Developmental Psychology, 13,* 29–38.

Waber, D. (1979a). Cognitive abilities and sex-related variations in the maturation of cerebral cortical functions. In M. Wittig and A. Petersen (Eds.), *Sex-related differences in cognitive functioning.* New York: Academic Press.

Waber, D. (1979b). Neuropsychological aspects of Turner syndrome. *Developmental Medicine and Child Neurology, 231,* 58–70.

Wada, J., Clarke, R., and Hamm, A. (1975). Cerebral hemispheric asymmetry in humans. *Archives of Neurology, 32,* 239–246.

Walzer, S., Wolff, P., Bowman, D., Silbert, A., Bashir, A., Gerald, P., and Richmond, J. (1978). A method for the longitudinal study of behavioral development in infants and children: The early development of XXY children. *Journal of Child Psychology and Psychiatry, 19,* 213–230.

Witelson, S. (1976). Sex and the single hemisphere: Right hemisphere specialization for spatial processing. *Science, 193,* 425–427.

Witelson, S. (1977). Early hemispheric specialization and interhemisphere plasticity: An empirical and theoretical review. In S. J. Segalowitz and F. A. Gruber (Eds.), *Language development and neurological theory.* New York: Academic Press.

Witelson, S., and Paillie, W. (1973). Left hemisphere specialization for language in the newborn: neuroanatomical evidence of asymmetry. *Brain, 96,* 641–646.

Witkin, H., Mednick, S., Schulsinger, F., Bakkestrom, P., Christiansen, K., Goodenough, D., Hirshborn, K., Lundsteen, C., Owen, D., Philip, J., Rubin, D., and Stocking, M. (1976). Criminality in XYY and XXY men. *Science, 193,* 547–555.

CHAPTER 12

Language and Brain Dysfunction in Dementia

Loraine K. Obler

INTRODUCTION

As we look to the classic syndromes of aphasia to teach us about the distribution of function within the language zone of the left hemisphere, so we may look to the dementias to learn about the interactions (1) between cortical and subcortical mechanisms in language and (2) between the form of language and semantics. At first glance, it would appear more difficult to derive knowledge about language–brain relationships from the relatively diffuse lesions of the dementias than it has been from study of the more localized lesions associated with the classic aphasias. And perhaps that explains the paucity of work done on the language disturbances of the dementias. However, as our localizing instruments become more sensitive, and as our neurolinguistic strategies become more sophisticated, it becomes clear that the dementias will provide a rich realm for expanding our knowledge of brain systems or mechanisms underlying language processes.

Dementia, it must be stressed, is a broad term, encompassing a number

267

LANGUAGE FUNCTIONS
AND BRAIN ORGANIZATION

of diseases which have in common the deterioration of various cognitive or intellectual abilities. As a rule the disabilities range from attentional and memory deficits to deterioration of visuospatial abilities and logical manipulations. Language abilities, some argue, may be entirely spared in dementia. Within a narrow definition of linguistic abilities, this may be true. Thus it is generally held that phonological and syntactic systems *are* spared in dementia while semantic and pragmatic systems show deficits (i.e., Irigaray, 1973; Schwartz, Marin, and Saffran, 1979). To the extent that semantic and pragmatic aspects of communication are deemed "linguistic," then, there is language loss in dementia. To the extent that they rely on cognitive skills distinct from language, we may conclude that apparent language deficits in dementia are secondary to these cognitive deficits. If the latter is the case, one may explore the relations between the specific cognitive abilities implicated and the resulting communicative behaviors. Before we turn to specific issues concerning language vis-à-vis cognition, let us briefly review the various dementias in terms of their localizing potential, and let us consider the indications that basic capabilities, such as memory and attention, interact with language processes.

CORTICAL AND SUBCORTICAL DEMENTIAS

Localization of brain damage in the dementias is not as clear-cut as it is in the aphasias. Yet while the standard belief is that dementias result from diffuse lesions, unlike the focal lesions of stroke or tumor, in fact one must speak of brain regions *relatively* deteriorated and *relatively* spared by a dementing disease.

Among those dementias which evidence predominantly subcortical lesions are such diseases as Parkinson's disease and progressive supranuclear palsy. Neurobehaviorally, these diseases typically present slowed behavior on the part of the patients, memory disorders, personality changes (generally toward apathy), and a loss in "ability to manipulate acquired knowledge" (Albert, 1978). This inability to manipulate acquired knowledge may be seen, for example, in difficulty with listing the months of the year backward. Language disturbance in these diseases is less florid than in the cortical diseases, and is most easily detected in free oral discourse and in free writing, and less easily seen with more restricted tasks. What is striking, rather, are speech disturbances (dysarthria) which in severe cases may mask language disturbances. Yet word-finding difficulties are also evidenced, by hesitations or by employment of verbal paraphasias. And in the writing of patients with subcortical dementias we see lack of morphological agreement (in addition to nonlanguage disturbances, such as micrographia and omission or addition of letters). For example, one patient wrote about the Cookie Theft Picture of the Boston Diagnostic Aphasia Examination:

> *the boy on the stoll tripp and the girl laugh at the boy and then she spill water on the floor.*

Such examples of "language disturbance," I would argue, are only tangentially linguistic disturbances. That is to say, they may reflect the patient's inattention to errors, and reduced self-monitoring. As normals, of course, we make such errors but we usually self-correct in writing, and, somewhat less often, in speech. However, one might claim that inattention alone is not at fault in the dementing patient, since were that the case, agraphic disturbances should be randomly distributed to any letter or part of a word; in fact morphological errors such as those in the example above are relatively common. Also, the errors made are not random errors but rather involve substitution of one inflectional affix for a related one.

One might likewise explain the paraphasias of subcortical dementias as due to inattention on the part of the patient, who makes the same sort of semantically close verbal substitutions most of us healthy people will self-correct or inhibit before production. Presumably one might test this hypothesis by analysis of speech produced before and after administration of an alerting medication to one of these patients. In our study of Dr. T. (Obler, Cummings, and Albert, 1979), a 64-year-old ophthalmologist with progressive supranuclear palsy, we noted that levodopa served generally to speed the patient up. This made his speech more intelligible; he became able to produce four to five words in one breath as compared to two to three words before medication, so he no longer needed to pause within constituent phrases. His paraphasias in discourse were still evident, however (e.g., in describing the Cookie Theft Picture of the BDAE, he now said, "The young lady he is *hang up* in his mischief,") so either he was not sufficiently alerted to correct the paraphasias, or alertness alone cannot override the anomic disorder. Patients with subcortical dementias, we believe at this point, evidence language disturbance in two realms: lexical selection, and paragrammatism (Obler and Albert, 1981). Since it is possible that attentional deficits may account for the disturbed language production, one would want to correlate the degree of attentional disturbance with the degree of anomia or agraphia in patients with subcortical dementias. To the extent that there was a positive correlation between nonlanguage measures of attention and the frequence of anomic and agraphic errors, one might conclude that impaired attentional mechanisms underlay the apparent language disturbance. To the extent that there was no correlation, we could assume that as for aphasics, language abilities themselves were impaired.

As to the dementias stemming from predominantly cortical disorders, language disturbances resembling those of Wernicke's anomic, transcortical sensory, or transcortical mixed aphasias are not infrequent. In fact, making the differential diagnosis between early or midstage Alzheimer's dementia and these aphasias is becoming a fairly frequent problem on aphasia wards as the human life span increases and the incidence of strokes decreases. Of course this problem of differential diagnosis is most severe if one is confronted with a patient who cannot report the medical history leading up to the picture of fluent empty speech or poor comprehension. A good history of progressive language deterioration combined with behavioral difficulties will lead to diagnosis of a cortical dementia, while sudden onset would more likely result in the diagnosis of aphasia.

For purposes of this chapter we may sketch a common pattern of linguistic decline in the cortical dementias. In the early stages of the disease, patients will digress somewhat in their discourse, but may return to their earlier points. Verbal paraphasias, if produced in speech, may be corrected by the patient, who is aware of word-finding difficulties at this stage and will comment on them. Naming to confrontation may still be fairly good, except for low-frequency items, and comprehension tasks involving single sentences are performed without error. In natural conversation the patient may respond to questions closely related to the ones which are asked. On more specific tasks the patient will perform somewhat better, although he or she may ask for repetition of instructions. On repetition tasks, patients will have no trouble with long sentences composed of high-frequency items, but when sentences are composed of low-frequency words, the patient will ask for a second chance, and may omit words, or produce phonetically related jargon, suggesting that he or she has not comprehended the stimulus sentences. This would appear to contrast with the loss of good communicative gesture in aphasia (see Foldi, Cicone, and Gardner, this volume).

By a later stage of the disease, patients will make more uncorrected verbal paraphasias in discourse and may eventually produce jargon. While a broad range of functors will be used, sometimes with disregard for their logical constraints, nouns will be substituted for by vague terms (*thing, anyone*). Naming of high-frequency items will diminish, and patients' responses will become wordy, or take the form of unhelpful circumlocutions. On comprehension tasks, patients will appear to free-associate to a single substantive in the stimulus, and may respond incorrectly to yes–no questions, or questions relying on morphosyntactic processing, rather than on word-order processing. So the patient may report that the tiger dies, when told "The lion was killed by the tiger." On repetition tasks patients will fail even on phrases with high-frequency items.

Pragmatic communication breaks down in late stages of Alzheimer's dementia, and the patient may be unresponsive or, conversely, may speak when no one is around. Responses may be entirely unrelated to linguistic context (Hutchinson and Jenson, 1980). Language may be used illogically; when Irigaray (1973) asked which month came first, April or May, one patient responded "It depends." Asked to repeat something he or she has just said for clarification ("What did you say?"), the patient will not do so. In final stages, palilalia and echolalia may predominate, as may muteness. Not all pragmatic behavior is lost, however, and it would be well worth studying the process of deterioration in order to discover the semiotic hierarchies of pragmatics. Gesture, for example, remains in the face of severe language loss; Schwartz, Marin, and Saffran (1979) report a patient who could name neither fork nor spoon but who differentially gestured their usage.

Although it is worthwhile to produce such a sketch in order to get some picture of the progressive decline of behaviors (see Obler and Albert, 1981, for a more detailed description), it is important to remember that there are marked individual differences among cortically dementing patients. We would maintain that these individual differences in which functions deteriorate, at what rate, and in what order or combination, relate to the "chance" of which areas are most affected at which stage. Thus, Gustafson, Hagberg, and Ingvar (1978) report their cerebral blood flow studies on 7 patients suffering Alzheimer's disease and 3 suffering Pick's disease. They were able to divide the patients into two different groups based on whether the reduction in blood flow was predominantly frontal or predominantly posterior. Patients with reduction of frontal blood flow, they reported, evidenced more signs of what they term *expressive aphasia,* such as stereotyped language, loquaciousness, and mutism or echolalia. By contrast, the patients with slowing predominantly in temporoparieto-occipital regions showed signs of receptive language disturbance.

This localizationist approach, however gross, we note, is more informative at present than more sophisticated measures of the general amount of brain atrophy when correlating neurological condition with dementing patients' specific behaviors. Computer tomography (CT) scan measures of cortical atrophy have shown a singular lack of correlation with the behavioral deteriorations of dementia. Individuals with extreme cortical deteriorations, as evident in CT scans, may be functioning perfectly normally, whereas individuals with severe dementias may evidence no cortical atrophy at all (Naeser, Gebhardt, and Levine, 1980). Measures of ventricular atrophy on the CT scans are somewhat more promising, as are computer measures of white matter density (Naeser, Gebhardt, and Levine, 1980), but it is not clear at present how we may associate these measures with specific lesion sites, or with the language areas presumably undermined. At this time, then, we may best rely on postmortem neuropathological reports of atrophy, in the attempt to distinguish the language behaviors of predominantly subcortical as opposed to cortical dementias, and on EEG and blood flow measures to distinguish predominantly anterior from posterior deterioration among patients with cortical dementias.

Just as we called upon inattention to explain some language disturbances of the subcortical dementias, so memory dysfunction may interact with the language disturbance of the cortical dementias. Memory is crucial for keeping to the point of conversation, for recalling lexical items for naming or list-generating tasks, and for repeating back sentences of any length. As with attention (which is also disturbed in some cortical dementias, of course) we would suggest that the next step in researching language disturbance in the cortical dementias is to determine which aspects of long-term and short-term memory underlie deficited language behaviors. Thus it may be possible to sort out the extent to which language-specific abilities are impaired, and the extent to which language disturbances may be secondary to memory disturbances.

LANGUAGE FORM AND SEMANTICS

Lexical Access

As mentioned earlier it is generally held that in the cortical dementias at least, lexicon deteriorates (as does pragmatics), while syntax and phonology do not. That lexical *access* deteriorates is unquestionable. Naming disturbances, especially semantic paraphasias, are reported in virtually all cases, from early in the disease, both in cortical and subcortical dementias. Whether or not lexicon "itself" deteriorates is a little more controversial, but Schwartz, Marin, and Saffran (1979), detail a convincing case.

WLP, a 62-year-old housewife, mother, and seamstress, was followed by Schwartz and her colleagues over a period of 30 months in the course of her dementing disease. Upon initial presentation, her severe anomia was manifested by an inability to name all but 2 out of 70 pictures of common objects presented to her. Given multiple-choice responses, however, she could correctly name more than 40 items, and 90% of her errors were due to selection of semantically related items (rather than phonologically–orthographically related, or unrelated items). Over the course of her decline, however, semantically related items accounted for fewer and fewer of her increasing errors, and the unrelated item was chosen more frequently. Having observed that WLP seemed to overextend the label *dog* to include cats, the authors developed a series of tests in which the patient was first given pictures of different dogs, cats, and birds to put beside the labels *dog, cat,* and *bird,* and then asked to match the same pictures with a representative exemplar of each of the 3 categories. With the first task, the authors demonstrated that the patient indeed extended the label *dog* to include many cats but virtually no birds. On the second task it became clear that rule-governed conceptual difficulties underlay the mislabeling of the first task. Thus, when asked to match a picture of either a dog or a cat with an exemplar picture of a dog, WLP invariably chose the cat. Again, birds were never confused with these two animals. Schwartz and her colleagues then neatly demonstrated what others merely claimed (Irigaray, 1973; de Ajuriaguerra and Tissot, 1975): in dementia it would seem that there is a breakdown of specific semantic features distinguishing two categories, but the semantic extension is not limitless or random. Rather, it would appear that recognition of superordinate categories is still maintained, at least in the early stages of the disease. Note that these superordinate categories are probably more strictly cognitive than linguistic, since in English we have no linguistic label that pairs only dogs and cats and excludes birds.

Syntax *versus* Lexicon?

Schwartz and her colleagues then went on to argue less convincingly, that whereas lexicon was impaired for WLP, syntax was not. They did this by means of an impressive demonstration that WLP could flawlessly perform a number of tasks that agrammatic patients cannot. Thus, after a lot of coaching, the authors were able to get WLP to perform such transformations on sentence stimuli as negative, question, and plural formation. An example follows:

Examiner:
   I have some bananas (negative required).
WLP:
   I don't have any bananas.

Irigaray (1973) as a rule was unable to get her demented patients to perform such tasks. She rightly asserted that her patients were unable to perform these tasks because those abilities tested by the task were metalinguistic abilities.

By contrast, de Ajuriaguerra and Tissot (1975) maintained that the inability of their dementing patient or patients (they never define their subject population) to perform on a similar task suggests breakdown of syntax per se in dementia. (One must read between the lines to figure out precisely what it was that they asked the subject[s] to do, but it would seem to be a task of explaining the order of events in a sentence spoken by the examiner such as, "We shall invite the gardener who did our garden [sic].") De Ajuriaguerra and Tissot themselves note that in conversation, the "errors" they report on their "syntactic" task do not occur in conversation. However, they did not recognize that it is the ability to perform metalinguistic tasks that is impaired, an ability that may not be universal among humans in any case—if we consider the difficulty some students have in introductory linguistics classes. The fact that WLP can perform the metalinguistic tasks of Schwartz and her colleagues, however surprising, can not conclusively demonstrate that WLP's syntax is unimpaired. Rather, it demonstrates that on certain tasks WLP evidenced unimpaired syntax.

In fact, the second syntax test that de Ajuriaguerra and Tissot employed suggests that had Schwartz and her colleagues used a different syntactic task, one that relied more heavily on the semantic nature of syntax, WLP might well have displayed deficits. In this second test de Ajuriaguerra and Tissot asked their patient(s) for repetition of anomalous sentences, and as most testers of dementing patients find, saw that certain sentences were unconsciously corrected. In the studies of Haiganoosh Whitaker (1976), two sorts of such stimuli were used—sentences with morphosyntactic errors, and sentences with semantic errors. In such a test, dementing patients spontaneously correct the sentences with morphosyntactic errors (e.g., *The boys is here.* becomes *The boy is here.*) but not the sentences with lexicosemantic errors (e.g., *The boy comes yesterday.* remains *The boy comes yesterday.*) De Ajuriaguerra and Tissot added a set of sentences whose errors rely on misuse of logicosyntactic functors (e.g., *It is raining for I cannot go*), and note that these sentences go uncorrected by their dementing patient(s). They conclude that syntax is disturbed in dementia. We would argue rather that *semantics* is disturbed in dementing patients; whether it be semantics expressed through lexicon or through syntax. Thus, lexical items that reflect semantic specificity may be difficult to access and may be misused, whereas lexical items that are vague, such as *thing* or *something,* may be overused. By the same token, formal syntactic elements that convey little semantic sense may be produced automatically, whereas syntactic items that bear semantic burden may be abused. That this latter holds true we have observed in the discourse of a fairly severely dementing patient with Alzheimer's disease, LA, who produced the following sentences:

1. *A museum doesn't get a museum **before** a museum gives you something.*
2. *I'd like a cup of coffee **but** I'd like one.*
3. *The mother's **neither** caring, the son's **neither** caring **or** can't help it.*
4. *She doesn't believe it **because** she's believing to see something that doesn't know I'm to do it.*

We have argued further (Obler 1979) that this behavior of the dementing patient, whereby he or she uses logical functors with disregard for their semantic constraints, may well distinguish Alzheimer's patients from Wernicke's aphasics. Wernicke's aphasics also produce fluent empty speech but their discourse contains a limited set of functors, heavily biased toward the emptier or less marked or less logical ones: the conjunction *and* is used especially frequently (Gleason, Goodglass, Green, Obler, Hyde, and Weintraub, 1980). One concludes that it is not the linguistic categories per se that count here, *lexicon* versus *syntax,* so much as the linguistic interface with cognition, namely *semantics,* which is disturbed in dementing patients.

Further evidence of the automaticity of *formal* syntax comes from Haiganoosh Whitaker's (1976) detailed case of transcortical aphasia in the dementing woman HCEM, whose speech production was effectively restricted to echolalia and reading aloud. As we mentioned previously this patient spontaneously corrected errors of morphosyntax in her echoing, in much the same way that she corrected the experimenter's deliberate mispronunciations of words (e.g., *pedcil→pencil*). Indeed, the patient's echoes converted the experimenter's British accent to her own American one.

When Whitaker goes on to argue that creative language is lost, however, I would take exception. In dementing patients, as in some aphasics, a certain creative morphologizing takes place, whereby recognizable roots are inflected innovatively, a behavior that Séglas (1892) termed *embolalia.* Thus one patient of ours spoke of *metropolity,* and of *predatorians,* and de Ajuriaguerra and Tissot report a patient calling a paintbrush (*pinceau*) an *epoussoir,* from the verb *epousseter,* "to dust." A second patient of ours, Dr. T., spoke of the boy in the Cookie Theft Picture as "stealing them *two-handedly.*" In searching for the word for the item of cowboy clothing "chaps," he said that they were "to guard your pants against thorny *outgrowings.*" AL assured us he was not *baloneying* us. So it is not CREATIVE language use per se that is diminished in dementia;

rather, it is MEANINGFUL PROPOSITIONAL language use. Indeed, Haiganoosh Whitaker's patient would make acceptable morphological transformations in her reading, thus producing:

> *clarify* for *clarification*
> *informative* for *information*
> *reminder* for *remind*
> *difference* for *different* and
> *honest* for *dishonest.*

Paragrammatism

Having discussed how formal syntax is relatively well preserved in the
dementias whereas meaning-bearing syntax may be abused, we can turn
to paragrammatism. Although not a terribly common event in the discourse of the dementing patient, it is of interest when it occurs. Several
kinds of paragrammatism can be distinguished in our dementing patients.
We may treat the less interesting ones first.

One patient in particular, at a fairly early state of the disease, would
regularly omit items, easily recoverable from the context of his discussion. When LB intended to refer to the name of the guy next door, he said
"The guy next name," clearly a case of omission for "the guy next door's
name." Later on, he said of his treatment by the VA, "I told the people
they tested me freed." Such responses might well be attributed to inattention, or to lack of self-monitoring or of self-correcting, rather than to any
breakdown in language per se, parallel to the writing errors noted in the
discussion of patients with subcortical aphasias.

A second set of paragrammatic errors occurs around inflectional endings and functors. Consider the following sentences produced by dementing patients:

(1) D.T. (written description of what happens at Christmas at his house):
    The grandchildren comes around. . . .

(2) EB (in attempting to name a pretzel): *. . . and they have little white
    peaks that comes up like that.*

(3) LB (describing a war experience): *Whenever they felt like it they
    would flip uh out machines guns and set them out and tell 'ems, us to
    lined up just to scare the devil but we didn't know when they were
    going to pull them.*

(4) Ak (re: his home's previous owners): *That was their affairs.*

(5) WS (writing to dictation *The President lives in Washington.*): *President live in Washington.*

What is particularly interesting here is that most of the instances we have collected (numbers 1–4) reveal an excessive use of inflection, rather than omission of it (number 5). This is in apparent contrast to Irigaray's (1973) finding that on the task of making a brief sentence out of two or three stimulus items, the dementing patients tended to use the verbs in infinitival form as they were given, rather than inflecting them. What might be argued here is that in spontaneous speech, but less so under test conditions, the morphologizing component is working uninhibitedly (not entirely independent of meaning, of course, since in the examples we found, an appropriate inflection is regularly appended, usually redundantly, to the wrong word). This suggests a stage in the process of sentence production at which the morphological endings or functors to be applied are available, but they are unassigned as yet to the relevant lexical items. Some disturbance in the course of production results in their assignment to the wrong item, or their double assignment. Perhaps the best example of double marking is LB's statement that he used to drive a truck to Boston "every day, or not every day; I guess it was twice times." A good example of a syntactic item being applied to the wrong phrase is WS's appreciation of patient's treatment at our hospital: "There's nothing they don't get they want."

Those speech errors of normal subjects (Fromkin 1973) that suggest the independence of grammatical morphemes and morphophonemic rules, it should be noted, tend to occur somewhat differently. The bulk of the examples in this section of Fromkin's appendix (pp. 258, 259) are of the type wherein *I cooked a roast* becomes *I roasted a cook;* inflections occur in the right place, but the wrong lexical items have been inserted. Only one of Fromkin's 30 examples is of the type common in our dementing patient: *ministers in the church* became *ministers in the churches.* It is tempting to suggest that the cases of paragrammatism resulting from redundant marking are due to perseveration. That is to say, the patient intended to mark for plural, or for past tense, and succeeded in this marking in more places than the rules of English call for. Perseverative behavior is seen in the dementing patient in severe form in *palilalia*, the multiple spontaneous repetition of words and phrases. However, we are speaking of a more subtle form, relating to how a deep structure gets realized in surface form. Perseveration may account for two other sentences, but here the perseveration is at a higher level, inducing parallel syntactic forms inappropriately:

(1) LA ("What work did you do"?): *When I first laughed along I lived, uh worked in Boston. When I **second**, I worked the whole crew together.*

(2) LA (describing the Cookie Theft Picture): *and there's a **kid tipping** over and **mother spilling** over.*

A related process can be studied in another set of mildly paragrammatic sentences that illustrate the third type of paragrammatism, which involves the breaking of lexical cooccurrence constraints:

(1) ET (speaking about how the pitch of his speech has gotten higher under medication): *My **speech is, elevated tone,** oh yes, yes. **Talk is higher.***

(2) ET: *The wolf **rook a liking to** Red Riding Hood's **basket.***

(3) TG: *That will **expedite** an otherwise absent **clarity.***

(4) TG (re: animal listing task): *You wouldn't have that combination, that **mixture of animals.***

(5) TG (naming a spiral): *That looks like a **centrifugal drawing.***

In (1) and (2) it is clear enough that animateness is being violated; in (5), it is some like concept (more or less dynamic) that is at issue. In (3) and (4), however, we would need to get quite ad hoc to determine what specific "semantic features" were being violated. This is not to say that the notion of semantic features is not valid and that one must substitute for it the notion of fields of semantic association. Rather, the errors demonstrate that for certain semantic fields, distinctive features appear to operate, whereas for others they do not. Semantic features were operating in Irigaray's data from a test of relational terms (e.g., *The daughter of my brother is my* _____?). She found answers were regularly within the appropriate semantic fields, but off by one or two distinctive features (e.g., *The daughter of my brother is my nephew.*) This lines up nicely with the Schwartz, Marin, and Saffran (1979) analysis discussed earlier which demonstrated that lexicosemantic extension in dementia was rule-governed, and that more specific features for a word dissolve before higher-level features.

A fourth sort of paragrammatism arises through idiom breaking, which is surprising because one expects a certain integrity of automatic, over-learned forms in the dementias. Consider the following:

(1) LB ("Is your wife older than you?"): *Oh yeah, **one year low,** uh, I am one year older than my wife.*

(2) LA ("Are you married?"): *No, I don't* **have any married thing** *at all. . . .*

(3) LB (*Omitting* **out of 'em,** *see fuller context p. 276,* [3]) *. . . just to scare the devil, but we don't know. . . .*

(4) LB: *Then I* **made small pittance.**

(5) ET: *I still don't know whether Little Red Riding Hood* **comes into mischief.**

This idiom breaking would indicate that idioms are *not* represented in the brain exclusively as units, but rather must be composed to some extent out of their constituent lexical items. In instances (1–5), the patient commits a verbal paraphasia for one of the lexical items, and thus throws the entire idiom off. A similar explanation may be made for broken verb–particle units:

(1) ET: *I think that he* **took a linking towards** *pastry.*

(2) LB: *I* **landed as** *as* [sic] *a salesman for one year.*

(3) ML (re: logs in the water): *Big boats can* **fend** *them* **out.**

(4) AK (referring to his scar): *It still* **hurts around** *a bit, you know.*

(5) LA: *I've got a sister; she* **died away** *from a certain. . . .*

In each instance, the patient's intonation indicates a verb–particle construct is intended. In all but (2), the appropriate semantic realm is selected; some anticipatory production error would easily explain the peculiar construct of (2). What is of interest then is that we must now think of verb–particle *units,* like idioms, as being generated as if they were composed of two independent lexical items. (Actually in the fourth and fifth examples, no particle is necessary; it might even be seen as an instance of creative morphology.)

Yet in a further set of items, verb–particle compounds have been selected where both verb and particle are ill-chosen:

(1) ET: *The boy is* **hanging** *the girl* **up** *in his mischief.* (Intends **involving.**)

(2) LB (re: carrying out tests): *I wonder how many people can* **come through with** *these things.*

(3) LB (re: his story structure): *I'm sorry, I'm* **going behind** *myself.* (Intends **getting ahead of.**)

In some sense, one could argue that lexical cooccurrence constraints are broken in all instances of paragrammatism. To what extent they indicate lexical or semantic deterioration per se, and to what extent they result from inattention, disinhibition, and paraphasia, is still unclear. However, it seems patently clear that the earlier belief that the lexicon itself is impaired and that syntax remains entirely normal is not quite true for the dementias. Rather, semantics is impaired, as it is realized in both lexicon and syntax. In addition, the paragrammatic errors suggest that there are production difficulties in realizing surface forms from deep intent.

CONCLUSION: LANGUAGE
AND NEUROPSYCHOLOGICAL BEHAVIOR

Jakobson's regression hypothesis was gently laid in its grave with respect to aphasics by the contributions to Caramazza and Zurif (1978). Yet we have argued elsewhere (Obler, 1981) that perhaps the aphasic was Jakobson's straw man and that in fact it is the dementing patient whose language behavior "should" parallel in reverse the language development of the child. That is to say, to the extent that attention, memory, and other aspects of cognition must interact with language, one might expect that their dissolution in dementia could be the reverse of their development in the child. Given the complexity of human development, however, it is probably the case that neither is true, that neither language nor other cognitive and basic psychological behaviors deteriorate in precisely the reverse fashion to their development. For example, Schwartz, Marin, and Saffram (1979) pointed out that WLP did not confuse semantic opposites like *in–under, old–young,* whereas children continue to confuse these items long after they stop making semantic overextensions. Likewise, de Ajuriaguerra and Tissot asserted that the language of childhood and that of senile dementia are not very similar. However, dementing patients do get treated like children, and spoken to like children, so the analysis holds as far as the lay person is concerned. And of course, the symmetry of the regression hypothesis is elegant, and thus attractive.

In any event, we are convinced that memory, attention, perseveration, disinhibition, and perception of the real world do interact in language performance in ways that the dementias are only beginning to tell us. Certainly a number of the behaviors of dementia discussed previously such as paragrammatism and verbal paraphasias, also obtain in the aphasias, in ways that have been less obviously attributable to deficits of underlying "psychological" mechanisms. Are we to conclude then that there is cognitive decline in the aphasias? That debate has raged for a century. Certainly, if there is necessary intellectual decline in the aphasias, it is minimal compared to the decline of the dementias. Must we then presume that similar language behaviors result from different lesions—from language area lesions in the aphasias, but from other lesions in the dementias? Although theoretically possible, that is probably not the case, if only because the diffuse lesions of the cortical dementias do touch language areas, particularly the superior temporal gyrus (Brody, 1976). In sum, in order to work out the roles that inattention, memory loss, perseveration, and deficits of semantic cognition and disinhibition play in the language disturbances of the dementias, what we must now do is to look for relations and correlations between the development of each of these deficits, and the appropriate language behaviors.

ACKNOWLEDGMENTS

Years of work with Martin Albert on the language of healthy and dementing elders have informed this chapter. Critical readings by him and by Margaret Fearey, Lise Menn, Sid Segalowitz, and Steve Woodward have sharpened each succeeding draft.

REFERENCES

Albert, M. (1978). Subcortical dementia. (1978). In R. Katzman, R. Terry, and K. Bick (Eds.), *Alzheimer's disease, senile dementia, and related disorders*. New York: Raven.

Brody, H. (1976). An examination of cerebral cortex and brain-stem aging. In R. Terry and S. Gershon (Eds.), *Neurobiology of Aging*. New York: Raven.

Caramazza, A., and Zurif, E. (Eds.). (1978). *Language acquisition and language breakdown: Parallels and divergencies*. Baltimore: Johns Hopkins Univ. Press.

de Ajuriaguerra, J., and Tissot, R. (1975). Some aspects of language in various forms of senile dementia (comparisons with language in childhood). In E. and E. Lenneberg (Eds.), *Foundations of language development: A multidisciplinary approach*, Vol. 1. New York: Academic Press. Pp. 323–339.

Fromkin, V. (Ed.). (1973). *Speech errors as linguistic evidence*. The Hague: Mouton.

Gleason, J., Goodglass, H., Green, E., Obler, L., Hyde, M., and Weintraub, S. (1980). Narrative strategies in aphasic and normal subjects. *Journal of Speech and Hearing Research, 23*, 370–382.

Goodglass, H., and Kaplan, E. (1972). *The assessment of aphasia and related disorders*. Philadelphia: Lea and Febiger.

Gustafson, L., Hagberg, B., and Ingvar, D. (1978). Speech disturbances in presenile dementia related to local cerebral blood flow abnormalities in the dominant hemisphere. *Brain and language, 5*, 103–118.

Hutchinson, J. and M. Jensen, (1980). A Pragmatic evaluation of discourse communication in a normal and senile elderly in a nursing home, in L. K. Obler and M. L. Albert, Eds., *Language and communication in the elderly; Clinical, therapeutic, and experimental issues*, Lexington, Mass.: D. C. Heath and Co.

Irigaray, L. (1973). *Le langage des déments*. The Hague: Mouton.

Jakobson, R. (1941/1968). *Child language, aphasia, and phonological universals*. Trans. A. Keiler. The Hague: Mouton. (German 1941.)

Naeser, M. A., Gebhardt, C., and Levine, H. (1980). Decreased computerized tomography numbers in patients with presenile dementia: Detection in patients with otherwise normal scans. *Archives of Neurology, 37,* 401–409.

Obler, L. (1979). Psycholinguistic aspects of language in dementia. Paper presented at Academy of Aphasia, San Diego.

Obler, L. (1981). Review of *Le langage des déments. Brain and Language, 12,* 375–386.

Obler, L., and Albert, M. (1981). Language in the elderly aphasic and in the dementing patient. In M. Taylor Sarno (Ed.), *Acquired aphasia.* New York: Academic Press.

Obler, L., Cummings, J., and Albert, M. L. (1979). Subcortical dementia: Speech and language functions. Paper presented at American Geriatrics Society Meeting, Washington, D.C., 1979.

Schwartz, M., Marin, O., and Saffran, E. (1979). Dissociations of language functions in dementia: A case study. *Brain and Language, 7,* 277–306.

Séglas, J. (1892). *Des troubles du langage chez les aliénés.* Paris: Rueff et Cie.

Whitaker, Haiganoosh (1976). A case of the isolation of the language function. In H. and H. Whitaker (Eds.), *Studies in neuroloinguistics,* Vol. 2. New York: Academic Press. Pp. 1–58.

PART IV

Is Brain Lateralization
a Single Construct?

INTRODUCTION

Kimura's original linking of the results of a dichotic listening task to those of sodium amobarbital testing produced great hope of a single behavioral measure that would reflect hemispheric specialization. Unfortunately, other behavioral tasks that reliably produce expected asymmetries, such as the visual half-field technique, do not necessarily reflect identical asymmetric processes. For example, Bryden (1965) and Fennell, Bowers, and Satz (1977) find a relatively low correlation between the asymmetry scores from individuals on the two tasks. This divergence may be due to modality differences or to differing processing requirements. The same diversity exists with regard to body laterality; although the intercorrelations among hand, eye, and ear preference are all positive and statistically significant with a large sample, they are depressingly low (R ≤ .24) (Porac and Coren, 1981). A single construct underlying bodily lateral preferences may be inadequate.

There are two factors contributing to the nonconvergence of the various measures of laterality. The first involves the problem of low test–retest reliability and is relevant more for the discussion of hemispheric specialization than for bodily lateral preferences. The test–retest reliability of the dichotic listening and visual half-field tasks is relatively high for a group of right-handers, but low for individuals (Blumstein, Goodglass, and Tartter, 1975; Fennell, Bowers, and Satz, 1977). Similarly, asymmetric interference on a finger-tapping task also provides a good group effect while simultaneously not being reliable for individuals (Hiscock and Kinsbourne, 1980). This means, of course, that attempts to find consistency across modalities (and therefore, measures) will be hampered by the unreliability problem since a single measure does not correlate well with itself.

The second factor speaks to both cerebral specialization and bodily lateral preferences. Although researchers in this field have spoken of cerebral dominance for this function or that, the complexity of the nervous system probably requires a more sophisticated approach. There are, of course, many cognitive and motor skills that we, as researchers, investigate. It is probably the case that the pattern of cerebral specialization for all these functions is rather complex. Recently, this has been shown empirically. Porac, Coren, Steiger, and Duncan (1980) found three orthogonal factors in bodily lateral preferences, one for the limbs and one each for the eye and the ear. It is harder to demonstrate similar complexity with cerebral specialization for speech or cognitive functions in a single study. Some suggestions, however, have been made. Molfese (1978) reported a right-hemisphere EEG differentiation of voice onset time and a left-hemisphere differentiation of place of articulation. On a different set of dimensions, Sidtis (1980) suggests that the harmonic structure of sound predisposes an ear advantage in dichotic listening. These are clearly very different, yet both auditory, dimensions on which to build a theory of cerebral specialization.

This multidimensionality has also been illustrated well by Samar (1981). He measured visual half-field (VHF) asymmetries in dealing with verbal (a lexical decision task) and nonverbal (same–different line slant judgment) stimuli and simultaneously recorded event related potentials (ERPs) to those visual stimuli. He extracted 13 orthogonal components from a principal components analysis of the ERPs averaged over each condition, accounting for almost 90% of the variance in the ERPs. These components represent orthogonal sources of variation in the brain waves. Of interest to the issue at hand are the attempts to predict VHF asymmetry scores from the asymmetry of the factor scores for each component (i.e., the difference in factor score values between the hemispheres). Four factors predicted the VHF linguistic asymmetry well, with R values of 0.90 for males and 0.98 for females. Thus the VHF asymmetry did reflect asymmetric brain functions that were also picked up by the EEG. However, the ERPs also reflected other orthogonal factors as well, missed by the VHF task. For example, the factor difference scores also predicted well the verbal reasoning skills of the subjects (as measured on a standardized test), with R's of 0.93 for males and 0.97 for females. The VHF asymmetry did not predict verbal reasoning significantly. The rest of ERP factors that predicted verbal skill only overlapped partially with the set that predicted the VHF asymmetries. Thus Samar has demonstrated that there are meaningful asymmetries in the EEG that are independent of one another.

Just as it is incorrect to expect that a single score will accurately reflect all lateral preferences, we should expect there to be multiple factors in hemispheric organization for language and for other cognitive functions. Now, this is taken for granted when, as in Part I of this book, we see a conscious division of language into various functions. However, perhaps we should go further yet, and allow that some people may have divergent enough early linguistic experiences that their neuropsychological organization for language may be different. For example, the cognitive requirements of sign languages for the deaf such as American Sign Language (ASL), may promote a different organization (see Ross's chapter, this volume); the bilingual's second language may not rest on the same neurophysiological basis as the first, if there is a different developmental history associated with it (see Vaid's chapter, this volume). Individual differences in hemispheric organization may be due to many factors, not all of which are amenable to measurement. An important next step in neurolinguistics is to recognize these factors, even if they cannot be measured, devise control tasks for them, and make use of appropriate statistical models to take them into account (see Segalowitz's and Bryden's chapter, this volume).

REFERENCES

Blumstein, S., Goodglass, H., and Tartter, V. (1975). The reliability of ear advantage in dichotic listening. *Brain and Language, 2,* 226–236.

Bryden, M. P. (1965). Tachistoscopic recognition, handedness, and cerebral dominance. *Neurospychologia, 3,* 1–8.

Fennell, E. B., Bowers, D., and Satz, P. (1977). Within-modal and cross-modal reliabilities of two laterality tests. *Brain and Language, 4,* 63–69.

Hiscock, M. & Kinsbourne, M. (1980). Asymmetry of verbal-manual time sharing in children: a follow-up study. *Neuropsychologia, 18,* 151–162.

Molfese, D. L. (1978). Neuroelectrical correlates of categorical speech perception in adults. *Brain and Language, 5,* 25–35.

Porac, C., and Coren, S. (1981). *Lateral preferences and human behavior*. New York: Springer-Verlag.

Porac, C., Coren, S., Steiger, J., and Duncan, P. (1980). Human laterality: A multidimensional approach. *Canadian Journal of Psychology, 34,* 91–96.

Samar, V. (1981). Multiple determination of visual half-field asymmetries and Differential Aptitude Test scores by independent components of cerebral specialization. Presented at International Neuropsychological Society, February 1981, Atlanta, Georgia.

Sidtis, J. J. (1980). On the nature of the cortical function underlying right hemisphere auditory perception. *Neuropsychologia, 18,* 321–330.

CHAPTER 13

Cerebral Specialization in Deaf Individuals

Phyllis Ross

INTRODUCTION

In hearing individuals, there is a division of labor between the two hemispheres. Traditionally, the left hemisphere has been known as the dominant, speech hemisphere, and the right hemisphere as the minor hemisphere specializing in nonverbal and visuospatial processing. Support for this view of hemispheric specialization comes from studies concerned with the effects of unilateral brain lesions (e.g., Milner, 1968, 1971), split-brain surgical procedures (e.g., Gazzaniga, 1970), intracarotid injection of sodium amytal (Wada and Rasmussen, 1960), and from studies in which material is presented differentially to the cerebral hemispheres of normal individuals (e.g., Geffen, Bradshaw, and Wallace, 1971; Kimura, 1967; Kimura, 1973a).

While it is clear that the two cerebral hemispheres have specialized functions, relatively little is known about the ontogeny of cerebral specialization. Both neuroanatomical and behavioral evidence point to the existence of certain cerebral asymmetries at or shortly after birth. Studies

287

indicate that the left planum temporale, which would appear by virtue of its location to include Wernicke's area, known to be crucial to normal language functioning, is larger than the right planum both in adults (Geschwind and Levitsky, 1968), and in newborns (Wada, Clarke, and Hamm, 1975; Witelson and Pallie, 1973). These asymmetries may provide a basis for left-hemisphere language dominance. Electrocortical hemispheric asymmetry has also been demonstrated in newborns (Molfese, Freeman, and Palermo, 1975). These investigators showed that left-hemisphere auditory evoked responses were larger in amplitude than right-hemisphere responses to speech stimuli for infants one week to ten months old, as well as for older adults and children, whereas nonspeech stimuli produced larger amplitude responses in the right hemisphere for all groups. Using a dichotic listening paradigm, Entus (1977) found a right-ear (left-hemisphere) advantage for processing verbal material and a left-ear (right-hemisphere) advantage for processing nonverbal material in infants a few weeks of age. In addition, asymmetrical head turning tendencies have been found in newborns (Turkewitz, Gordon, and Birch, 1965; Turkewitz, Moreau, and Birch, 1966), and such asymmetries may be related to functional asymmetries seen in older children and adults.

Although such studies suggest the presence of functional cerebral asymmetries at or shortly after birth, this does not imply that there are no changes in cerebral specialization during development. On the contrary, there is evidence suggesting that environmental factors, particularly those related to language experience, can influence the development of cerebral specialization. For example, aphasia has been noted to occur equally often in association with left- or right-hemisphere lesions in illiterate adults (Cameron, Currier, and Haerer, 1971: Wechsler, 1976). In addition, it has been found that extreme deprivation of exposure to and use of language until puberty, as in the case of Genie (Fromkin, Krashen, Curtiss, Rigler, and Rigler, 1974), is associated with lack of a left-hemisphere superiority on verbal dichotic listening tests. Finally, it has been shown that trilingual children may show a greater left-hemisphere superiority for verbal material than monolingual children (Starck, Genese, Lambert, and Seitz, 1977). All these results suggest that the degree of experience with language may be positively related to the degree of left-hemisphere language dominance.

CEREBRAL SPECIALIZATION
IN DEAF INDIVIDUALS: OVERVIEW

In this chapter, evidence relating to the nature and development of cerebral specialization in deaf individuals will be critically examined.

Such an examination serves a dual function. First, information concerning the pattern of cerebral specialization in deaf individuals may lead to a better understanding of cognitive and perceptual functioning in this population. Second, and more broadly, such information may contribute to an understanding of the factors which are important in the normal development of cerebral specialization.

Speculations on the Development of Cerebral Specialization in Deaf Individuals

Since the linguistic experience of deaf individuals is radically different from hearing individuals, cerebral specialization may not develop to the same extent or in the same manner in these two groups. Deaf individuals have little or no experience with speech perception, and more generally, auditory perception, and their verbal language development is both delayed and incomplete to varying degrees. This abnormal pattern of verbal language acquisition may be an impediment to the normal development of cerebral specialization. Furthermore, it is possible that the rate and pattern of cerebral specialization may be different for congenitally deaf and noncongenitally deaf individuals, since the latter group has had some degree of exposure to verbal language.

Although deaf individuals do not have a normal verbal linguistic environment, many deaf people do learn sign language, and its acquisition may be an important factor in the development of cerebral specialization. Thus, if learning some linguistic system, whether verbal or sign, is important in the development of lateralization, deaf individuals who have learned sign language may develop a normal pattern of cerebral specialization, in spite of their impoverished verbal language. Furthermore, if sign language is learned at an early age, in a naturalistic manner, it may foster the development of cerebral specialization to a greater extent than if it is learned at a later age.

Differences in Hemispheric Advantage for Deaf and Hearing Subjects: Reflection of Differences in Brain Organization or in Mode of Processing?

Many of the studies which will be reviewed in this chapter show differences in hemispheric advantage for processing linguistic and nonlinguistic material in deaf and hearing individuals. Such results may be due to differences between these groups in the specialized functions of the two cerebral hemispheres. However, it is also possible that these differences in hemispheric advantage may reflect differences in mode of information processing, rather than differences in brain organization. This latter possibility is consistent with the widely accepted view of hemispheric specialization which emphasizes differences between the hemispheres in processing strategy as opposed to the type of material processed.

One description of hemispheric specialization in terms of differences in processing which has been proposed is that the left hemisphere is specialized for analytic processing, whereas the right hemisphere is specialized for holistic processing. Support for this view of hemispheric specialization comes from studies showing that the processing of certain types of stimuli can result in either a right-hemisphere or a left-hemisphere advantage, depending upon the type of processing strategy used, that is, holistic or analytic. This has been demonstrated for the processing of music (Bever and Chiarello, 1974; Kellar and Bever, 1980; Johnson, 1977) and faces (Patterson and Bradshaw, 1975; Ross and Turkewitz, 1981). In addition to a distinction between analytic and holistic processing, a left-hemisphere superiority for sequential processing and a right-hemisphere superiority for simultaneous processing have been suggested (e.g., Cohen, 1973).

To the extent that the direction of hemispheric advantage on a task reflects the utilization of a particular processing strategy, differences between deaf and hearing individuals in hemispheric advantage may be due to differences in the mode of processing, rather than to differences in underlying brain organization. In the subsequent discussion and analysis of the literature, an attempt will be made to decide between these alternative explanations.

SPECULATIONS ON THE CEREBRAL LATERALIZATION OF SIGN LANGUAGE

Related to the question of whether the acquisition of sign language affects the development of cerebral specialization in deaf individuals is the question of the cerebral lateralization of sign language itself. Is sign language lateralized in the left hemisphere, as is the case for verbal language? Sign language and verbal language may both be processed in the left hemisphere since both are symbolic, linguistic systems. Furthermore, it has been suggested (Stokoe, 1972) that the linguistic structure of sign language, specifically American Sign Language (ASL), parallels that of spoken language. According to Stokoe, there are a finite number of visibly distinctive features which compose the signs of ASL. These formational features form three classes, designating the location of the sign, the type of hand configuration used, and the type of movement involved. Stokoe calls these formational features cheremes, and considers them analogous to the phonemes in spoken language. To the extent that the production and comprehension of sign language relies on an analysis of signs into distinctive features, the processing of sign language may be carried out more efficiently in the left hemisphere, which appears to be specialized for analytic processing.

Another reason why sign language may be processed more efficiently in the left hemisphere is that this hemisphere appears to be specialized for the control and analyses of skilled motor sequences (Kimura, 1973b,c; Kimura and Archibald, 1974; Kimura, Battison, and Lubert, 1976; Lomas and Kimura, 1976; Mateer and Kimura, 1977).

Kimura and Archibald (1974), for example, found that left hemisphere-damaged patients were impaired relative to right hemisphere-damaged patients in copying hand movement sequences, but not in copying static hand positions. The types of sequential hand movements used by Kimura and Archibald are similar to the types of hand configurations and movements used in sign language.

Although the production and comprehension of sign language may depend to a large extent upon analytic and sequential processing, and thus upon the specialized capacities of the left hemisphere, as a VISUAL mode of communication, it also relies heavily upon spatial and simultaneous processing. Therefore, sign language may be primarily controlled by the right hemisphere. On the most basic level, the formational elements composing signs have a simultaneous presence. A sign is characterized by the co-occurrence of a particular shape of the hand(s) articulated in a particular position with a particular movement. This differs from the spoken word, where the phonemes are sequentially ordered. In morphology too, sign language shows greater reliance than speech on simultaneous devices. Klima and Bellugi (1975) give the example of the concept *bluish,* which as an English word is composed of a sequence of *blue* plus *ish,* and in ASL is expressed through a modification in hand configuration and movement of the sign for *blue.* Signers also occasionally manifest creative and simultaneous combination of concepts, using one hand for one concept and the other for another concept. Klima and Bellugi (1976) give an example of a man who, when asked how he felt about leaving a city he loved for a good job in another city, signed simultaneously *excited* with the right hand and *depressed* with the left.

While English grammar relies heavily on the temporal ordering of words, the grammar of sign language seems to rely on spatial and simultaneous processes. Lane, Boyes-Braem, and Bellugi (1976) state that in signing the proposition *the dog bites the cat,* the signer

> may point to two locations in space which he labels by the signs for DOG and CAT. He can then perform the sign for BITE so that the direction of movement goes from the dog-designated location to the cat-designated location. To sign "The cat bites the dog", only the direction of the sign for BITE needs to be changed [p. 265].

This example shows how grammatical concepts, such as agent and object, which are commonly indicated in English by the order of words in a sentence, are indicated in ASL by spatial devices.

In sign language, it is possible to express an entire phrase in a single sign. Lane, Boyes-Braem, and Bellugi (1976, pp. 265–266) explain how the sentence *Three of them came over to me* can be expressed in ASL.

> The handshape parameter indicates that three people are involved; "of them" is shown by beginning the sign in the location in space which had been previously designated as a basis for people; the verb "came over", is communicated by the orientation of the hand and the pattern of movement; and "to me" is conveyed by the direction and endpoint of movement. The entire sign for this proposition is, then, a hand with three fingers extended which begins in a previously established location diagonal from the signer and moves in an arc toward the signer.

Since sign language relies heavily on spatial and simultaneous parameters at the cherological, morphological, and syntactic levels, sign language may be more efficiently processed in the right hemisphere.

Thus far, arguments for a right-hemisphere or left-hemisphere advantage in processing signs have been based on an analysis of the properties of sign language itself. It is also possible, however, that the hemispheric advantage for processing sign language may differ for individuals with different neurological and linguistic histories. Thus, if it is the case that the pattern of hemispheric specialization in deaf individuals depends to some extent on the age of onset of deafness, individuals who become deaf at different ages may differ with regard to the hemispheric lateralization of sign language. In addition, the age sign language is learned, as well as an individual's fluency in signing, may have an influence on its cerebral lateralization.

CLINICAL EVIDENCE CONCERNING THE CEREBRAL LATERALIZATION OF SIGN LANGUAGE

Several clinical case reports have described disturbances in signing following left-hemisphere dysfunction in deaf persons (Critchley, 1938; Grasset, 1896; Douglass and Richardson, 1959; Leischner, 1943; Sarno,

Swisher, and Sarno, 1969; Tureen, Smolik, and Tritt, 1951; Underwood and Paulson, 1981). However, a careful examination of the information provided in these reports prevents any definitive conclusions from being drawn concerning sign language lateralization.

First, in some of these reports, fingerspelling appears to be more generally impaired than does signing. Thus, in the case reported by Grasset (1896) and described by Critchley (1938), fingerspelling but not signing was impaired. In the patients described by Douglass and Richardson (1959) and by Sarno, Swisher, and Sarno (1969), the production of fingerspelling was more severely impaired than the production of signs, although the comprehension of fingerspelling and signing were equally impaired in the former patient. In the case reported by Tureen, Smolik, and Tritt (1951), while an impairment in fingerspelling is described, no mention is made of the patient's signing ability. Since fingerspelling is based on the conventional spoken alphabet rather than on sign language, the evidence from these reports does not provide compelling evidence directly concerning the question of the lateralization of sign language.

Second, other variables, which may affect the manner in which sign language is processed and thus its hemispheric lateralization as well, make interpretation of some of these reports difficult; one patient was postlingually deaf (Critchley, 1938), two were not proficient in their signing ability (Critchley, 1938; Douglass and Richardson, 1959), and one was left-handed (Underwood & Paulson, 1981).

The most convincing case report for clear implication of the left hemisphere in the processing of sign language is one cited by Kimura (1976). According to that report, a congenitally deaf man of deaf parents, who learned sign language long before using vocal speech, lost his signing ability following left-hemisphere damage (Leischner, 1943). However, in the absence of reports involving the effects of right-hemisphere damage in deaf individuals on signing ability, no definitive conclusions can be drawn from these reports concerning sign language lateralization. Thus, it is possible that damage to the right hemisphere may also impair signing ability, which would suggest greater bilateral control for sign language than for verbal language.

EXPERIMENTAL EVIDENCE CONCERNING CEREBRAL SPECIALIZATION IN DEAF INDIVIDUALS: TACHISTOSCOPIC STUDIES

There is a small but growing body of experimental literature concerning the cerebral lateralization of linguistic (both English and sign) and nonlin-

guistic material in deaf individuals. The majority of these studies have
used the technique of briefly presenting a visual stimulus to the left or
right of a central fixation point for an exposure duration sufficiently brief
to prevent eye movement. Using this technique, stimuli presented to the
right visual field (RVF) are projected to the left hemisphere and those presented to the left visual field (LVF) are projected to the right hemisphere.
In some of the studies, a stimulus is presented to either the RVF or LVF
on each trial (unilateral presentation) and in other studies, two different
stimuli are presented simultaneously on each trial, one to each visual field
(bilateral presentation).

In the following sections, experimental evidence concerning the cerebral lateralization of verbal, nonlinguistic, and sign language stimuli will
be examined.

Tachistoscopic Presentation of Verbal Material

LETTER RECOGNITION In a study by Phippard (1977), a letter was
tachistoscopically presented to the RVF or LVF, and the subject's task
was to find the letter on a response card. Hearing subjects were more accurate on RVF presentations. Prelingually deaf subjects attending an oral
school, concentrating on the training of speech, speechreading, and the
use of residual hearing, were more accurate on LVF presentations. Prelingually deaf subjects attending a total-communication school, whose
communication mode involved a combination of sign language, fingerspelling, speech, and speechreading, showed no difference between the
two visual fields in accuracy of recognition. The absence of a visual field
advantage for the total-communication group on the letter task but not for
the oral group may suggest that the emphasis on the production and comprehension of speech, such as that which occurs in an oral school, is necessary for the development of hemispheric specialization for processing
verbal material. On the other hand, it is possible that the different results
were due to differences between the groups on other variables which may
have an influence on the development of cerebral lateralization, such as
age of onset of deafness or fluency in signing.

The results of the deaf subjects attending the oral school may suggest
that the right hemisphere of some deaf individuals is specialized for linguistic processing, rather than the left hemisphere, as is the case for hearing individuals. However, an alternative interpretation of the results is
that the use of different types of strategies in this task was responsible
for the opposing directions of hemispheric advantage found for the hearing and the oral deaf subjects. The multiple-choice recognition paradigm
may have induced a spatial strategy for coding the letters in the deaf sub-

jects, which would predominantly involve holistic processing and presumably result in a right-hemisphere advantage. However, hearing subjects may have based their decisions on the letter names, which would presumably result in a left-hemisphere advantage.

It has been previously shown (Conrad, 1972) that deaf subjects may code letters on a spatial basis. Conrad found that in recalling letters after they had been presented visually, deaf subjects made more errors on the basis of visual similarities between letters than on the basis of phonetic similarities, whereas the opposite was found for hearing subjects. Thus, deaf subjects may not treat letters as linguistic material in situations that do not require a verbal analysis. In some tasks, even hearing subjects may show better performance in the right hemisphere in letter recognition tasks when responses can be made on the basis of the visual similarity between letters, rather than on the basis of the letter name (Cohen, 1972; Davis and Schmit, 1973; Geffen, Bradshaw, and Nettelton, 1972; Wilkins & Stewart, 1974).

Ross, Pergament, and Anisfeld (1979) had subjects judge whether two letters presented simultaneously were the same or different. The two letters were tachistoscopically presented to the same visual field. The letters in a pair were always physically different, consisting of one uppercase and one lowercase letter (e.g., *Aa*). Hearing subjects showed faster reaction times on RVF presentations, but there was no difference in reaction time between the visual fields for the congenitally deaf subjects tested.

As was the case for the results of the task by Phippard (1977), these results may be interpreted in two ways. On the one hand, the results may reflect differences in the strategy used by deaf and hearing subjects for processing letters. In hearing individuals, the naming response to letters is so automatic that it was almost certainly elicited in the letter task used, and mediated the decision concerning sameness or difference of the two letters presented. The name, being a speech entity, would require left-hemisphere processing. The deaf, however, cannot be assumed to have responded with speech names to the letters in this task. It is possible that they performed the task on the basis of direct associations between individual uppercase and lowercase letters. That is, they had specifically learned that "A" and "a" belong together, and responded "same" when they could retrieve a specific connection and "different" when they could not. Such a strategy would not clearly require the predominant use of a particular cerebral hemisphere, and thus processing in this task for the deaf may have been under bilateral control. Alternatively, different deaf subjects may have employed different strategies, with some processing verbally and others spatially.

Finally, the results of a letter recognition task by McKeever, Hoemann, Florian, and VanDeventer (1976) were inconclusive, since neither the deaf nor hearing subjects showed a significant visual field asymmetry.

WORD RECOGNITION. McKeever, Hoemann, Florian, and VanDeventer (1976) tested hearing and congenitally deaf adults on a unilateral and a bilateral word recognition task. The subjects' task was to report the word(s) tachistoscopically presented on each trial. Hearing subjects reported the words vocally, and deaf subjects used sign language. Hearing subjects were more accurate on RVF presentations on both of these tasks. The results of the deaf subjects went in the same direction, but were significant only for the unilateral word task, and even here, the difference in favor of the RVF was much smaller than for the hearing subjects.

Manning, Goble, Markman, and LaBreche (1977) tested several groups of deaf subjects, differing with respect to the onset of deafness and degree of hearing loss, as well as a group of hearing individuals, on a bilateral word recognition task. The subjects' task was to report the words tachistoscopically presented on each trial. Hearing subjects reported the words vocally and deaf subjects fingerspelled their responses. Hearing subjects were more accurate on RVF presentations. As a whole, deaf subjects also showed a RVF superiority. However, although all groups of deaf subjects showed greater accuracy on RVF presentations, statistically reliable differences were found only for the group of subjects where the onset of deafness occurred between the ages of 0 to 2 years, and for the hard-of-hearing group with a 65–75 db hearing loss. The RVF superiority was not significant for congenitally deaf subjects, for subjects who became deaf between the ages of 3 and 8 years, and for a hard-of-hearing group with a 50–60 db hearing loss. The results for the different groups of deaf subjects do not form a pattern that can be easily interpreted. However, they clearly do not provide strong support for a relationship between age of onset of deafness or degree of hearing loss and left-hemisphere specialization for verbal processing.

Manning, Goble, Markman, and LaBreche (1977) also tested congenitally deaf and hearing individuals on a second bilateral word recognition task. On this task, subjects had to choose the pictures signified by the two tachistoscopically presented words from an array of pictures on a choice card. Both deaf and hearing subjects were more accurate for words presented in the RVF, although the differences between the visual fields for the deaf subjects was smaller than for the hearing subjects and only approached significance.

In a study by Poizner, Battison, and Lane (1979), congenitally deaf and hearing adults were given a task involving the tachistoscopic presentation of a word to the RVF or LVF, and the subject had to report the word presented on each trial. Hearing subjects vocally reported the words, whereas deaf subjects fingerspelled the words. Both deaf and hearing subjects were more accurate for words presented to the RVF.

The results of these word recognition tasks suggest that deaf individuals do have a left-hemisphere superiority for word recognition, although this superiority may not be as marked or as reliable as that shown by hearing individuals. If the left-hemisphere superiority shown by hearing individuals in the recognition of visually presented words is due, at least in part, to a phonetic analysis of these stimuli, then it is not surprising that deaf individuals show a less marked left-hemisphere advantage than hearing individuals in processing these stimuli. However, even when there is no phonetic recoding of visually presented words, simply processing words as lexical items may produce a left-hemisphere superiority. For example, deciding whether a string of letters is a word or a nonword is predominantly a left-hemisphere function (Lieber, 1976). In addition, the decision concerning whether two words belong to the same semantic category is also carried out more efficiently in the left hemisphere (Gross, 1972). Furthermore, the fact that English words are composed of sequentially ordered letters may contribute to a left-hemisphere advantage in lexical processing. Thus, the left-hemisphere advantage shown by deaf individuals for word recognition may reflect the superiority of the left hemisphere in such individuals for the semantic and sequential processing of lexical items.

NUMBER RECOGNITION Poizner and Lane (1979) tested congenitally deaf and hearing adults on a task involving the recognition of tachistoscopically presented numbers. The subjects' task was to press a key as quickly as possible when presented with target numbers and refrain from responding when other numbers were presented. Both deaf and hearing subjects had faster reaction times on RVF presentations, although this difference was significant only for the hearing subjects.

If the left-hemisphere superiority for the hearing subjects in recognizing numbers is due primarily to an auditory recoding of these stimuli, then it is not surprising that deaf individuals do not show a left-hemisphere advantage in number recognition. As in the processing of visually presented letters, deaf individuals may recognize visually presented numbers primarily on the basis of spatial rather than auditory characteristics.

**COMPARISON OF WORD OR LETTER WITH PREVIOUSLY PRE-
SENTED STIMULI** Scholes and Fischler (1979) tested congenitally deaf
and hearing adults on a task involving a comparison of a tachistoscopically presented letter with a previously presented picture. In this task,
subjects were first shown a picture of an object in the center of the visual
field followed by the tachistoscopic presentation of a probe letter to the
RVF or LVF. For the deaf subjects, either a manually signed or an orthographic letter was presented. For the hearing subjects, only orthographic
letters were presented. The subjects' task was to judge whether the letter
was present in the spelling of the object's name. Hearing subjects showed
faster reaction times on RVF presentations. Deaf subjects showed faster
reaction times on LVF presentations, although the effect just failed to
reach significance. On the basis of their performance on a sentence comprehension task, deaf subjects were divided into linguistically skilled and
linguistically unskilled groups. When these two groups of deaf subjects
were examined separately, skilled subjects showed more of a trend
toward a LVF advantage than unskilled subjects, and a significant LVF
advantage was produced by skilled subjects for manual letters.

Ross, Pergament, and Anisfeld (1979) tested congenitally deaf and hearing adults on a task involving a comparison of a tachistoscopically presented word with a previously presented sign. In this task, subjects were
first shown a videotaped sign in the center of the visual field, followed by
the tachistoscopic presentation of a word to the RVF or LVF, and the
subject had to judge whether the word corresponded to the sign. Hearing
subjects had faster reaction times when the words were presented to the
RVF, whereas deaf subjects had faster reaction times on LVF presentations.

In the picture–letter task employed by Scholes and Fischler (1979),
subjects were asked to make a comparison between a tachistoscopically
presented verbal stimulus (i.e., letter) and the name of a previously presented concept (i.e., picture). Similarly, in the sign–word task employed
by Ross, Pergament, and Anisfeld (1979), subjects were asked to make a
comparison between a tachistoscopically presented verbal stimulus (i.e.,
word) and a previously presented concept (i.e., sign). In both of these
studies, hearing and deaf subjects showed opposite directions of hemispheric advantage: hearing subjects showed a left-hemisphere advantage
and deaf subjects a right-hemisphere advantage (although the right-hemisphere advantage only tended toward significance in the Scholes and
Fischler study).

As was the case for the letter recognition tasks used by Phippard (1977)
and Ross, Pergament, and Anisfeld (1979), the results of these studies
may be interpreted in two ways. In the study by Scholes and Fischler

(1979), in which subjects were required to match a test letter to the letters in the name of a picture, the left-hemisphere advantage shown by hearing subjects most likely reflects the superiority of this hemisphere for phonetic processing. One possible interpretation of the tendency towards a right-hemisphere advantage for deaf subjects on this task is that these individuals make such judgments, that is, matching letters, more efficiently in the right hemisphere, implying that linguistic cerebral dominance does not develop normally in deaf individuals. However, an alternative interpretation is that deaf subjects compared letters on the basis of spatial rather than linguistic characteristics, and that such spatial processing resulted in a right-hemisphere advantage.

Similarly, in the study by Ross, Pergament, and Anisfeld (1979), the results may be interpreted as reflecting a difference between deaf and hearing individuals in cerebral specialization for linguistic processing. The left-hemisphere superiority in hearing subjects was most likely due to a comparison of signs and words on a semantic basis. In addition, hearing subjects may have converted the sign to a word and compared words on a phonetic basis, which would also contribute to left-hemisphere superiority. Deaf subjects, on the other hand, may not have treated the presented words as linguistic stimuli. Instead, they may have converted the signs to words and compared words primarily on the basis of their spatial configurations rather than comparing linguistic items. Such a strategy would presumably involve holistic processing of the material and result in a right-hemisphere rather than a left-hemisphere advantage.

DISCUSSION The results of studies involving the tachistoscopic presentation of verbal material to deaf and hearing subjects indicates that these groups of subjects differ in the direction and/or degree of hemispheric advantage in the processing of such material. Deaf subjects generally show a left-hemisphere advantage in word recognition tasks, although it is relatively weak when compared with hearing subjects. In other tasks involving the processing of letters, words, or numbers, they show either no hemispheric advantage or a right-hemisphere advantage.

The different patterns of hemispheric advantage shown by deaf subjects on word recognition tasks, as opposed to other tasks involving the processing of verbal material, may be attributable to differences in task requirements. In the word recognition tasks described, subjects either had to report the word(s) presented by signing, or had to choose a picture matching the presented words. Such task requirements force subjects to process the words as linguistically meaningful items. Assuming left-hemisphere dominance for lexical processing in deaf individuals, such tasks would result in a left-hemisphere superiority. The other tasks discussed did not require the tachistoscopically presented material to be processed as linguistic stimuli. In these studies, subjects either had to judge whether a stimulus corresponded with a previously presented stimulus or set of stimuli, or had to identify the stimulus using a multiple-choice recognition procedure. In such tasks, subjects can compare stimuli on a spatial rather than linguistic basis, and assuming right-hemisphere dominance for processing spatial information in deaf individuals, such a strategy would result in a right-hemisphere advantage in these tasks. Given the fact that most deaf individuals are not as experienced as hearing individuals in dealing with letters and words, it is not unreasonable to propose that they would rely more heavily than hearing individuals on nonlinguistic, that is, spatial processing of linguistic material where possible.

Since both deaf and hearing individuals show a left-hemisphere advantage on some verbal tasks, an interpretation of the different patterns of hemispheric advantage found for these groups on other tasks (in terms of differences in processing strategy) seems to be more parsimonious than an interpretation in terms of differences in cerebral organization. Although it is possible (as some authors have concluded on the basis of the results of their individual studies McKeever, Hoemann, Florian, and VanDeventer, 1976; Scholes and Fischler, 1979) that deaf individuals do not develop a left-hemisphere specialization for linguistic processing, such a conclusion seems unwarranted on the basis of the entire body of available data.

Tachistoscopic Presentation of Nonlinguistic Stimuli

RECOGNITION OF LINE ORIENTATION Phippard (1977) tested prelingually deaf and hearing adolescents on a task involving the recognition of the spatial orientation of short lines. On each trial, a line in a particular orientation was tachistoscopically presented to the RVF or LVF, and the subjects' task was to choose the line on a response card corresponding to the one just seen. Hearing subjects, as well as deaf subjects attending an oral school, were more accurate on LVF presentations. This suggests a right-hemisphere specialization in these deaf individuals, as well as in hearing individuals for processing this type of visuospatial material. However, deaf subjects attending a total-communication school showed no difference between the two visual fields in accuracy of recognition. The absence of a visual-field asymmetry for the total-communication group on this task, as well as on the letter recognition task previously described, suggests that such individuals may not have developed a cerebral specialization for processing either verbal or visuospatial material.

PICTURE RECOGNITION   Neville (1977) tested congenitally deaf and hearing children on a task involving the recognition of line drawings of common objects. The deaf children attended a school which did not encourage the use of sign language, although some of the children used sign language at home. None of the deaf children tested had the ability to speak. On each trial, a drawing was tachistoscopically presented to the RVF or LVF, and the subjects' task was to choose the picture corresponding to the one he had just seen. Although the objects depicted on the choice card were the same as those tachistoscopically presented, the choice could not be made on the basis of a simple physical match, since the pictures were different (e.g., an open umbrella was to be matched to a closed one). There was no difference between the visual fields in accuracy of recognition for either the deaf or hearing groups. However, visual evoked potentials for the hearing group showed greater activity in the right hemisphere. For the deaf group as a whole, the visual evoked potentials did not reveal any differences between the hemispheres. However, when the deaf subjects were separated into those who used sign language at home and those who used pantomime, differences between these groups became apparent. The evoked potentials did not indicate any differences between the hemispheres for the pantomime group, but for the sign group, the evoked potentials indicated greater left-hemisphere involvement.

The opposing directions of hemispheric advantage found for the hearing subjects and for the deaf subjects using sign language at home can be interpreted as reflecting either differences in cerebral organization between deaf and hearing individuals or differences in processing strategy. Thus, it is possible that, in contrast to hearing individuals, the left hemispheres of deaf individuals who know sign language develop as the specialists in processing visuospatial material. Alternatively, it is possible that on this task, hearing and deaf subjects used different strategies in comparing pictures. Although the tachistoscopically presented picture of an object was always somewhat different from the corresponding picture on the choice card (so that a choice could not be made on the basis of a simple physical match), it is possible that hearing subjects compared pictures on the basis of global physical similarity, which would presumably involve holistic processing and result in a right-hemisphere advantage. However, deaf subjects may have compared the NAMES of the objects represented in the pictures, resulting in a left-hemisphere advantage. Both of these strategies seem appropriate for the task, and it is not obvious why one or the other would be used.

Whatever the reason for the different patterns of hemispheric advantage found for deaf and hearing subjects, the results do suggest the importance of the acquisition of some linguistic system (whether verbal or sign) in the development of lateralization. However, this conclusion must be viewed as tentative, since separation into pantomime and sign language groups was posthoc, and the groups may have differed along dimensions other than the language–no language dimension. For example, Neville (1977) states that most of the signers had deaf parents, while those in the pantomime group had hearing parents. This may have resulted in rather different emotional and intellectual histories for the two groups.

DOT LOCALIZATION   Neville and Bellugi (1978) tested congenitally deaf and hearing subjects on a dot localization task. Deaf subjects were tested in a unilateral and a bilateral version of this task. However, hearing subjects were only tested in the unilateral condition, since they found the bilateral version of the task too difficult. On each trial, a dot was tachistoscopically presented to the RVF and/or LVF, and the subject had to choose the location of the dot(s) just seen from a four-by-five matrix of dots. Hearing subjects showed a LVF advantage for the unilateral dot localization task. In contrast, deaf subjects showed a RVF advantage for unilateral presentations, but no difference between the visual fields for bilateral presentations.

The results of the dot localization task again raise the possibility that the left rather than the right hemisphere of congenitally deaf individuals develops as the specialist for visuospatial functions. However, the results may also be interpreted as reflecting a holistic approach to the task by hearing subjects and an analytic approach by deaf subjects. Thus, hearing subjects may have perceived the tachistoscopically presented dot as an integral part of the entire matrix of dots, and made judgments on the basis of the spatial relationship between the dot and the entire matrix. This type of holistic processing strategy would presumably result in a right-hemisphere advantage. However, deaf subjects may have perceived the dot as an independent component of the matrix, and then judged the spatial position of this component (e.g., "lower right"). This type of analytic and verbal processing strategy would presumably result in a left-hemisphere advantage. The fact that hearing subjects could not be tested on the bilateral version of the task because they found it too difficult suggests that deaf and hearing individuals may, in fact, use different strategies in this task, with one being more efficient than the other.

RECOGNITION OF MEANINGLESS HANDSHAPES   Poizner and Lane (1979) tested congenitally deaf and hearing adults on a task involving the recognition of hand configurations that never occur in ASL. Half of the hand configurations were designated as target items and the other half were nontarget items. On each trial, a hand configuration was tachistoscopically presented to the RVF or LVF, and the subject had to press a key as rapidly as possible to target items and refrain from responding to other stimuli. Both deaf and hearing subjects had faster reaction times on LVF presentations, suggesting a right-hemisphere specialization for processing this type of visuospatial material.

RECOGNITION OF RANDOM SHAPES Two attempts to detect asymmetries in deaf and hearing subjects for recognition of random shapes failed to find a visual field advantage for either group (Poizner and Lane, 1979; Manning, Goble, Markman, and VanDeVenter, 1977). Therefore, these studies are inconclusive with regard to cerebral specialization in deaf individuals.

FACIAL RECOGNITION In a study by Phippard (1977), neither deaf children or age-matched hearing children showed a visual field advantage on a facial recognition task, again providing inconclusive evidence regarding cerebral specialization in the deaf.

DISCUSSION As was the case for studies involving the processing of verbal material, the results of some studies involving the processing of nonlinguistic stimuli suggest that at least certain subgroups of deaf individuals show a right-hemisphere advantage in processing these stimuli, as do hearing individuals. However, the results of other studies show that certain subgroups of deaf individuals show a left-hemisphere advantage in processing nonlinguistic visual material, a pattern opposite to that shown by hearing individuals.

The left-hemisphere advantage shown by deaf individuals on some tasks involving the processing of nonlinguistic visual material may reflect a left-hemisphere specialization for certain types of visuospatial functions. Alternatively, it is possible that opposite patterns of hemispheric advantage for deaf and hearing subjects in processing certain types of visuospatial material reflect differences in processing strategy; hearing subjects may use a holistic mode of processing on certain visuospatial tasks, resulting in a right-hemisphere advantage, whereas deaf subjects may use a more analytic or linguistic mode of processing, resulting in a left-hemisphere advantage.

Why would deaf individuals sometimes use an analytic mode of processing nonlinguistic material while hearing individuals use a holistic mode of processing? There is evidence which suggests that with certain types of nonlinguistic material, the use of a left-hemisphere mode of processing may be associated with better performance than the use of a right-hemisphere mode (Gordon, 1975). In addition, there is evidence that individuals who are relatively experienced with certain types of nonlinguistic material may show a left-hemisphere advantage in processing such material, whereas less experienced individuals show a right-hemisphere advantage (Bever and Chiarello, 1974; Johnson, 1977; and see Goldberg and Costa, 1981). Since it is possible that the greater dependence of deaf individuals on visual processing makes them more proficient than hearing individuals in processing at least certain types of visuospatial material, deaf individuals may use an analytic left-hemisphere mode for processing such material, whereas hearing individuals may use a less advanced holistic right-hemisphere processing mode.

In sum, both deaf and hearing individuals show a right-hemisphere advantage on certain visuospatial tasks, but opposing patterns of hemispheric advantage on other visuospatial tasks. An interpretation of the latter results in terms of a different pattern of cerebral organization in deaf and hearing individuals cannot explain the results of studies in which both groups show a similar pattern of cerebral asymmetry. Therefore, differences in hemispheric advantage between deaf and hearing individuals in visuospatial tasks seems to be best explained in terms of differences in processing strategy rather than differences in cerebral organization.

Tachistoscopic Presentation of Signs and Letters of the Manual Alphabet

DESCRIPTIONS AND RESULTS OF STUDIES McKeever, Hoemann, Florian, and VanDeventer (1977) tested congenitally deaf and hearing adults, both groups fluent in sign language, on a task involving the recognition of outline drawings of signs and letters of the manual alphabet. On each trial, either letters or signs were bilaterally presented, and the subject had to report the stimuli presented. Signs were selected which involved little or no movement, so that they could be recognized in an outline drawing from their hand configuration. Hearing subjects reported the stimuli vocally whereas deaf subjects reported the stimuli by signing. When the results for these two types of stimuli were combined, hearing subjects were significantly more accurate on LVF presentations. The results for the deaf subjects went in the same direction but were not significant. The results of this study suggest that sign language stimuli, which require substantial spatial processing, are processed more efficiently in the right hemisphere than in the left, at least in hearing individuals. The results also suggest that deaf individuals do not have a hemispheric advantage in processing sign language.

Poizner and Lane (1979) tested congenitally deaf and hearing adults on a task involving the recognition of photographs of ASL signs which require no movement and therefore could be recognized from a static photograph. On each trial, a sign was tachistoscopically presented to the RVF or LVF, and the subject had to press a key as rapidly as possible to target items and refrain from responding to other stimuli. Both deaf and hearing subjects had faster reaction times on LVF presentations, suggesting that the sign language stimuli were processed more efficiently in the right hemisphere for both deaf and hearing subjects.

Manning, Goble, Markman, and VanDeventer (1977) also tested deaf subjects on a task involving the recognition of outline drawings of signs. On each trial, signs were presented bilaterally, and the subject had to report by signing the stimuli which had been presented. There was no visual field asymmetry shown for this task, but again the absence of a hearing comparison group for this task makes interpretation of the negative results for the deaf subjects difficult.

In a second experiment, Manning, Goble, Markman, and LaBreche (1977) tested congenitally deaf subjects on a somewhat different task involving the recognition of signs. In contrast to the outline drawings used in the first experiment, the stimuli in this experiment were black and white photographs of an individual, with the hands held in positions representing ASL signs. On each trial, signs were presented bilaterally. The subjects' task was to choose the pictures of objects on a choice card signifying the signs presented. Responses were more accurate for signs presented to the LVF, although the difference between the visual fields just failed to reach significance. The tendency toward better recognition on LVF presentations again suggests better right-hemisphere processing of signs in deaf individuals.

Neville and Bellugi (1978) also tested congenitally deaf subjects on a sign recognition task. The stimuli consisted of line drawings of a person making various signs of ASL. Signs were chosen which were easily recognizable from line drawings. Signs were presented both unilaterally and bilaterally. There was a RVF advantage for the unilateral presentations, but no difference between the visual fields for bilateral presentations. Thus, in contrast to the results of the previous studies, these results suggest superior left-hemisphere processing of signs in deaf individuals.

The opposite directions of hemispheric advantage found by Neville and Bellugi (1978), and by Poizner and Lane (1979), McKeever, Hoemann, Florian, and VanDeventer (1977), and Manning, Goble, Markman, and LaBreche (1977) for the recognition of signs may have been due to differences in the nature of the sign language stimuli used. In the stimuli used by Neville and Bellugi (1978), the hand configuration is not more prominantly displayed than the remaining portion of the body, but rather is imbedded in the entire perceptual stimulus. Therefore, in order to recognize the sign, the subject must first perceptually separate the hands from the remainder of the body (similar to separating foreground from background) and only then can the sign be identified. Thus, the task seems to require an analysis of the stimulus into separate elements, and the left-hemisphere advantage found by Neville and Bellugi may have been due, at least in part, to the substantial analytic processing requirements of the task.

In contrast, the stimuli used by Poizner and Lane (1979) consisted only of handshapes, with no other body parts represented. Similarly, in most of the stimuli used by McKeever, Hoemann, Florian, and VanDeventer (1977) only hand configurations were presented. In order to recognize the signs presented in these studies, a perceptual analysis of the stimulus is not required prior to the identification of the hand configuration. Therefore, the right-hemisphere advantage found in these studies may have been due to the fact that the task predominantly involved the use of configurational or holistic processing.

Although the stimuli presented by Manning, Goble, Markman, and LaBreche (1977) in their second experiment were photographs of a person signing, so that both the hand configuration as well as some portion of the remainder of the body were presented, the handshape sharply contrasted with the remainder of the body, since the signer was photographed wearing dark clothing. Thus, most of the sign stimuli for this study show light hands held against a dark background. Therefore, the processing of these stimuli also does not seem to demand the degree of perceptual analysis required by the stimuli used by Neville and Bellugi (1978), and the right-hemisphere advantage may have been due to the predominant use of holistic processing.

All of the experimental studies examining cerebral lateralization for signs discussed thus far have inferred hemispheric advantage on the basis of the recognition of signs presented statically. This restricted the choice of signs to those that could be identified without movement—only a small fraction of the signs in ASL. Since movement is an integral part of sign language, any definitive evaluation of sign language lateralization should in some way incorporate the movement component. If the left hemisphere is specialized for the control and analysis of skilled motor sequences (as suggested by Kimura, 1973b,c; Kimura and Archibald, 1974; Kimura, Battison, and Lubert, 1976; Lomas and Kimura, 1976; Mateer & Kimura, 1977), a left-hemisphere advantage might be expected for moving signs.

In an innovative study by Poizner, Battison, and Lane (1979), movement was incorporated into tachistoscopically presented signs by sequentially presenting still photographs of a person signing, taken at different points during the execution of the sign, but keeping the entire duration for the presentation of the three photographs below the latency required for eye movement. A second task in this study involved the presentation of static signs, which were identifiable without movement cues. As in the first task, the stimuli consisted of photographs of a person signing, but only one static image was presented. Congenitally deaf adults fluent in ASL, all of whom had deaf parents, were tested. On each trial, a sign was presented to the RVF or LVF and the subject had to report by signing the stimulus presented. Subjects were more accurate in the recognition of static signs presented to the LVF, but there was no difference between the visual fields in the recognition of the moving signs.

The right-hemisphere advantage for statically presented signs may have been due to the predominant use of holistic processing for these stimuli. The fact that there was no hemispheric advantage in the recognition of signs which incorporated movement may be interpreted as a shift towards greater left-hemisphere processing for these stimuli. Thus, the processing of signs in which movement is portrayed may be under greater bilateral control than is the case for static signs.

DISCUSSION It has been demonstrated that the direction of hemispheric advantage for the recognition of complex perceptual material may depend upon somewhat subtle changes in the nature of the stimuli presented, which may affect the type of processing utilized (e.g., Patterson and Bradshaw, 1975). Similarly, the hemispheric advantage shown for the recognition of sign language may differ with various types of stimuli and depend upon the type of processing required for the recognition of the stimuli presented. Thus, individuals may show a right-hemisphere advantage when the signs can be recognized predominantly on the basis of the configuration of the perceptual stimulus, but may show a left-hemisphere advantage when analytic processing is required for recognition of the signs. Therefore, the hemisphere which shows an advantage in processing an individual sign may depend upon particular properties of the sign itself, such as the type of movement involved and the type of visual information presented for recognition.

The finding of a right-hemisphere advantage in some of the studies examining sign language lateralization does not necessarily indicate that the signs were not treated as linguistic elements. Hearing subjects may also show a right-hemisphere advantage to written language stimuli when the spatial processing requirements are high. For example, Bryden and Allard (1976) have found that the hemispheric advantage shown for letter recognition depends on the typeface of the lettering. They found a left-hemisphere advantage for the recognition of printlike letters but a right-hemisphere advantage with certain scriptlike lettering. In addition, a right-hemisphere advantage has been found for the recognition of nonphonetic written Japanese linguistic symbols in individuals who show the typical left-hemisphere advantage for phonetically based script. (Sasanuma, Itoh, Mori, and Kobayashi, 1977; Hatta, 1977). Thus, the right-hemisphere advantage for certain types of sign language stimuli may indicate that processing the spatial properties of such signs predominates over their linguistic properties in determining cerebral asymmetry.

It must be kept in mind that the studies that have been discussed involve an examination of the lateralization of individual signs. However, processing of sign language in a more naturalistic context necessarily involves an analysis of the movement which is incorporated in a sequence of signs, as well as an analysis of sign language grammar. Such processing requirements may produce a more consistent hemispheric advantage than has been demonstrated for the processing of individual signs. Techniques other than tachistoscopic presentation would be required for an examination of the processing of sign language in a more naturalistic context. For example, procedures have been developed for measuring EEG alpha asymmetry during the performance of cognitive tasks. Using this procedure, greater left-hemisphere involvement has been observed during tasks requiring verbal analysis, such as writing, speaking, reading, and making syntactic and semantic discriminations. Greater right-hemisphere involvement has been observed during spatial tasks which seem to require holistic processing (e.g., Ornstein and Galin, 1976; Ehrlichman and Wiener, 1979; Ornstein, Herron, Johnstone, and Swencionis, 1979; Ornstein, Herron, Johnstone, and Swenciones, 1980). Since the EEG can be used to measure ongoing activity without the restrictive time constraints imposed by the tachistoscopic procedure, this measure might be used to determine hemispheric involvement during the production or comprehension of sign language discourse.

GENERAL DISCUSSION

Many of the studies discussed show that tasks which produce significant hemispheric asymmetries in hearing individuals also produce significant hemispheric asymmetries in deaf individuals (Phippard, 1977; Ross, Pergament, and Anisfeld, 1979; McKeever, Hoemann, Florian, and Van-Deventer 1976; Poizner, Battison, and Lane, 1979; Neville and Bellugi, 1978; Poizner and Lane, 1979); such results suggest that normal auditory experience is not necessary for the development of cerebral specialization.

Although the majority of evidence suggests that deaf individuals as a group develop cerebral asymmetries, the results of some studies do suggest that at least certain subgroups of the deaf population may not develop cerebral specialization. Thus, Phippard (1977) did not find hemispheric asymmetries for the processing of either verbal or nonverbal material for deaf children attending a total-communication school, whereas deaf children attending an oral school showed hemispheric advantages for both types of material. Further evidence for different patterns of cerebral organization in different subgroups of deaf individuals was found in the study by Neville (1977). In a group of deaf children lacking oral language skills, Neville found that those children who signed at home showed a hemispheric advantage in processing nonverbal material, whereas those who did not use sign language did not show a hemispheric asymmetry. Finally, Scholes and Fischler (1979) found that deaf individuals who were grammatically skilled showed a hemispheric asymmetry in processing linguistic material, while those who were grammatically unskilled did not show a hemispheric asymmetry for such processing.

These results suggest that deaf individuals cannot be regarded as a homogeneous group. The types of educational and linguistic environments to which a deaf individual has been exposed may be important variables in the development of cerebral specialization. In addition, competency in verbal language and in sign language may be important in determining the type of strategy used in a particular task for processing linguistic or nonlinguistic material. The findings in some studies of a smaller degree of cerebral asymmetry in deaf individuals than in hearing individuals may be due to the fact that particular subgroups of deaf individuals may exhibit different patterns of lateralization which are obscured when the group is analyzed as a whole. Thus, in studies examining cerebral asymmetries in deaf individuals, special attention should be paid to variables which might have an effect on the development of cerebral specialization and/or on the types of processing strategies utilized, such as age of onset of deafness, verbal fluency, and fluency in sign language.

In many of the studies reviewed, deaf and hearing individuals showed opposite directions of hemispheric advantage. Since in some studies involving both linguistic and nonlinguistic material, deaf and hearing subjects showed the same direction of hemispheric advantage, the results of studies in which opposite directions of hemispheric advantage were found seem to be best explained in terms of differences in processing strategy rather than in terms of differences in cerebral organization. Since deaf individuals have little or no auditory experience, and are more dependent than hearing individuals on visual experience, it is not unlikely that deaf and hearing individuals may use different modes of processing linguistic as well as nonlinguistic material when the circumstances permit such flexibility.

ACKNOWLEDGMENTS

I wish to thank Gerald Turkewitz for his comments on an earlier draft of this chapter.

REFERENCES

Bever, T. G., and Chiarello, R. (1974). Cerebral dominance in musicians and nonmusicians. *Science, 185,* 537–539.
Bryden, M. P., and Allard, F. (1976). Visual hemifield differences depend on typeface. *Brain and Language, 3,* 191–200.
Cameron, R. F., Currier, R. D., and Haerer, A. F. (1971). Aphasia and literacy. *British Journal of Disorders of Communication, 6,* 161–163.
Cohen, B. (1972). Hemispheric differences in a letter classification task. *Perception and Psychophysics, 11,* 139–142.
Cohen, G. (1973). Hemispheric differences in serial vs. parallel processing. *Journal of Experimental Psychology, 97,* 349–356.
Conrad, R. (1972). Short-term memory in the deaf: A test for speech coding. *British Journal of Psychology, 63,* 173–180.
Critchley, M. (1938). "Aphasia" in a partial deaf-mute. *Brain, 61,* 163–169.
Davis, R., and Schmit, V. (1973). Visual and verbal coding in the interhemispheric transfer of information. *Acta Psychologia, 37,* 229–240.
Douglass, E., and Richardson, J. C. (1959). Aphasia in a congenitally deaf mute. *Brain, 82,* 68–80.
Ehrlichman, H., and Wiener, M. S. (1979). Consistency of task-related EEG asymmetries. *Psychophysiology, 16,* 247–252.
Entus, A. K. (1977). Hemispheric asymmetry in processing of dichotically presented speech and nonspeech stimuli by infants. In S. J. Segalowitz and F. A. Gruber (Eds.), *Language development and neurological theory.* New York: Academic Press.
Fromkin, V. A., Krashen, S., Curtiss, S., Rigler, D., and Rigler, M. (1974). The development of language in Genie: A case of language acquisition beyond the "Critical Period". *Brain and Language, 1,* 81–107.
Gazzaniga, M. S. (1970). *The bisected brain.* New York: Appleton.
Geffen, G., Bradshaw, J. L., and Nettleton, N. C. (1972). Hemispheric asymmetry: Verbal and spatial coding of visual stimuli. *Journal of Experimental Psychology, 95,* 25–31.
Geffen, G., Bradshaw, J. L., and Wallace, G. (1971). Interhemispheric effects on reaction time to verbal and nonverbal visual stimuli. *Journal of Experimental Psychology, 87,* 415–422.
Geschwind, N., and Levitsky, W. (1968). Human brain: Left–right asymmetries in temporal speech region. *Science, 161,* 186–187.

Goldberg, E., and Costa, L. D. (1981). Hemisphere differences in the acquisition and use of descriptive systems. *Brain and Language, 14,* 144–173.

Gordon, H. W. (1975). Hemispheric asymmetry and musical performance. *Science, 189,* 68–69.

Grasset, J. (1896). Aphasie de la main droite chez un sourd-muet. *Le Progres Medical, 4,* 281.

Gross, M. (1972). Hemispheric specialization for processing of visually presented verbal and spatial stimuli. *Perception and Psychophysics, 12,* 357–363.

Hatta, T. (1977). Recognition of Japanese Kanji in the left and right visual fields. *Neuropsychologia, 15,* 685–688.

Johnson, P. R. (1977). Dichotically stimulated ear differences in musicians and non-musicians. *Cortex, 13,* 385–389.

Kellar, L. A., and Bever, T. G. (1980). Hemispheric asymmetries in the perception of musical intervals as a function of musical experience. *Brain and Language, 3,* 24–38.

Kimura, D. (1967). Functional asymmetry of the brain in dichotic listening. *Cortex, 111,* 163–178.

Kimura, D. (1973a). The asymmetry of the human brain. *Scientific American, 228,* 70–78.

Kimura, D. (1973b). Manual activity during speaking I. Right-handers. *Neuropsychologia, 11,* 45–50.

Kimura, D. (1973c). Manual activity during speaking II. Left-handers. *Neuropsychologia, 11,* 51–55.

Kimura, D. (1976). The neural basis of language qua gesture. In H. Whitaker and H. A. Whitaker (Eds.), *Studies in neurolinguistics,* Vol. 2. New York: Academic Press.

Kimura, D., and Archibald, Y. (1974). Motor functions of the left hemisphere. *Brain, 97,* 337–350.

Kimura, D., Battison, R., and Lubert, B. (1976). Impairment of non-linguistic hand movements in a deaf aphasic. *Brain and Language, 3,* 566–571.

Klima, E. S., and Bellugi, U. (1975). Perception and production in a visually based language. *Annals of the New York Academy of Sciences, 263,* 244–250.

Klima, E. S., and Bellugi, U. (1976). Poetry and song in a language without sound. *Cognition, 4,* 45–97.

Lane, H., Boyes-Braem P., and Bellugi, U. (1976). Preliminaries to a distinctive feature analysis of handshapes in American Sign Language. *Cognitive Psychology, 8,* 263–289.

Leischner, A. (1943). Die "Aphasie" der Taubstummen. *Archiv für Psychiatrie und Nervenkrankheiten, 115,* 469–548.

Lieber, L. (1976). Lexical decisions in the right and left cerebral hemispheres. *Brain and Language, 3,* 443–450.

Lomas, J., and Kimura, D. (1976). Interhemispheric interaction between speaking and sequential manual activity. *Neuropsychologia, 14,* 23–33.

McKeever, W. F., Hoemann, H. W., Florian, V. A., and VanDeventer, A. D. (1976). Evidence of minimal cerebral asymmetries for the processing of English words and American Sign Language stimuli in the congenitally deaf. *Neuropsychologia, 14,* 413–423.

Manning, A. A., Goble, W., Markman, R., and LaBreche, T. (1977). Lateral cerebral differences in the deaf in response to linguistic and nonlinguistic stimuli. *Brain and Language, 4,* 309–321.

Mateer, C., and Kimura, D. (1977). Impairment of nonverbal oral movements in aphasia. *Brain and Language, 4,* 262–276.

Milner, B. (1968). Visual recognition and recall after right temporal-lobe excision in man. *Neuropsychologia, 6,* 191–209.

Milner, B. (1971). Interhemispheric differences in the localization of psychological processes in man. *British Medical Bulletin, 27*, 272–277.

Molfese, D. L., Freeman, R. B., and Palermo, D. S. (1975). The ontogeny of brain lateralization for speech and nonspeech stimuli. *Brain and Language, 2*, 356–368.

Neville, H. J. (1977). Electroencephalographic testing of cerebral specialization in normal and congenitally deaf children: A preliminary report. In S. J. Segalowitz and F. A. Gruber (Eds.), *Language development and neurological theory*. New York: Academic Press.

Neville, H. J., and Bellugi, U. (1978). Patterns of cerebral specialization in congenitally deaf adults. In P. Siple (Ed.), *Understanding language through sign language research*. New York: Academic Press.

Ornstein, R., and Galin, D. (1976). Physiological studies of consciousness. In R. Ornstein and P. Lee (Eds.), *Symposium on consciousness*. New York: Viking Press.

Ornstein, R., Herron, J., Johnstone, J., and Swencionis, C. (1979). Differential right hemisphere involvement in two reading tasks. *Psychophysiology, 16*, 398–401.

Ornstein, R., Herron, J., Johnstone, J., and Swencionis, C. (1980). Differential right hemisphere engagement in visuospatial tasks. *Neuropsychologia, 18*, 49–64.

Patterson, K., and Bradshaw, J. L. (1975). Differential hemispheric mediation of nonverbal visual stimuli. *Journal of Experimental Psychology: Human Perception and Performance, 1*, 246–252.

Phippard, D. (1977). Hemifield differences in visual perception in deaf and hearing subjects. *Neuropsychologia, 15*, 555–561.

Poizner, H., Battison, R., and Lane, H. (1979). Cerebral asymmetry for American Sign Language: The effects of moving stimuli. *Brain and Language, 7*, 351–362.

Poizner, H., and Lane, H. (1979). Cerebral asymmetry in the perception of American Sign Language. *Brain and Language, 7*, 210–226.

Ross, P., Pergament, L., and Anisfeld, M. (1979). Cerebral lateralization of deaf and hearing subjects for linguistic comparison judgments. *Brain and Language, 8*, 69–80.

Ross, P., and Turkewitz, G. (1981). Individual differences in cerebral asymmetries for facial recognition. *Cortex, 17*, 199–214.

Sarno, J. E., Swisher, L. P., and Sarno, M. T. (1969). Aphasia in a congenitally deaf man. *Cortex, 5*, 398–414.

Sasanuma, S., Itoh, M., Mori, K., and Kobayashi, Y. (1977). Tachistoscopic recognition of Kana and Kanji words. *Neuropsychologia, 15*, 547–553.

Scholes, R. J., and Fischler, I. (1979). Hemispheric function and linguistic skill in the deaf. *Brain and Language, 7*, 336–350.

Stokoe, W. C., Jr. (1972). *Semiotics and human sign languages*. The Hague: Mouton.

Starck, R., Genese, F., Lambert, W. E., and Seitz, M. (1977). Multiple language experience and the development of cerebral dominance. In S. J. Segalowitz and F. A. Gruber (Eds.), *Language development and neurological theory*. New York: Academic Press.

Tureen, L. I., Smolik, E. A., and Tritt, J. H. (1951). Aphasia in a deaf-mute. *Neurology, 1*, 237–244.

Turkewitz, G., Gordon, E., and Birch, H. G. (1965). Head-turning in the human neonate: Spontaneous patterns. *Journal of Genetic Psychology, 107*, 143–158.

Turkewitz, G., Moreau, T., and Birch, H. G. (1966). Head position and receptor organization in the human neonate. *Journal of Experimental Child Psychology, 4*, 169–177.

Underwood, J. K., and Paulson, C. J., (1981). Aphasia and congenital deafness: A case study. *Brain and Language, 12*, 285–291.

Wada, J. A., Clarke, R., and Hamm, A. (1975). Cerebral hemispheric asymmetry in humans: Cortical speech zones in 100 adult and 100 infant brains. *Archives of Neurology, 32*, 239–246.

Wada, J., and Rasmussen, T. (1960). Intracarotid injection of sodium amytal for the lateralization of cerebral speech dominance. *Journal of Neurosurgery, 17,* 266–282.

Wechsler, A. F. (1976). Crossed aphasia in an illiterate dextral. *Brain and Language, 3,* 164–172.

Wilkins, A., and Stewart, A. (1974). The time course of lateral asymmetries in visual perception of letters. *Journal of Experimental Psychology, 102,* 905–908.

Witelson, S. F., and Pallie, W. (1973). Left hemisphere specialization for language in the newborn: Neuroanatomical evidence of asymmetry. *Brain, 96,* 641–646.

CHAPTER 14

Bilingualism and Brain Lateralization

Jyotsna Vaid

INTRODUCTION

Research on the brain organization of language functions has, until recently, ignored the question of the neuropsychological repercussions of bilingualism. However, insights derived from a long-standing literature on polyglot aphasia (summarized by Albert and Obler, 1978; Paradis, 1977) have generated a growing experimental literature on cerebral lateralization of language in normal bilinguals (see Galloway, 1982, and Vaid and Genesee, 1980, for reviews). Taken together, the available clinical and experimental studies suggest that competence in more than one language may influence brain functioning so that it differs from that characterizing speakers of a single language.

Unfortunately, several experimental studies and a majority of the clinical reports are methodologically less than adequate. Since methodological issues have already been addressed elsewhere (Obler, Zatorre, Galloway, and Vaid, 1982), they will not be considered in detail in the present review which will, instead, focus on theoretical issues bearing on the problem of

315

cerebral hemispheric correlates of language processing in bilinguals. Following a general summary of clinical and experimental findings, specific studies will be discussed with a view to addressing whether and why factors intrinsic to bilingualism might influence the extent to which the left and right cerebral hemispheres partake differentially in language processing.

CLINICAL EVIDENCE

The notion that the right hemisphere may participate in the relearning of language(s) or certain language functions following left-hemisphere damage has been suggested in several case studies of polyglot aphasia (Anastosopolous, 1959; Bychowski, 1919; Minkowski, 1963). It has also been proposed (on the basis of clinical studies) that the right hemisphere in bilinguals may, even premorbidly, share language functions with the left hemisphere to a greater extent than is the case in unilinguals (Gloning and Gloning, 1965; Ovcharova, Raichev, and Geliva, 1968; Vildomec, 1963).

Support for the latter possibility obtains from an extensive survey of the early polyglot aphasia literature, which revealed that the incidence of crossed aphasia (i.e., language deficits following right-sided lesions) in right-handers was 14% among the polyglots, as compared to an estimated 2% among unilinguals; the corresponding values among left-handers were 71% and 29%, respectively (Galloway and Krashen, 1980). However, a recent survey of a large population of Chinese–English aphasics did not find a higher incidence of crossed aphasia among the bilinguals as compared to unilinguals (April and Han, 1980a). Thus, Galloway and Krashen's findings, as Galloway (1981) has herself conceded, may reflect a tendency in the early clinical literature to report only the unusual and thereby the more interesting cases.

The suggestion of greater right-hemisphere mediation of language in bilinguals than in unilinguals, or in one of the bilingual's languages versus the other, has also failed to draw support from sodium amytal studies of bilingual and multilingual epileptics (Ojemann and Whitaker, 1978; Rapport, Tan, and Whitaker, 1980). However, electrical stimulation of different sites on the exposed cortex of the language dominant hemisphere conducted on the two patients described in Ojemann and Whitaker (1978) and on three of the five patients discussed in Rapport, Tan, and Whitaker (1980) suggested the existence of differential intrahemispheric organization of the patients' languages. Specifically, the effects of electrical stimulation on object-naming ability in each language were of three kinds: areas were found where stimulation (1) always produced naming errors in both of the languages, (2) usually produced naming errors in one of the languages and occasionally produced errors in the other, and (3) only produced naming errors in one language or the other. According to Ojemann and Whitaker (1978), these findings suggest a "partially overlapping" form of organization of the bilingual's languages and provide an anatomical explanation for the variation in language loss and recovery patterns observed in the polyglot aphasia literature.

In summary, evidence from the clinical literature supports the possibility of differential intrahemispheric organization of the bilingual's languages but is equivocal with respect to the role of the right hemisphere in language mediation prior to brain damage. Because of the lack of systematic and reliable information available from early case reports and the limited range of linguistic skills so far examined in empirical studies, further research is needed to clarify and extend the available clinical observations.

EXPERIMENTAL EVIDENCE

Research on cerebral lateralization of language in normal bilinguals presents an initially confusing picture. Some studies report differences, either in the direction of greater right-hemisphere involvement (e.g., Hardyck, 1980; Sussman, Franklin, and Simon, 1980) or greater left-hemisphere involvement (e.g., Carroll, Exp. 2, 1980; Kotik, 1975), while others (e.g., Galloway and Scarcella, 1979; Piazza and Zatorre, 1981; Soares, 1982b) report no differences in the lateralization pattern for one or both of the languages spoken by bilinguals relative to that characterizing unilingual or bilingual comparison groups. This diversity of results may in part be attributed to differences across studies in the criteria used to screen subjects on second-language proficiency and to an unsystematic inclusion of appropriate control groups, whether unilingual or bilingual (see Obler, Zatorre, Galloway and Vaid, 1982). It may, additionally, reflect differences across studies in the particular hypotheses being investigated (see Vaid and Genesee, 1980). In order to interpret the available laterality findings with bilinguals, it is necessary to refer to a standard conceptual framework of the nature of lateralization differences.

Cerebral Lateralization of Function

Evidence from the experimental literature on unilinguals supports the notion that the two hemispheres are specialized not so much for particular types of stimuli (e.g., verbal versus nonverbal) as for the types of processing required of any stimulus to satisfy particular task demands (Cohen, 1977; Sergent and Bindra, 1981). Various processing-style dichotomies have been hypothesized, including analytic versus holistic, and serial versus parallel processing (see Bradshaw and Nettleton, 1981, and Moscovitch, 1979, for a review of relevant experiments). Since the different dichotomies proposed have generally not been defined so as to be mutually exclusive, it is difficult at present to distinguish among them.

However the differences between the two hemispheres are characterized, an implication of the information-processing approach for language processing is that both hemispheres contribute to it; they differ only in the nature and degree of their contribution. Thus, for example, if serial processing of linguistic input is required by a given task, one would expect greater left-hemisphere (LH) participation than when the same input is to be processed holistically (Geffen, Bradshaw, and Nettleton, 1972; Van-Lancker and Fromkin, 1973).

Apart from the influence of task-related processing demands, hemisphere asymmetries may also be affected by individual differences in processing strategy and/or in the resources brought to bear on a given task (Bryden, 1978). To the extent that bilingualism influences the strategies and resources used to process linguistic input (Ben-Zeev, 1977; Duncan and DeAvila, 1979; Kessler and Quinn, 1980; Lambert, 1981), it may be considered a source of individual differences in hemispheric processing of language as well.

For the purpose of this review, bilingualism may be defined as functional knowledge and regular use of two languages. Two types of factors inherent to bilingualism will be examined with regard to their neuropsychological implications. These concern characteristics of the bilingual's languages and characteristics of the contexts in which the languages are learned. Discussion of each factor will proceed with a rationale for expecting it to influence the pattern of cerebral lateralization and a review of relevant studies. When all of the studies have been discussed, their implications for a model of hemispheric processing of language in bilinguals will be summarized.

LANGUAGE-SPECIFIC FACTORS

These factors refer to properties unique to particular languages that might give rise to differences in patterns of interhemispheric and/or intrahemispheric activity. Language-specific factors that have been investigated in the clinical and experimental bilingual literature include (1) language-related modes of thinking, (2) the salience of vowels, (3) the linguistic significance of tones, (4) directional scanning effects related to reading habits, and (5) the sound–symbol correspondence of different writing systems. Since the first four of these factors have previously been discussed in some detail (Vaid and Genesee, 1980) they will only briefly be presented below.

Modes of Thinking

It has been hypothesized that languages such as Navajo and Hopi, whose structure is presumed to give rise to an "appositional" mode of thinking, rely more on the right hemisphere (RH) than languages, such as English, that induce a "propositional" mode of thinking (Rogers, Ten-Houten, Kaplan, and Gardiner, 1977). Evidence bearing on this hypothesis is equivocal; while three studies report differential RH participation among speakers of Navajo (Hynd and Scott, 1980; Scott, Hynd, Hunt, and Weed, 1979) and Hopi (Rogers, TenHouten, Kaplan, and Gardiner, 1977), two others do not (Carroll, Exp. 1, 1980; Hynd, Teeter, and Stewart, 1980). Moreover, the results of the former set of studies lend themselves to other interpretations besides the mode of thinking one, since such factors as second-language proficiency and age of second language acquisition were not controlled.

Vowels

It is reasonable to expect that vowels will be perceived more analytically in languages in which they often form meaningful words as compared to languages in which consonants are more salient. Languages in the former category might, thus, be expected to rely more on the LH in processing vowels as compared to those of the latter kind. This hypothesis is supported for Japanese, Korean, and Samoan, languages with several words consisting of monosyllabic vowels, irrespective of whether speakers of these languages are also fluent in less vowel-dependent languages such as English or French (Shimizu, 1975; Tsunoda, 1971, 1978).

Tones

When tonal changes carry changes in meaning, one might expect tones, like words, to be processed more efficiently in the LH. Evidence in support of this hypothesis has been presented for Thai–English (VanLancker and Fromkin, 1973, 1978), Chinese–English (Naeser and Chan, 1980; but see Benson, Smith, and Arreaga, 1974) and Vietnamese–French bilinguals (Hécaen, Mazaro, Rannier, Goldblum, and Merienne, 1971).

Direction of Script

The issue of whether laterality effects in the visual modality are influenced by left-to-right versus right-to-left reading habits has been investigated in a number of tachistoscopic studies comparing Yiddish (Mishkin and Forgays, 1952; Orbach, 1953) or Hebrew (Barton, Goodglass, and Shai, 1965; Gaziel, Obler, and Albert, 1978; Orbach, 1967) with English and, recently, Urdu with Hindi (Vaid, 1981a), and Persian with Spanish (Sahibzada, 1982). Although some studies indicate the presence of a scanning effect (i.e., a left-visual-field (LVF) preference under unilateral presentation for languages read from right-to-left), others demonstrate an overriding cerebral laterality effect (i.e., a right visual field (RVF) superiority), especially under conditions where scanning effects are minimized, as when words are presented vertically, for shorter exposure durations, or with a central fixation control.

Type of Script

There appear to be neuropsychological repercussions of differences across languages in degree of sound–symbol correspondence. Studies of unilingual aphasics have provided evidence for the existence of intra-hemispheric specialization for auditory and visual aspects of language, the auditory being associated with left temporal and the visual with parieto-occipital activity. Since readers of a language with a phonetic script may rely more on auditory than visual processing, one would expect them to experience more severe alexia or agraphia than readers of an ideographic script, following damage of the temporal cortex. Conversely, greater impairment in reading and writing should accompany left occipital damage among readers of an ideographic script relative to readers of a phonetic script (see deAgostini, 1977).

In bilingual speakers of languages whose scripts differ in their degree of sound–symbol correspondence it may, accordingly, be hypothesized that, WITHIN THE LEFT HEMISPHERE, DAMAGE TO A PARTICULAR CORTICAL AREA (TEMPORAL VERSUS PARIETO-OCCIPITAL) WILL PRODUCE A DIFFERENTIAL PATTERN OF READING AND WRITING DEFICITS ACROSS THE TWO LANGUAGES.

In support of this hypothesis, Luria (1960) reported greater impairment in writing French (which has a less direct sound–symbol correspondence) than Russian following a left inferior parietal lesion. Lyman, Kwan, and Chao (1938) reported greater impairment for written Chinese than English after a left parieto-occipital injury, also in support of the hypothesis. Similarly, several case studies of Japanese aphasics have reported selective impairment in reading or writing *kanji,* the ideographic script, following left parieto-occipital damage, and *kana,* the phonetic script, following left temporal damage (Sasanuma, 1975).

The hypothesis of differential processing of phonetic and ideographic writing systems would predict, AT THE INTERHEMISPHERIC LEVEL, GREATER LEFT LATERALIZATION OF THE MORE PHONETIC SCRIPT. In support of this form of the hypothesis, Sugishita, Iwata, Toyokura, Yoshida, and Yamada (1978) reported that reading of *kanji* presented in the LVF (RH) was preserved in a Japanese commissurotomy patient, while reading of *kana* was impaired. Similarly, a right-handed commissurotomy patient was reported to be agraphic in *kana* but not in *kanji* when writing with his left hand (Sugishita, Toyokura, Yoshioko, and Yamada, 1980).

In normal Japanese subjects, hemisphere differences between *kana* and *kanji* processing have been observed in an evoked potentials study (Hink, Kaga, and Suzuki, 1980) and in several tachistoscopic studies. In the latter, a reliable RVF superiority is typically found for words written in *kana* (Hirata and Osaka, 1967; Endo, Shimizu, and Hori, 1978) and an LVF superiority for words written in *kanji* (Hatta, 1977, 1981).

Right hemisphere involvement has also been reported in the processing of Chinese characters, from which *kanji* are derived.[1] In a case study of a tridialectal aphasic, comprehension of isolated Chinese characters was preserved in contrast to a marked impairment in nearly all other language modalities, including tone perception (Naeser & Chan, 1980). Two case reports of crossed aphasia in Chinese–English bilinguals have also implicated the RH in the processing of Chinese characters (April and Han, 1980a; April and Tse, 1977), especially if they are pictographic (Nguy, Allard, and Bryden, 1980). However, presentation of character pairs produces either a RVF superiority (Feustel and Tsao, 1978) comparable to that noted for English in the same individuals (Kershner and Jeng, 1972), or no significant visual field asymmetries (Hardyck, Tzeng, and Wang, 1978). An analogous effect of single versus double character presentation is evident for *kanji* characters (Hatta, 1978; Sasanuma, Itoh, Mori, & Kobayashi, 1977). Moreover, task parameters, such as whether phonetic or visual judgments are required, appear to influence the direction and de-

---

[1] Purported RH involvement in the processing of *kanji* or of Chinese characters is apparently restricted to the visual modality, for LH superiority has been reported in dichotic studies of Japanese and Chinese unilinguals (Nagafuchi, 1970; and Bryson, Mononen, and Yu, 1980, respectively).

gree of visual field asymmetries in the processing of both Chinese ideograms (Huang and Jones, 1980) and *kanji* (Sasanuma, Itoh, Kobayashi, and Mori, 1980).

In summary, evidence from clinical and experimental investigations supports the existence of differential intrahemispheric and interhemispheric organization of reading and writing in bilingual users of phonetic and ideographic scripts. However, in these studies, as in studies of other language-specific effects, bilingualism has generally been considered to be incidental to the phenomenon under investigation serving, at most, to highlight cross-linguistic differences that are presumably comparable, though less striking, in unilingual speakers of the languages in question. Nevertheless, it is of theoretical interest to determine whether, or under what circumstances, language-specific effects are also bilingual-specific, that is, are found only among bilinguals. Bilingual-specific effects might arise from particular language combinations, for example, French–Spanish versus French–Vietnamese, or from an interaction of linguistic and language-acquisitional parameters, such as sequence of language acquisition, for example, Chinese–English versus English–Chinese; from modality, for example, written versus spoken (Hinshelwood, 1902; Minkowski, 1963), or second-language proficiency.

LANGUAGE-ACQUISITIONAL FACTORS

The role of three language acquisitional factors, namely, manner, stage, and age of second language acquisition, will be discussed below. Although, as will soon become evident, these factors are interrelated, they will nevertheless be presented separately at first for clarity of exposition.

Manner of Second Language Acquisition

In the second-language literature, Krashen (1977) has proposed a distinction between formal and informal modes of language acquisition which may have neuropsychological implications. Formal language acquisition, or what Krashen terms language learning, may be characterized by contexts in which there is an emphasis on the structure of the language through, for example, error correction and rule isolation. Informal language acquisition, on the other hand, occurs in naturalistic, communicative contexts in which the user's attention is directed not so much to the form as to the content of the utterance. While formal acquisition engenders in the learner a metalinguistic awareness of language as an abstract,

rule-governed system, informal acquisition entails a relatively uncon-
scious internalization of linguistic rules.

Language acquisition in unilinguals is initially informal but becomes
more formal with increased cognitive maturity (Rosansky, 1975). This
shift is reflected on the neurological level by an increasing reliance on the
LH following an initial stage of bilateral (though not necessarily equal)
mediation of language (see Witelson, 1977, pp. 269–271).

With respect to the manner of second language acquisition, the follow-
ing hypothesis may be proposed: THERE WILL BE GREATER RH INVOLVE-
MENT IN THE SECOND AS COMPARED TO THE FIRST LANGUAGE OF ADULT
BILINGUALS IF THE SECOND LANGUAGE IS LEARNED INFORMALLY. CON-
VERSELY, THERE WILL BE GREATER LH INVOLVEMENT IN THE SECOND
THAN IN THE FIRST LANGUAGE IF THE FORMER IS LEARNED FORMALLY.

Five experimental studies are relevant to the manner hypothesis and all
support it. Using a conjugate lateral eye movement paradigm, Hartnett
(1976) found evidence for greater LH participation among English-speaking students learning Spanish in a deductive teaching method than in students learning Spanish with an inductive method. Interestingly, this finding was statistically significant only for successful learners in each group.

Kotik (1975) reported that Russian–Estonian bilinguals who had
learned their second language in a naturalistic context evinced an equivalent right-ear advantage (REA) in their two languages on a dichotic listening task. However, Estonian–Russian bilinguals who had learned their
second language formally, particularly those who were proficient in it,
showed a significantly larger REA in their second than in their first language. The fact that the former group had acquired their second language
in early childhood while the latter had acquired it after puberty may also
have contributed to the observed group difference.

Carroll (Exp. 2, 1980) similarly found a significantly larger REA for the
second than for the first language of native English speakers learning
Spanish in a formal, classroom setting. In contrast, a group of English
speakers who had had informal exposure to Spanish (in an immersion program) showed no significant differences in ear advantage between the two
languages.

Albert and Obler (1978) and Gordon (1980) reported a significantly
greater REA in the second than in the first language among Hebrew–English adult bilinguals in Israel (i.e., those for whom Hebrew was the first
language), but not among English–Hebrew bilinguals (i.e., those for
whom English was the first language). This finding is compatible with the
manner hypothesis since, in Israel, English is learned primarily in a formal, academic setting whereas far greater opportunities exist for naturalistic acquisition of Hebrew. Interestingly, Gordon (1980) also reported a

trend for an overall smaller REA among English–Hebrew bilinguals in Israel and among Hebrew–English bilinguals in the United States.

In summary, all of the studies bearing on the manner-of-language-acquisition hypothesis support it. However, the criteria used to define formal and informal modes of second language acquisition (e.g., manner of instruction, context of exposure) have varied across studies. Moreover, in each study, other language acquisitional parameters, in particular, age and proficiency, have also been operating. Thus, it remains unclear precisely what conditions are necessary and sufficient for the emergence of manner-of-acquisition effects.

Stage of Second Language Acquisition

In the beginning stages of second language acquisition, both child and adult learners tend to rely more on content than function words, prosodic rather than syntactic cues, and linguistic information in context rather than in isolation (Krashen, 1977). Beginning second language learners also make extensive use of verbal routines and formulaic utterances (Scarcella, 1979; Wong-Fillmore, 1979). There is evidence from the unilingual laterality literature of RH participation in automatized, concrete, and melodic aspects of language (Searleman, 1977; see Foldi, Cicone, and Gardner, Chapter 3 and Millar and Whitaker, Chapter 4, this volume, for reviews). The compatibility between language functions apparently mediated by the RH, and aspects of language salient for beginning learners, leads to the following hypothesis: RH INVOLVEMENT IN SECOND LANGUAGE PROCESSING WILL BE MORE EVIDENT IN THE INITIAL THAN IN THE FINAL STAGES OF SECOND LANGUAGE ACQUISITION (Galloway and Krashen, 1980; Obler, 1977).

An ideal test of the stage hypothesis would require a longitudinal assessment of the same learners at increasingly advanced levels of second language proficiency. Such a test has, so far, not been undertaken. The hypothesis has usually been tested by comparing differential hemispheric involvement in the first and second language processing of nonproficient bilinguals relative to that of proficient bilingual and/or unilingual controls, or by comparing first and second language performance of nonproficient bilinguals. As such, the hypothesis gives rise to two predictions: (1) the LH is dominant in the processing of both languages of proficient bilinguals, as it is in unilinguals, and (2) there is less LH involvement in nonproficient bilinguals in the processing of their second language when compared to their first language, and to that of the same language in proficient bilinguals or in unilinguals.

Support for the first prediction derives from laterality studies that have reported an equivalent REA (Schönle & Breuninger, 1977) or RVF superiority (Hamers & Lambert, 1977) for both languages of bilinguals who acquired their languages in early childhood. Unfortunately, several other studies of proficient bilinguals in which an overall LH superiority was reported did not control for age of exposure to the second language (Albert and Obler, 1978; Gaziel, Obler, and Albert, 1978; Hardyck, 1980; Obler, Albert, and Gordon, 1975; Walters and Zatorre, 1978). However, studies of proficient bilinguals who learned their second language in adolescence report less LH participation in the second than in the first language (Hynd, Teeter, and Stewart, 1980; Sussman, Franklin, and Simon, 1980; see also next section), a finding that is inconsistent with the first prediction of the stage hypothesis.

The second prediction of the stage hypothesis, addressed in some 12 studies, has received even less support than the first.

Among studies that have examined adults, Obler, Albert, and Gordon (1975) in a pilot dichotic listening study reported a smaller REA in the second than in the first language of nonproficient English–Hebrew bilinguals, a finding compatible with the second prediction of the stage hypothesis. However, this effect was statistically not significant (Obler, personal communication) and was subsequently not replicated (Albert & Obler, 1978).

The results of another dichotic listening experiment (Maitre, 1974) provide partial support for the second prediction, insofar as a smaller REA was obtained for words in the second than in the first language of native English-speaking students enrolled in college-level Spanish courses. However, a similar effect was not obtained for sentences, nor was the size of the REA for Spanish words or sentences statistically different from that observed among native Spanish-speaking controls.

Using words similar to those employed by Maitre, Galloway and Scarcella (1982) found no significant differences in the size of the REA across the two languages of male native Spanish-speaking adults acquiring English informally. Moreover, the size of the REA of the bilinguals was equivalent to that obtained in English-speaking and in Spanish-speaking unilingual controls.

In yet another study providing evidence contradictory to the stage hypothesis, proficient and nonproficient speakers of Hebrew and English showed an equivalent REA in their first and second languages (Gordon, 1980). Subjects' proficiency was inferred from their self-ratings and from their length of exposure to the second language.[2]

[2] Gordon's (1980) results refer to the combined data of the Hebrew–English and English–Hebrew subgroups. However, these groups, according to Gordon, may have used different criteria in their self-ratings of proficiency in the second language. Thus, possible hemisphere differences related to proficiency may have been obscured.

   In apparent support of the hypothesis, Schneiderman and Wesche (1980) reported a significant REA for words in the first language (English) but no significant ear differences for words in the second language (French) of nonproficient bilinguals. However, it is not possible to rule out an alternative explanation of the results in terms of a language-specific effect arising from characteristics of the French stimuli (see Bellisle, 1975), since French unilingual controls were not tested. Moreover, the authors failed to obtain a significant correlation between degree of second language proficiency (as measured by a test of written and aural comprehension) and strength of ear asymmetry for French, thereby undermining the stage hypothesis further.

   In an electroencephalographic (EEG) study of nonproficient English–Italian bilinguals and proficient speakers of standard Italian and a regional dialect, Hardyck (1980) reported overall LH participation on a task assessing recognition of language of input. However, greater RH activation for the second than for the first language was noted only among nonproficient bilinguals. Although compatible with the second prediction of the stage hypothesis, this finding was interpreted by Hardyck as evidence for RH superiority in tasks requiring perceptual effort, of which processing a language in which one is not proficient is only one example (see also Mohr and Costa, 1981).

   Laterality studies with nonproficient child bilinguals have also produced negative evidence with respect to the second prediction of the stage hypothesis. Piazza and Zatorre (1981) found an equivalent REA in the first and second language of 11½- and 13½-year-old Spanish–English speakers. Moreover, a post hoc estimate of the subjects' proficiency in their second language (English) did not correlate with their laterality measures on English.

   Rupp (1980) reported a significantly greater REA for the second as compared to the first language of 6–13-year-old native Vietnamese speakers who had had two years of exposure to English. Greater LH activation in the second than in the first language was also noted in an EEG study of Hopi–English children in grades 4 to 6 (Rogers, TenHouten, Kaplan, and Gardiner, 1977).

   The findings of a tachistoscopic study of Hebrew schoolchildren with 2, 4, and 6 years of formal exposure to English appear to support the stage hypothesis (Silverberg, Bentin, Gaziel, Obler, and Albert, 1979): in contrast to a consistent RVF superiority for Hebrew words obtained across all three grade levels, visual field asymmetries for English differed across grades, ranging from a significant LVF preference to a slight RVF preference with increasing proficiency in English. However, findings of two subsequent laterality studies cast a different light on the Silverberg, *et al.*

(1979) results. In the first study, a LVF to RVF shift was observed in native Hebrew speakers learning to read Hebrew, although a consistent REA was noted in the same individuals for dichotically presented Hebrew words (Silverberg, Gordon, Pollack, and Bentin, 1980). Thus, the observed shift in the direction of LH specialization may simply reflect a learning-to-read effect specific to the visual modality. In the second study, a significant RVF preference for Hebrew and no significant visual field differences for English were noted in a group of native Hebrew speakers with two years of instruction in English (Bentin, 1981). However, when the English stimuli were initially presented for longer exposure durations (thereby making them easier to perceive during the experiment) and when the Hebrew stimuli were presented out of focus (making them difficult to perceive) the same subjects now showed a significant RVF superiority for English and no significant visual field differences for Hebrew (Bentin, 1981). This finding suggests that the RH is involved when information, whether in the first or second language, is perceptually taxing. Taken together, the findings of Silverberg, Gordon, Pollack, and Bentin (1980) and Bentin (1981) argue against an interpretation of the Silverberg, Bertin, Gaziel, Obler, and Albert (1979) results in terms of the stage hypothesis and suggest, instead, that RH involvement is neither unique to the second language nor is it necessarily generalizable to language modalities other than reading.

To summarize, of the available studies bearing on the first prediction of the stage hypothesis, that, in proficient bilinguals, the LH is dominant for language functioning, most provide supportive evidence. However, the prediction was not supported in studies of proficient bilinguals who acquired their second language after first language acquisition (Hynd, Teeter, and Stewart, 1980; Sussman, Franklin, and Simon, 1982; and see next section).

With regard to the second prediction, that in nonproficient bilinguals there is a greater likelihood of RH participation in the second than first language, a majority of the relevant studies do not provide supportive evidence. Of the five that appear to support it, the results of two offer equivocal evidence (Maitre, 1974; Obler, Albert, and Gordon, 1975) and the remaining three lend themselves to alternative interpretations that are not unique to the process of second language acquisition (Hardyck, 1980; Schneiderman and Wesche, 1980; Silverberg, Bentin, Gaziel, Obler, and Albert, 1979).

Two types of findings characterize studies that are incompatible with the second prediction of the stage hypothesis: (1) equivalent LH involvement in the first and second language of nonproficient bilinguals (Albert and Obler, 1978; Galloway and Scarcella, 1982; Gordon, 1980; Piazza and

Zatorre, 1981) and (2) greater LH participation in the less proficient as compared to the more proficient language (Rogers, TenHouten, Kaplan, and Gardiner, 1977; Rupp, 1980).

In light of these negative findings, the stage hypothesis would appear to be invalidated. However, certain issues need to be considered before such a conclusion can be reached. First, the fact that subjects within and across studies varied widely in levels of second language proficiency may have increased the variance in the results, thereby obscuring possible effects due to the stage factor. Second, although the stage hypothesis rests on the assumption that different aspects of language are salient for beginning versus advanced learners, none of the studies just reviewed, with the exception of Maitre (1974), compared hemispheric specialization across different linguistic units. Finally, it is questionable whether, even if properly tested, the predicted differences in the extent of hemispheric involvement in the two languages of bilingual subgroups can be reliably detected by current testing procedures, especially since the size of ear or visual field asymmetries may be influenced by factors other than degree of cerebral lateralization (Bryden, 1978; Colbourn, 1978; White, 1972). This consideration applies, of course, to any laterality study that involves group differences.

AGE OF SECOND LANGUAGE ACQUISITION

Differences in the state of brain maturation during first versus second language acquisition may give rise to different patterns of cerebral lateralization in bilinguals who acquired both their languages in infancy and those who acquired their second language around puberty. Evidence from psychological studies of semantic satiation (Jakobovits and Lambert, 1961), concept formation (Lambert and Rawlings, 1969), verbal memory (Lambert, 1969), and semantic interference (Preseon and Lambert, 1969) points to differences between "early" and "late" bilinguals in language processing. According to Lambert (1981), the psychological literature suggests a greater functional segregation of the two languages among late as compared to early bilinguals, as late bilinguals are less susceptible to either facilitative or disruptive effects of linguistically mixed input presentation.

Since language-of-input is signalled more by surface than by semantic features, one might expect late bilinguals to be more sensitive than early bilinguals to surface features of verbal input (see Vaid, 1981b). Moreover, since the RH is thought to partake in the processing of orthographic (Bryden and Allard, 1976) and prosodic aspects of linguistic material (Blumstein and Cooper, 1974), both of which may be considered surface-level components of language, one might expect greater reliance on the RH among late than among early bilinguals. More generally, it may be hypothesized that THE PATTERN OF HEMISPHERIC ASYMMETRY IN BILINGUALS WILL MORE CLOSELY RESEMBLE THAT OF UNILINGUALS THE EARLIER SECOND LANGUAGE ACQUISITION OCCURS AND WILL DIFFER FROM THAT OF UNILINGUALS THE LATER THE SECOND LANGUAGE IS ACQUIRED.

There are over a dozen studies of relevance to the age-of-acquisition hypothesis, including two with children. Consistent with the hypothesis, a dichotic listening study of 4-year-old French–English bilinguals revealed an equivalent REA in English among bilinguals and unilingual controls, but no significant ear differences in French among the bilinguals and unilinguals alike (Bellisle, 1975). Similarly, in another dichotic listening experiment, an REA found for English in 6 to 8-year-old unilinguals was equivalent to that noted in native English speakers of the same age who were also competent in French and Hebrew; unfortunately, the trilinguals' dichotic listening performance on the other two languages was not assessed (Starck, Genesee, Hamers, Lambert, and Seitz, 1977).

Among studies with adults, a variety of paradigms have been used. In an auditory Stroop test presented monaurally, significant right-ear semantic interference was obtained in male French–English early bilinguals and in English-speaking unilingual controls (Vaid and Lambert, 1979). In a visual version of the Stroop test (Vaid, 1979b), French–English early bilinguals and English unilinguals again performed similarly. By contrast, male and female late bilinguals in both studies showed significant interference in the left ear (Vaid, and Lambert, 1979) or the left visual field (Vaid, 1979b), suggesting a tendency among late bilinguals to encode words superficially, even when the words are directed to the LH.

A concurrent verbal/manual task was used in three studies to examine hemisphere differences in speech production. A study by Sussman, Franklin and Simon (1982) found that early bilinguals experienced more interference in tapping with their right than with their left hand while concurrently speaking in either their first or second language; however, late bilinguals experienced greater right-hand than left-hand disruption during concurrent speech in their first language only—in their second language, they showed equivalent disruption for right-hand *and* left-hand tapping. Hynd, Teeter, and Stewart (1980) similarly reported greater right-hand than left-hand tapping interference during concurrent speech in the first as compared to the second language of Navajo–English late bilinguals. The hypothesis suggested by Sussman of bilateral participation in the second language of late bilinguals has recently been called into question by the results of a tapping study using Portuguese–English late bilinguals

(Soares, 1982a), as well as by the results of a tachistoscopic study with the same group of subjects (Soares and Grosjean, 1981). However, as Soares did not include an early bilingual subgroup in either of his studies, his findings cannot strictly be taken as evidence against the age-of-acquisition hypothesis.

In two dichotic listening studies, support was obtained for *greater* left hemisphere participation in the second as compared to the first language among late bilinguals, but equivalent ear asymmetries in the two languages of early bilinguals (Gordon, 1980; Kotik, 1975). This finding may reflect the fact that the second language in both cases was taught and used in a formal manner. Thus, it would appear that in certain contexts of language acquisition, age-related effects may be superseded.

Two studies employed an evoked potentials procedure to study language lateralization. Genesee, Hamers, Lambert, Mononen, Seitz and Starck (1978) reported that latencies to the N1 and P2 components of the evoked potentials were shorter in the LH for both languages of early French–English bilinguals, but in the RH for both languages of late bilinguals in a task of language recognition. Using a similar task, Kotik (1980) also reported a faster neural response in the RH than in the LH among a group of Polish–Russian late bilinguals. Although these two studies support the age-of-acquisition hypothesis, some further results reported by Kotik indicate that another variable must also be considered, for, on a task requiring judgments of word animacy, evoked responses were found to be faster in the LH (Kotik, 1980). Kotik, regrettably, did not include a group of early bilinguals in her study, thereby leaving open the question of whether the two groups may have performed differently in either task.

The relative contribution of task and group parameters was subsequently investigated by Vaid (1981b) in a series of tachistoscopic experiments measuring response latencies of French and English early and late bilinguals for different types of word pair comparisons. The results pointed to the importance of task demands in determining the pattern of visual field asymmetries: in early *and* late bilinguals, a significant RVF preference in response time was found for phonetic and syntactic (i.e., parts-of-speech) comparisons (Vaid, 1981b, 1981c, 1982), a significant LVF superiority was found for orthographically based judgments (Vaid, 1979a), and no significant visual field differences were obtained for semantic comparisons (Vaid, 1979a; 1981c). Moreover, overall differences were observed between the groups in speed of processing different aspects of language; that is, late bilinguals were faster than early bilinguals (and monolinguals) on orthographic and phonetic comparisons (Vaid, 1979a, 1981c, 1982) while early bilinguals were faster on semantic comparisons (Vaid, 1981c).

In summary, studies pertaining to the age of second language acquisition suggest that hemispheric processing in early bilinguals resembles what is found in unilinguals and that, in proficient late bilinguals, hemispheric specialization may shift to the right for both languages, unless the second language was learned formally (in which case there is greater LH control of the second language). Some evidence has also accumulated to suggest that group differences relate to differences in speed of processing various aspects of language which themselves may be differentially lateralized (Vaid, 1981b). It will be important for further research to compare proficient and nonproficient late bilinguals on various linguistic tasks in a laterality context.

DISCUSSION

On the basis of the studies reviewed above there appears to be little evidence to suggest that RH involvement is more likely in the beginning than in the advanced stages of second language acquisition. The evidence does suggest that RH participation is more likely the later the second language is acquired relative to the first and the more informal the exposure to the second language. On the other hand, LH involvement in second language processing appears to be more likely the more advanced the stage of second language acquisition, the earlier the second language is acquired relative to the first and the more formal the exposure to the second language.

A general principle that emerges from the research findings is that bilinguals are more likely to show a comparable pattern of hemispheric involvement across their two languages the more similar the language acquisition conditions. Conversely, the less similar the language acquisition conditions the greater the likelihood that the pattern of hemispheric involvement will differ across the two languages of bilinguals. The exact nature of this difference will depend on the outcome of interaction effects of a variety of factors, as discussed subsequently.

Interaction Effects

Interactions among the factors of manner, stage, and age of second language acquisition are not only possible but probable, although few studies to date have directly addressed them. Findings from available studies suggest that manner effects are observed only among advanced second language learners (Carroll, Exp. 2, 1980; Gordon, 1980; Hartnett, 1976; Kotik, 1975). Nonproficient bilinguals do not appear to behave in accordance with the predictions of the manner hypothesis (Galloway and Scarcella, 1979). Whether this is an actual or only an apparent effect, and whether other types of interactions are operating, can only be determined through further research in which the factors of age, stage, and manner are systematically manipulated to the extent possible.

Interaction effects may also obtain between language-acquisitional and language-specific factors, as previously suggested. For example, would the hemispheric processing of tones in Thai differ in native versus nonnative speakers of the language? Would the observed differences in hemispheric specialization for phonetic versus ideographic scripts be influenced by the context in which the scripts were learned? At present there is no evidence relating to such questions.

A third class of interaction effects that potentially contributes to differences in cerebral lateralization involves factors specific to bilingualism, that is, those discussed in this review, and those shared by bilinguals and unilinguals. The latter include subject variables such as handedness and sex (Andrews, 1977; McGlone, 1980; Piazza, 1980) and task variables such as phonetic versus visual processing (Levy and Trevarthen, 1977; Sasanuma, Itoh, Kobayashi, and Mori, 1980).

Laterality studies of bilinguals, with two exceptions, (Gaziel, Obler, and Albert, 1978; Orbach, 1967) have only tested right-handers. However, it may be of interest to examine possible interactions between handedness and parameters of bilingualism, as it has been suggested that left-handedness is associated with difficulty in foreign language learning (Fraser, 1980).

Unlike the variable of handedness, that of sex has not been adequately controlled in the bilingual laterality literature. Nearly a third of the studies have not mentioned the sex composition of their subject samples. Several other studies have employed more females than males without, however, analyzing for sex differences (Genesee, Hamers, Lambert, Mononen, Seitz, and Starck, 1978; Obler, Albert, and Gordon, 1975). Sex differences in lateralization in bilinguals have been reported in the direction of greater LH lateralization among males than females and may either override (Gordon, 1980) or interact with (Vaid and Lambert, 1979) parameters of bilingualism. Further studies should seek to isolate the locus and nature of sex differences in lateralization and relate them to observations of sex differences in language learning aptitude.

The importance of task variables has also been demonstrated in a few recent studies in the bilingual laterality literature (Bentin, 1981; Kotik, 1980; Vaid, 1981b). In order that effects due to task-related processing demands not be associated with bilingualism per se, it is necessary that unilingual and/or bilingual controls be routinely tested under identical experimental conditions.

Taken together, findings from neuropsychological studies of bilinguals suggest that what distinguishes bilinguals from unilinguals, and particular bilingual subgroups from others, when differences do occur, is their use of strategies that deploy the specialized processes of the two hemispheres differently. As Genesee (1980) has noted, the functional competencies of the two hemispheres of early and late bilinguals may be no different from one another, nor from those of unilinguals. Rather, it may be the extent to which late bilinguals tend to use RH-based strategies in language processing that differs.

Further theorizing on cerebral lateralization of language in bilinguals will, ultimately, have to address the issue of just what it is about bilingualism that may give rise to observed differences in processing strategy. To this end, psycholinguistic and cognitive research on bilingualism should provide some needed insights.

ACKNOWLEDGMENTS

I would like to thank Fred Genesee, Loraine Obler, E. Schneiderman, Sid Segalowitz, and Carlos Soares for providing useful comments on earlier drafts of this chapter.

REFERENCES

Albert, M., and Obler, L. (1978). *The bilingual brain.* New York: Academic Press.

Anastosopoulos, G. (1959). Linksseitige Hemiplegie mit Alexie, Agraphie und Aphasie bei einem polyglotten Rechtshänder. *Deutsche Zeitschrift für Nervenheilkunde, 179,* 120–144.

Andrews, R. (1977). Aspects of language lateralization correlated with familial handedness. *Neuropsychologia, 15,* 769–778.

April, R. S., and Han, N. (1980a). Crossed aphasia in a right-handed bilingual Chinese man: A second case. *Archives of Neurology, 6,* 342–346.

April, R. S., and Han, M. (1980b). Current observations of aphasia in Chinese subjects. Paper presented at Third INS European Conference, Chianciano-Terme, Italy.

April, R. S., and Tse, P. (1977). Crossed aphasia in a Chinese bilingual dextral. *Archives of Neurology, 34,* 766–770.

Barton, M., Goodglass, H., and Shai, A. (1965). Differential recognition of tachistoscopically presented English and Hebrew words in right and left visual fields. *Perceptual and Motor Skills, 21,* 431–437.

Bellisle, F. (1975). *Early bilingualism and cerebral dominance.* Unpublished manuscript, McGill University.

Benson, P., Smith, T., and Arreaga, L. (1974). Lateralization of linguistic tone: Evidence from Cantonese. *Conference on Cerebral Dominance* (Brain Information Service Report No. 34). Los Angeles: Univ. of California Press, 19–20.

Bentin, S. (1981). On the representation of a second language in the cerebral hemispheres of right-handed people. *Neuropsychologia, 19,* 599–603.

Ben-Zeev, S. (1977). Mechanisms by which childhood bilingualism affects understanding of language and cognitive structures. In P. Hornby (Ed.), *Bilingualism: Psychological, social, and educational implications.* New York: Academic Press.

Blumstein, S., and Cooper, W. E. (1974). Hemispheric processing of intonation contours. *Cortex, 10,* 146–158.

Bradshaw, J. and Nettleton, N. (1981). The nature of hemispheric specialization in man. *The Behavioral and Brain Sciences, 4,* 51–91.

Branch, C., Milner, B., and Rasmussen, T. (1964). Intracarotid amytal for the lateralization of cerebral speech dominance. *Journal of Neurosurgery, 21,* 399–405.

Bryden, M. (1978). Strategy effects in the assessment of hemispheric asymmetry. In G. Underwood (Ed.), *Strategies of information processing.* New York: Academic Press.

Bryden, M. P., and Allard, F. (1976). Visual hemifield differences depend on typeface. *Brain and Language, 3,* 191–200.

Bryson, S., Mononen, L., and Yu, L. (1980). Procedural constraints on the measurement of laterality in young children. *Neuropsychologia, 18,* 243–246.

Bychowski, Z. (1919). Über die Restitution der nach einem Schädelschuss verlorenen Umgangssprache bei einem Polyglotten. *Monatschrift für Psychologie und Neurologie, 45,* 183–201.

Carroll, F. (1980). Neurolinguistic processing of a second language: Experimental evidence. In R. Scarcella and S. Krashen (Eds.), *Research in second language acquisition.* Rowley, Mass.: Newbury House.

Cohen, G. (1977). *The psychology of cognition.* New York: Academic Press.

Colbourn, C. J. (1978). Can laterality be measured? *Neuropsychologia, 16,* 283–289.

DeAgostini, M. (1977). A propos de l'agraphie des aphasiques sensoriels: Etude comparative Italien-Français. *Langages, 47,* 120–130.

Duncan, S., and DeAvila, E. (1979). Bilingualism and cognition: Some recent findings. *Journal of the National Association for Bilingual Education 4,* 15–50.

Endo, M., Shimizu, A., and Hori, T. (1978). Functional asymmetry of visual fields for Japanese words in *kana* (syllable-based) writing and Japanese shape-recognition in Japanese subjects. *Neuropsychologia, 16,* 291–297.

Feustel, T., and Tsao, Y.-C. (1978). Differences in reading latencies of Chinese characters in the right and left visual fields. Paper presented at the meeting of the Eastern Psychological Association, New York.

Fraser, B. (1980). A sinister problem in second language learning. Paper presented at the 14th Annual TESOL Convention, San Francisco.

Galloway, L. (1982). Bilingualism: Neuropsychological considerations. In G. Hynd (Ed.), *Journal of Research and Development in Education, 15,* 12–28.

Galloway, L. (1981). *Contributions of the right cerebral hemisphere to language and communication: Issues in cerebral dominance with special emphasis on bilingualism, second language acquisition, sex differences, and certain ethnic groups.* Unpublished doctoral dissertation, U.C.L.A., Los Angeles.

Galloway, L., and Krashen, S. (1980). Cerebral organization in bilingualism and second language. In R. Scarcella and S. Krashen (Eds.), *Research in second language acquisition.* Rowley, Mass.: Newbury House. .

Galloway, L., and Scarcella, R. (1982). Cerebral organization in adult second language acquisition: Is the right hemisphere more involved? *Brain and Language, 16,* 56–60.

Gaziel, T., Obler, L., and Albert, M. (1978). A tachistoscopic study of Hebrew–English bilinguals. In M. Albert and L. Obler (Eds.), *The bilingual brain*. New York: Academic Press.

Geffen, G., Bradshaw, J., and Nettleton, N. (1972). Hemispheric asymmetry: Verbal and spatial encoding of visual stimuli. *Journal of Experimental Psychology*, 95, 25–31.

Genesee, F. (1980). Bilingual brains? Paper presented at Symposium on Neurolinguistics and bilingualism: The Question of Individual Differences. Albuquerque, New Mexico.

Genesee, F., Hamers, J., Lambert, W. E., Mononen, L., Seitz, M., and Starck, R. (1978). Language processing in bilinguals. *Brain and Language*, 5, 1–12.

Gloning, I., and Gloning, K. (1965). Aphasien bei Polyglotten. Beitrag zur Dynamik des Sprachabbaus sowie zur Lokalisationsfrage dieser Störungen. *Wiener Zeitschrift für Nervenheilkunde*, 22, 362–397.

Gordon, H. W. (1980). Cerebral organization in bilinguals: I. Lateralization. *Brain and Language*, 9, 255–268.

Hamers, J., and Lambert, W. E. (1977). Visual field and cerebral hemisphere preferences in bilinguals. In S. J. Segalowitz and F. A. Gruber (Eds.), *Language development and neurological theory*. New York: Academic Press.

Hardyck, C. (1980). Hemispheric differences and language ability. Paper presented at Symposium on Neurolinguistics and Bilingualism. The Question of Individual Differences. Albuquerque, New Mexico.

Hardyck, C., Tzeng, O., J.-L., and Wang, W. S.-Y. (1978). Cerebral lateralization of function and bilingual decision processes: Is thinking lateralized? *Brain and Language*, 5, 56–71.

Hartnett, D. (1976). The relation of cognitive style and hemispheric preference to deductive and inductive second language learning. *Conference on Cerebral Dominance* (Brain Information Service Report No. 42). Los Angeles: Univ. of California Press, 41.

Hatta, T. (1977). Recognition of Japanese *kanji* in the left and right visual fields. *Neuropsychologia*, 15, 685–688.

Hatta, T. (1978). Recognition of Japanese kanji and hirakana in the left and right visual fields. *Japanese Psychological Research*, 20, 51–59.

Hatta, T. (1981). Differential processing of *kanji* and *kana* stimuli in Japanese people: Some implications from Stroop test results. *Neuropsychologia*, 19, 87–93.

Hécaen, H., Mazaro, G., Rannier, A., Goldblum, M., and Merienne, L. (1971). Aphasie croisée chez un sujet droitier bilingue. *Revue Neurologique*, 124, 319–323.

Hellige, J., and Webster, R. (1979). Right hemisphere superiority for initial stages of letter processing. *Neuropsychologia*, 17, 653–660.

Hink, R., Kaga, K., and Suzuki, J. (1980). An evoked potential correlate of reading ideographic and phonetic Japanese scripts. *Neuropsychologia*, 18, 455–464.

Hinshelwood, J. (1902). Four cases of word blindness. *Lancet*, 1, 358–363.

Hirata, K., and Osaka, R. (1967). Tachistoscopic recognition of Japanese letter materials in left and right visual fields. *Psychologia*, 10, 7–18.

Huang, Y. L., and Jones, B. (1980). Naming and discrimination of Chinese ideograms presented in the right and left visual fields. *Neuropsychologia*, 18, 703–706.

Hynd, G., and Scott, S. (1980). Propositional and appositional modes of thought and differential speech lateralization in Navajo Indian and Anglo Children. *Child Development*, 51, 909–911.

Hynd, G., Teeter, A., and Stewart, A. (1980). Acculturation and the lateralization of speech in the bilingual native American. *International Journal of Neuroscience*, 11, 1–7.

Jakobovits, L., and Lambert, W. E. (1961). Semantic satiation among bilinguals. *Journal of Experimental Psychology*, 62, 576–582.

Kershner, J., and Jeng, A. G.-R. (1972). Dual functional asymmetry in visual perception: Effects of ocular dominance and postexposural processes. *Neuropsychologia, 10,* 437–445.

Kessler, C., and Quinn, M. (1980). Cognitive development in bilingual environments. Paper presented at International Symposium on Standard Language–Vernacular Relationship in Bilingual Communities, Racine, Wisconsin.

Kotik, B. (1975). *Investigation of speech lateralization in multilinguals.* Unpublished doctoral dissertation, Moscow State University.

Kotik, B. (1980). An evoked potential study of Polish–Russian bilinguals. Personal communication.

Krashen, S. (1977). The monitor model for adult second language performance. In M. Burt, H. Dulay, and M. Finocchiaro (Eds.), *Viewpoints on English as a second language.* New York: Regents.

Lambert, W. E. (1969). Psychological studies of interdependencies of the bilingual's two languages. In J. Puhvel (Ed.), *Substance and structure of language.* Los Angeles: Univ. of California Press.

Lambert, W. E. (1981). Bilingualism and language acquisition. In H. Winitz (Ed.), *Native language and foreign language acquisition.* New York: The New York Academy of Sciences.

Lambert, W. E., and Rawlings, C. (1969). Bilingual processing of mixed language associative networks. *Journal of Verbal Learning and Verbal Behavior, 8,* 604–609.

Lamendella, J. (1977). General principles of neurofunctional organization and their manifestation in primary and nonprimary language acquisition. *Language Learning, 27,* 155–196.

Levy, J., and Trevarthen, C. (1977). Perceptual, semantic and phonetic aspects of elementary language processes in split-brain patients. *Brain, 100,* 105–118.

Luria, A. R. (1960). Differences between disturbance of speech and writing in Russian and French. *International Journal of Slavic Linguistics and Poetics, 3,* 13–22.

Lyman, R., Kwan, S. T., and Chao, W. H. (1938). Left occipito-parietal brain tumor with observations on alexia and agraphia in Chinese and English. *Chinese Medical Journal, 54,* 491–516.

McGlone, J. (1980). Sex differences in human brain asymmetry: A critical survey. *The Behavioral and Brain Sciences, 3,* 215–264.

Maitre, S. (1974). *On the representation of second languages in the brain.* Unpublished M.A. thesis, University of California, Los Angeles.

Minkowski, M. (1963). On aphasia in polyglots. In L. Halpern (Ed.), *Problems of dynamic neurology.* Jerusalem: Hebrew University.

Mishkin, M., and Forgays, D. (1952). Word recognition as a function of retinal locus. *Journal of Experimental Psychology, 43,* 43–48.

Mohr, E., and Costa, L. (1981). Right versus left performance on dichotic stimulation tasks which increase in difficulty. Poster presented at Fourth INS European Conference, Bergen, Norway.

Moscovitch, M. (1979). Information processing and the cerebral hemispheres. In M. Gazzaniga (Ed.), *The handbook of behavioral neurobiology: Neuropsychology,* Vol. 2. New York: Plenum.

Naeser, M., and Chan, S. W-C. (1980). Case study of a Chinese aphasic with the Boston Diagnostic Aphasia Exam. *Neuropsychologia, 18,* 389–410.

Nagafuchi, M. (1970). Development of dichotic and monaural hearing abilities of young children. *Acta Otolaryngyologica, 69,* 409–414.

Nguy, T. V.-H, Allard, F., and Bryden, M. P. (1980). Laterality effects for Chinese characters: Differences between pictorial and nonpictorial characters. *Canadian Journal of Psychology, 34,* 270–273.

Obler, L. (1977). Right hemisphere participation in second language acquisition. Paper presented at Conference on Individual Differences and Universals in Language Learning Aptitude, Durham, New Hampshire.

Obler, L., Albert, M., and Gordon, H. W. (1975). Asymmetry of cerebral dominance in Hebrew–English bilinguals. Paper presented at the Thirteenth Annual Meeting of the Academy of Aphasia, Victoria, British Columbia.

Obler, L., Zatorre, R., Galloway, L., and Vaid, J. (1982). Cerebral lateralization in bilinguals: Methodological issues. *Brain and Language, 15,* 40–54.

Ojemann, G., and Whitaker, H. A. (1978). The bilingual brain. *Archives of Neurology, 35,* 409–412.

Orbach, J. (1953). Retinal locus as a factor in the recognition of visually perceived words. *American Journal of Psychology, 65,* 555–562.

Orbach, J. (1967). Differential recognition of Hebrew and English words in right and left visual fields as a function of cerebral dominance and reading habits. *Neuropsychologia, 50,* 127–134.

Ovcharova, P., Raichev, R., and Geliva, T. (1968). Afaziia u Poligloti. *Nevrologiia, Psikhiatriia i Nevrochirurgiia, 7,* 183–190.

Paradis, M. (1977). Bilingualism and aphasia. In H. A. Whitaker and H. A. Whitaker (Eds.), *Studies in neurolinguistics,* Vol. 3. New York: Academic Press.

Piazza, D. (1980). The influence of sex and handedness in the hemispheric specialization of verbal and nonverbal tasks. *Neuropsychologia, 18,* 163–176.

Piazza, D., and Zatorre, R. (1981). Right ear advantage for dichotic listening in bilingual children. *Brain and Language, 13,* 389–396.

Preston, M., and Lambert, W. E. (1969). Interlingual interference in a bilingual version of the Stroop Color–Word Task. *Journal of Verbal Learning and Verbal Behavior, 8,* 295–301.

Rapport, R. L., Tan, C., and Whitaker, H. A. Language function and dysfunction among Chinese and English-speaking polyglots: Cortical stimulation, Wada testing and clinical studies. In Paradis, M., and Lebrun, Y. (Eds.), La neurolinguistique du bilinguisme, a special issue of *Langages.* Paris: Larousse (in press).

Rogers, L., TenHouten, W., Kaplan, C., and Gardiner, M. (1977). Hemispheric specialization of language: An EEG study of bilingual Hopi children. *International Journal of Neuroscience, 8,* 1–6.

Rosansky, E. (1975). The critical period for the acquisition of language: Some cognitive developmental considerations. *Working Papers on Bilingualism, 6,* 92–102.

Rupp, J. (1980). *Cerebral language dominance in Vietnamese–English bilingual children.* Unpublished doctoral dissertation, The University of New Mexico, Albuquerque.

Sahibzada, N. (1982). Effect of directional reading habits on recognition ability in different fields. Paper presented at Second Annual Meeting of the South Asian Language Analysis Roundtable, Syracuse, New York.

Sasanuma, S. (1975). *Kana* and *kanji* processing in Japanese aphasics. *Brain and Language, 2,* 369–383.

Sasanuma, S., Itoh, M., Kobayashi, Y., and Mori, K. (1980). The nature of the task–stimulus interaction in the tachistoscopic recognition of *kana* and *kanji* words. *Brain and Language, 9,* 298–306.

Sasanuma, S., Itoh, M., Mori, K., and Kobayashi, Y. (1977). Tachistoscopic recognition of *kana* and *kanji* words. *Neuropsychologia, 15,* 547–553.

Scarcella, R. (1979). "Watch up!" A study of verbal routines in adult second language performance. *Working Papers on Bilingualism, 19,* 79–88.

Schneiderman, E., and Wesche, M. (1980). Right hemisphere participation in second language acquisition. In K. Bailey, M. Long, and S. Peck (Eds.), *Issues in second language acquisition: Selected papers of the Los Angeles Second Language Forum.* Rowley, Mass.: Newbury House.

Schönle, P., and Breuninger, H. (1977). Untersuchungen mit dichotischen Hörtesten bei Zweisprachigkeit. *Archiv für Ohren-, Nasen-und Kehlkopfheilkunde, 216,* 574.

Scott, S., Hynd, G., Hunt, L., and Weed, W. (1979). Cerebral speech lateralization in the native American Navajo. *Neuropsychologia, 17,* 89–92.

Searleman, A. (1977). A review of right hemisphere linguistic capabilities. *Psychological Bulletin, 84,* 503–528.

Sergent, J., and Bindra, D. (1981). Differential hemispheric processing of faces: Methodological considerations and reinterpretation. *Psychological Bulletin, 89,* 544–554.

Shimizu, K. (1975). A comparative study of hemispheric specialization for speech perception in Japanese and English speakers. *Studia Phonologica, 9,* 13–24.

Silverberg, R., Bentin, S., Gaziel, T., Obler, L., and Albert, M. (1979). Shift of visual field preference for English words in native Hebrew speakers. *Brain and Language, 8,* 184–190.

Silverberg, R., Gordon, H. W., Pollack, S., and Bentin, S. (1980). Shift of visual field preference for Hebrew words in native speakers learning to read. *Brain and Language, 11,* 99–105.

Soares, C. (1982a). Converging evidence for left hemisphere language lateralization in bilinguals: use of the concurrent activities paradigm. Unpublished Manuscript, Northeastern University.

Soares, C. (1982b). *Language processing in bilinguals: Neurolinguistic and psycholinguistic studies.* Unpublished doctoral dissertation, Northeastern University, Boston.

Soares, C., and Grosjean, F. (1981). Left hemisphere language lateralization in bilinguals and monolinguals. *Perception and Psychophysics, 29,* 599–604.

Starck, R., Genesee, F., Lambert, W. E., and Seitz, M. (1977). Multiple language experience and the development of cerebral dominance. In S. J. Segalowitz and F. A. Gruber (Eds.), *Language development and neurological theory.* New York: Academic Press.

Sugishita, M., Iwata, M., Toyokura, Y., Yoshida, M., and Yamada, R. (1978). Reading of ideograms and phonograms in Japanese patients after partial commissurotomy. *Neuropsychologia, 16,* 417–426.

Sugishita, M., Toyokura, Y., Yoshioko, M., and Yamada, R. (1980). Unilateral agraphia after section of the posterior half of the truncus of the corpus callosum. *Brain and Language, 9,* 215–225.

Sussman, H., Franklin, P., and Simon, T. (1982). Bilingual speech: Bilateral control? *Brain and Language, 15,* 125–142.

Tsunoda, T. (1971). The difference of the cerebral dominance of vowel sounds among different languages. *Journal of Auditory Research, 11,* 305–314.

Tsunoda, T. (1978). *Some observations on race and language.* Unpublished manuscript, University of Tokyo Medical School.

Tzeng, O. J-L., Hung, D., Cotton, B., and Wang, W. S-Y. (1979). Visual lateralization effect in reading Chinese characters. *Nature, 202,* 299–301.

Vaid, J. (1979a). On the neurolinguistic processing of rhyme. Paper presented at the Winter Meeting of the Linguistic Society of America, Los Angeles.

Vaid, J. (1979b). *Visual field asymmetries on a bilingual Stroop test.* Unpublished manuscript, McGill University.

Vaid, J. (1981a). Cerebral lateralization of Hindi and Urdu: A pilot tachistoscopic Stroop study. Paper presented at the First Annual Meeting of the South Asian Language Analysis Roundtable, Stony Brook, New York.

Vaid, J. (1981b). *Hemisphere differences in bilingual language processing: A task analysis.* Unpublished doctoral dissertation, McGill University, Montreal.

Vaid, J. (1981c). On the lateralization of rhyme and meaning in bilinguals and unilinguals. Paper presented at the annual meeting of the Canadian Psychological Association, Toronto.

Vaid, J. (1982). Hemisphere differences in language processing: A task analysis. Paper presented at the annual meeting of the International Neuropsychological Society, Pittsburgh.

Vaid, J., and Genesee, F. (1980). Neuropsychological approaches to bilingualism: A critical review. *Canadian Journal of Psychology, 34,* 417–445.

Vaid, J., and Lambert, W. E. (1979). Differential cerebral involvement in the cognitive functioning of bilinguals. *Brain and Language, 8,* 92–110.

VanLancker, D., and Fromkin, V. (1973). Hemispheric specialization for pitch and "tone": Evidence from Thai. *Journal of Phonetics, 1,* 101–109.

VanLancker, D., and Fromkin, V. (1978). Cerebral dominance for pitch contrasts in tone language speakers and in musically untrained and trained English speakers. *Journal of Phonetics, 6,* 19–23.

Vildomec, V. (1963). *Multilingualism.* Leyden: A. W. Sythoff.

Walters, J., and Zatorre, R. (1978). Laterality differences for word identification in bilinguals. *Brain and Language, 6,* 158–167.

Whitaker, H. A. (1978). Bilingualism: A neurolinguistics perspective. In W. Ritchie (Ed.), *Second language acquisition research: Issues and implications.* New York: Academic Press.

White, M. (1972). Hemispheric asymmetries in tachistoscopic information processing. *British Journal of Psychology, 62,* 497–508.

Witelson, S. (1977). Early hemisphere specialization and interhemisphere plasticity: An empirical and theoretical review: In S. J. Segalowitz and F. A. Gruber (Eds.), *Language development and neurological theory.* New York: Academic Press.

Wong-Fillmore, L. (1979). Individual differences in second language acquisition. In C. Fillmore, D. Kempler, and W. S.-Y. Wang (Eds.), *Individual differences in language ability and language behavior.* New York: Academic Press.

CHAPTER 15

Individual Differences in Hemispheric Representation of Language

Sidney J. Segalowitz and M. P. Bryden

INTRODUCTION

There is a comfortable generalization in neurolinguistics: that in the vast majority of people, cortical representation of the skills underlying language is in the left hemisphere. But individual differences plague neurolinguistic research, just as much as any other field involving live organisms. These differences are due to individual variation in task performance, differences among individuals in brain morphology, and wide differences in developmental experiences that affect characteristics of the language function, which perhaps in turn affect the way language comes to be organized in the brain. In this chapter, we outline a number of sources of neurolinguistic variation that serve to complicate any simple model of language representation in the brain, and we present some research techniques designed to accommodate these individual differences.

DIFFERENCES IN BRAIN MORPHOLOGY

The degree of structural similarity between the two cerebral hemispheres is large compared to the vast divergence in linguistic functioning.

341

Yet, there have been isolated reports of systematic differences in size between portions of the left and right temporal lobes since the end of the last century (see Witelson, this volume, for a more complete review). Ever since Geschwind and Levitsky's (1968) systematic study of the planum temporale, an area undoubtedly involved in language functions, the possibility of some direct link between brain morphology and neuropsychological functions has been held tantalizingly close. They found that of 100 brains, the planum temporale was larger on the left in 68 cases, on the right in 11 cases and about equal in the remainder. Similar findings have been reported for infants and fetuses and for some primates as well (Witelson, 1980, and this volume; LeMay and Geschwind, 1975; Yeni-Komshian and Benson, 1976). One could argue that this asymmetry cannot account for the asymmetry in language representation since less than 11% of the population is right-dominant for language (not that it would "account" for the asymmetry without the caveat). What is of interest here is the question of what makes the planum temporale larger or smaller. As Rubens (1977) has shown with 36 cases, the angle and length of the Sylvian fissure is quite different in the two hemispheres: more vertical and shorter on the right side in 27 of the cases. What is impressive is the degree of individual variation in the structure of this neurolinguistically critical area. There is similar variation in asymmetries in the frontal and occipital areas as well (Witelson, 1980). It is not surprising that variations in functional representation exist, considering the variation in the structures that uphold these functions. It is clear that any attempt to generalize across people and to talk of "the brain" as opposed to individual brains will be bedeviled by such variation.

Besides such gross morphological variation, there is a growing list of physical and functional brain differences attributable entirely to experience. For example, early stimulative experiences have been shown to produce a number of changes in brain chemistry (Rosenzweig, 1979), in neuronal morphology (Greenough, 1975), in neuropsychological organization of the visual system (Greenough and Volkmar, 1973; Rose, 1979), and perhaps even in the emotional system (Dennenberg, 1981). Given this scope for inconsistencies across individuals, it is not surprising that attempts to provide relatively direct measures of linguistic representation often produce nonunamimous results (Galambos, Benson, Smith, Schulman-Galambos and Osier, 1975; Ojemann and Whitaker, 1978).

Of course, there is also extensive literature on the effects of early brain damage in promoting some reorganization of skills in the brain (Thorne, 1974; St. James-Roberts, 1981). Without going into a complete review here, it is clear that any early damage creates permanent changes in a brain (Isaacson, 1975; Schneider, 1979), and we do not know the extent to which language skills "relocate" after anything but the most severe brain damage (Annett, 1975; Dennis & Whitaker, 1977). The problem here is that a "relocation" may reflect a change in the cognitive strategies used in communication as much as a true representation of the same language skills in another brain site, and it may be very difficult to detect the difference.

EARLY EXPERIENCES AND PSYCHOLINGUISTIC STRATEGY

Neurological differences among individuals undoubtedly add variance to the "neuro" side of the equation linking the brain to language knowledge and behavior. There are other constitutional factors that also correlate with neuropsychological differences in organization—handedness and sex—which are dealt with below. However, language consists of a number of complex behaviors. Presumably, there are some experiences that influence the cognitive framework of the language. If the cognitive underpinnings of language can vary, presumably so can the neuropsychological representation. Whether or not differences in cognitive linguistic style do in fact contribute to the variation found in neurolinguistic organization has yet to be clearly and consistently demonstrated. It may be instructive to review here a number of possible factors that may influence psycholinguistic strategy in such a way as to affect the results of neurolinguistic studies.

Literacy and Style

Literacy brings great changes to the communicative style of an individual and a culture. Could it be that language as a skill is dealt with by an illiterate person differently from a literate person? Written language imposes a certain form on communication: an increase in propositionalizing and an emphasis on grammar and form, with a relative decrease in emphasis on global and emotional communication, which of course is reintroduced in poetry. Indeed, Cameron, Currier, and Haerer (1971) report relatively less left-hemisphere dominance for language among illiterates. The rate of aphasia after left-hemisphere damage was 78% for literates, 64% for semiliterates, and only 36% for illiterates of the same social and racial group.

Similarly, it is claimed by some that there are languages (or perhaps cultures that are associated with certain languages) that de-emphasize the propositional form that English traditionally takes. Thus, Rogers, Ten-Houten, Kaplan, and Gardiner (1977) report that when Hopi schoolchildren listen to stories in Hopi and in English, the right hemisphere shows significantly more alpha desynchrony (implying greater involvement in the task) during the Hopi story. Similarly, Scott, Hynd, Hunt, and Weed (1979) reported that Navajo subjects show a reversed dominance on a dichotic listening task (a left-ear advantage) compared to Anglo subjects. If these results hold up, they may be due to true differences in neurolinguistic representation or to some alternate strategy during the testing procedure. In either case, individual differences in measures of hemispheric representation of speech can certainly be found.

Deafness and Bilingualism

The vast majority of studies in neurolinguistics consider only monolinguals who have learned an Indo-European language, primarily English. Yet, clearly options exist for other linguistic experiences. For example, when a first language is learned through the visual medium of hand signs, it would be odd indeed if the cognitive requirements of the skill were identical to those of a vocal language (see Ross, this volume, for a review). Similarly, when more than one language is learned, a number of acquisition factors can influence the intellectual niche of each language: the age of acquisition, the stage of competence in the language, the context of communicative style (informal–colloquial versus formal–schooled), and so on (see Vaid, this volume, for a review). The issue here is whether any of these differences in the cognitive aspects of the skill have neurolinguistic significance. From what is known about brain lateralization, we would expect that visual-spatial processing (in the case of sign language) or the person's cognitive use of the particular language to have some bearing on neurolinguistic representation. More generally, the varying cognitive processes involved in language use may have varying intrahemispheric and interhemispheric relationships, perhaps accounting for some of the reports of selective loss of one language in bilinguals and multilinguals (Albert and Obler, 1978; Paradis, 1980).

THE PROBLEM OF TASK DEMANDS

Not all variation in scores on measures of neurolinguistic representation (e.g., aphasia battery subtests, dichotic listening, visual half-field tasks) reflects actual differences in brain organization for language. An elegant demonstration of the possibility of varying neuropsychological de-

mands for language has been shown in the literature on Japanese aphasics, where often the patient retains some reading skill in the ideographic
script, *kanji,* compared to the syllabic sound-based script, *kana* (Sasanuma, 1975). Clearly this change in the task varies the neuropsychological
requirements sufficiently to allow different brain substrates for this aspect
of language function. Differences in apparent neurolinguistic representation, then, may be due to differing task demands in the test situation. For
example, McCusker, Hillinger, and Bias (1981) suggest that English
words of high frequency are read with a visual strategy while those of low
frequency with a phonological strategy. Any differences across studies in
such task demands would confound any differences in results. Similarly,
as Bryden (1978) has shown, allowing subjects in a dichotic listening test
to report freely permits an attentional bias independent of cerebral dominance for language. Asking subjects to attend to only one ear for a series
of trials controls for this bias and produces more modest right-ear advantages.

Also under this category fall individual differences due to selective
hemisphere activation. For example, personality differences among subjects have been found to correlate with direction of lateral eye movements
(Bakan, 1969; Gur and Reivich, 1980; Day, 1977). The neuropsychological
basis for such differences is not known, although several hypotheses have
been offered (e.g., Tucker, 1981). If we accept that lateral eye movements
do indeed reflect hemispheric activation of some sort, then we must accept that there are wide individual differences in hemispheric activation.
This is the notion of hemisphericity: that for some people, one hemisphere tends to be more active, and for others, vice versa. When a range
of scores is produced in a dichotic listening or visual half-field study with
language stimuli, is this due to true varying neurolinguistic representation
or to intersubject variability in hemispheric activation? That is, the pattern of brain organization may be common across subjects, yet the neurolinguistic ''scores'' will show marked variation due to differences in hemisphere activation. Some techniques for measuring, and thereby
controlling for, such differences in hemisphere activation are discussed in
the last section of this chapter.

EVIDENCE FOR THE IMPORTANCE
OF HEMISPHERICITY

Assuming that brain activity reflects mental states, individual differences in mental states will be reflected in neurolinguistic scores. If some
mental states do promote asymmetric brain functions, then of course in
any group with similar neurolinguistic organization there will be variation

due to the range of relevant mental states, as well as that due to measurement error. This will be so despite the care put into constructing the test materials; the background activity cannot be removed, although it can be controlled for.

The strongest evidence of the strength (or weakness) of asymmetric activation comes from studies where differences in mental states are manipulated, although incidental support can be found in naturally occurring differences in hemisphericity. For example, Dabbs (1980) found group differences in resting blood flow to the left and right sides of the head (measured by temperature change at the inside corner of each eye): English majors had greater left-hemisphere flow, architecture majors greater right-hemisphere flow. He did not, however, give a neurolinguistic test to see if general activation differences could be reflected in ear or visual field asymmetries for verbal material. Similarly, lateral eye movements are reputed to reflect general activation, left-hemisphere activity inducing movements to the right and vice versa for the other side. Group differences, as measured by choice of university major chosen (Kocel, Galin, Ornstein, and Merrin, 1972) or by clinical diagnosis (Smokler and Shevron, 1979), have been reported using this measure. Whether or not the lateral eye movements truly reflect the psychological construct suggested is partly irrelevant for the argument here. Rather, it is important that systematic activation differences do occur.

More directly related to the issue of measuring neurolinguistic organization are the studies that find mental states affecting verbal asymmetry scores. The first set involves direct priming: by having subjects first attend to a verbal stimulus, Kinsbourne (1975), and Carter and Kinsbourne (1979) report that subjects show a displaced visual field asymmetry, in favor of the left-hemisphere score. A nonverbal stimulus induces the opposite. The idea here is that the initial, although brief, exposure to the verbal or nonverbal stimulus primes the appropriate hemisphere for action. The empirical reliability of this effect, however, is not clear (Allard and Bryden, 1979; Boles, 1979).

More impressive is a series of studies by Ley (1980), where he administered a lateralization task in three separate experiments (a verbal dichotic listening task, a musical dichotic listening task, and a visual half-field face recognition task) before and after subjects were exposed to a list of words they were to remember. The word lists were either high or low in imagery and either high or low in emotionality. He consistently got a relative increase in left-ear or left visual field reporting for the high imagery and high emotionality groups. Presumably the memory tasks induced different mental states, however subtle, that resulted in a shift in favor of increased right-hemisphere dominance.

Note that in all these studies, no one is suggesting that the pattern of lateralization for mental functions differs across groups. Rather, only the asymmetry SCORES differ as a function of group, of course, increasing the variance. To our knowledge, there is only one personality trait said to be related to actual differences in functional organization. Oltman, Ehrlichman, and Cox (1977), Hoffman and Kagan (1979), and Zoccolotti, Passafiume, and Pizzamiglio (1979) report that high field-dependent subjects produce lower asymmetry scores, which may of course be due to differing attentional and perceptual strategies in these subjects. Note, however, that even here the suggestion is only that the degree of asymmetry is reduced, not that a different pattern is present.

Variance due to individual differences in scores designed to reflect the functional organization of the brain can be used to account for two annoying aspects of this research field. The first is the low amount of variance in asymmetry scores that is accounted for by the lateralization effects under study. For example, Segalowitz and Orr (1981) reported a hemisphere by task (verbal–nonverbal) interaction at the $p < .025$ level, yet accounted for only 4% of the variance. Perusal of the literature indicates that this figure is not uncommon. The rest of the variance may be due to other main factors, such as sex or handedness, but the major share is due simply to subject variation. This subject variation cannot be analysed in the traditional experimental paradigm (some suggestions for how to do so are given below), and this is due in some part to experimental measurement error, and in some part to individual differences of the type discussed so far.

The other annoying aspect of neurolinguistic research with normal subjects is the low test–retest reliability of the standard tasks. For example, Blumstein, Goodglass, and Tartter (1975) report that 29% of subjects switched ear preference on a dichotic listening task with consonants from the first to the second session one week apart (see also Fennell, Bowers, and Satz, 1977; and Porter, Troendle, and Berlin, 1976). Hiscock and Kinsborne (1980) report low reliability on asymmetric tapping tasks in a one-year followup with children. Segalowitz and Orr (1981) found wide fluctuations in visual half-field asymmetries on six testings. The intrasubject variability is alarming and, given no method for stabilizing the scores (possibly due to intrasubject variation in hemisphere activation), it would be useful to be able at least to measure it.

HANDEDNESS AND LANGUAGE
LATERALIZATION

The most widely studied constitutional correlate of cerebral dominance for language is handedness. In this section, we review some of the literature indicating the extent of neurolinguistic variation associated with this variable. We go into some detail here to underscore the difficulty still apparent in determining the differences precisely, despite the large amount of data available.

Clinical Studies

It has long been clear that aphasic disturbances sometimes result from damage to the right hemisphere, and therefore that the right hemisphere is involved in language in some individuals. Furthermore, there seems to be an association between handedness and speech lateralization: disturbances of speech following right-hemispheric damage are more common in left-handers than in right-handers. However, this relationship is not a particularly close one, since the majority of left-handers, like the majority of right-handers, are left-hemispheric for speech. The incidence of right-hemisphere speech in left-handers has not been clearly established, nor do we know what factors predict which left-handers will show left-hemispheric speech representation and which ones will show right-hemispheric speech. A variety of suggestions have been offered in recent years, such as the degree of handedness, the familial history of left-handedness, and the position of the hand in writing (Bradshaw, 1980).

There have been at least two major attempts to integrate the literature on the relation between handedness and aphasia following unilateral brain damage, those of Annett (1975) and Satz (1980). Both authors indicate that left-hemisphere speech representation is the more common state in left-handers, although Satz suggests that the incidence of bilateral speech involvement is very high in left-handers (40%). Annett's concern is more with showing the relation between speech lateralization and her theory of handedness than with establishing any fixed value for the incidence of various forms of lateralization in the left-handed individual.

Obviously, one's conclusions depend in part on the data one chooses to use. Both Satz (1980) and Annett (1975) have very valuable points to make, but Annett has not made provision for the possibility of bilateral speech representation. Satz's analysis also presents a difficulty, outlined below. Consequently, we offer here another examination of the literature on handedness, aphasia, and unilateral brain damage.

The major problem with Satz's (1980) analysis is that he has included data on unilateral damage in left-handers from a variety of different sources, but uses data from only a few major long-term series as a basis for estimating the incidence of aphasia in right-handers. In general, the incidence of aphasia in left-handers is higher in those studies that do not report data on right-handers, than in those that do, suggesting that the criteria for inclusion were rather different in the two types of report. As a result, Satz may well have overestimated the incidence of aphasia in left-handers. In order to have some assurance that the criteria for diagnosing aphasia were the same in left-handers and right-handers, we need studies in which detailed information is provided for both left-handed and right-handed subjects.

This criterion leaves us with six major studies. These are the German war injury study of Conrad (1949), the British war injury series of Newcombe and Ratcliffe (1973), and the lengthy series of patients reported by Penfield and Roberts (1959), Bingley (1958), Hécaen and Ajuriaguerra (1964), and Gloning and Quatember (1966). Data on the incidence of aphasia following unilateral brain damage as a function of handedness from these six series are shown in Table 1. In the case of the Hécaen and Ajuriaguerra series, the data are those for expressive disorders. Table 1 also shows the incidence of aphasia following unilateral damage for both left-handers and right-handers when lesion side is disregarded, the percentage of aphasic patients who were left-handed, and the percentage of all nonaphasic patients who were left-handed.

This table reveals two major problems. Gloning and Quatember (1966) do not indicate how many of their 650 right-handers had left-hemispheric damage and how many had right-hemispheric damage. Furthermore, in the Hécaen and Ajuriaguerra (1964) and the Gloning and Quatember series, the incidence of left-handedness among the aphasics is far higher than among the nonaphasics. While all series but that of Newcombe and Ratcliffe (1973) indicate that the incidence of aphasia is higher in left-handers than in right-handers, the excess of left-handers in the aphasic sample in these two is extremely high. We would agree with Annett (1975) in suggesting that this indicates a very generous criterion for classifying aphasic subjects as left-handed in these two series. We have therefore chosen to omit these studies from further consideration, and to concentrate on the remaining four series.

In their analyses, both Satz (1980) and Annett (1975) have chosen to sum cases across the studies they have considered. Such an approach gives a heavy weight to those studies that involve a large number of patients, such as that of Newcombe and Ratcliffe (1973). At the same time, minor differences in subject selection and criteria for diagnosis lead to differences in the frequency of left-hemispheric and right-hemispheric lesions within handedness groups in the four series. Because we have no reason to believe that any one series is more reliable than the others, we have chosen to average the incidence of aphasia for each hand–hemisphere combination across the four studies. Such a procedure leads to an

Table 1

INCIDENCE OF APHASIA IN MAJOR STUDIES OF UNILATERAL BRAIN DAMAGE

	Left-handers			Right-handers			% LH in aphasics	% LH in nonaphasics
	Left damage	Right damage	Combined	Left damage	Right damage	Combined		
Conrad (1949)	10/19 52.6%	7/18 38.9%	45.9%	175/338 51.8%	11/249 4.4%	31.7%	8.4%	4.8%
Bingley (1958)	2/4 50.0	3/10 30.0	35.7	68/101 61.3	1/99 1.0	34.5	6.8	6.4
Penfield and Roberts (1959)	13/18 72.2	1/15 6.7	42.4	115/157 73.2	1/196 0.5	32.9	10.8	7.4
Newcombe and Ratcliffe (1973)	11/30 36.7	8/33 24.8	30.2	218/338 56.2	19/136 6.0	36.7	7.4	9.5
Hécaen and Ajuriaguerra (1964)	22/37 59.5	11/22 50.0	55.9	55/163 33.7	0/136 0.0	18.4	37.5	9.6
Gloning and Quatember (1966)	28/32 87.5	19/25 76.0	82.4	------231/650 combined------		35.5	16.9	2.3
Average of first 4 studies	52.8%	25.1%		62.1%	3.0%			

estimate of aphasia following unilateral damage to the left hemisphere of 62.1% in right-handers, and 52.8% in left-handers. Comparable figures for damage to the right hemisphere are 3.0% for right-handers, and 25.1% for left-handers.

Now let us make a few assumptions. There is virtually no evidence for bilateral speech representation in right-handers (Rasmussen and Milner, 1977), and therefore let us assume that right-handers are either right-hemispheric or left-hemispheric for speech. The overall ratio of left-hemisphere speech to right-hemisphere speech, then, should be the same as the ratio of the two incidence values, that is, 62.1:3.0. This ratio is obtained when 95.5% of right-handers manifest left-hemispheric speech and 4.5% have right-hemispheric speech, which accords nicely with Rasmussen and Milner's estimate. Furthermore, for such data to arise, unilateral damage in a hemisphere involved in speech must produce an aphasic disturbance in 65% of the cases, given the criteria for aphasia, unilateral damage, and so forth involved in these four series.

Now, let us apply the same logic to the data from left-handers. Note first that the incidence of aphasia following unilateral damage, but disregarding side, is 39% in left-handers and only 32.8% in right-handers. To account for the increased incidence in left-handers, we must either admit the possibility of bilateral speech representation in this group, or assume that language representation in left-handers is somehow different from language representation in right-handers. To us, it seems plausible to assume that language representation in a left-hemispheric left-hander is the same as that in a left-hemispheric right-hander. If this is so, then one can argue that the aphasia in left-handers will also occur in 65% of the cases in which a speech-involved hemisphere is damaged. By this logic, the 52.8% of left-handers who show aphasia following unilateral damage represent 65% of those individuals with some left-hemispheric speech involvement: that is, the left-dominant and bilateral subjects. Similarly, the 25.1% who show aphasia following right-hemispheric damage represent 65% of those with right-hemispheric speech involvement: the right-dominant and bilateral subjects. Solving these two simultaneous equations, we arrive at estimates of 61.4% for left-hemisphere speech, 18.8% for right-hemisphere speech, and 19.8% for bilateral speech in left handers (see Table 2).

$$0.65 \, (L + B) = 0.528$$
$$0.65 \, (R + B) = 0.251$$

where L is the incidence of left-dominance in left-handers, R the incidence of right-dominance in left-handers, and B the incidence of bilateral representation $(1 - L - R)$.

Table 2

ESTIMATED DISTRIBUTION OF LANGUAGE LATERALIZATION

	Left hemisphere	Bilateral	Right hemisphere
Right-handers	95.5%	—	4.5%
Left-handers	61.4%	19.8%	18.8%

These figures suggest a far lower incidence of bilateral speech than the 40% estimated by Satz (1980). Our own confidence in our assumptions, however, is bolstered by the fact that Rasmussen and Milner (1977) have provided almost identical estimates for the relation between handedness and speech lateralization using a very different approach. They determined speech lateralization by sodium amytal testing in a series of patients about to undergo surgery for temporal lobe epilepsy. Excluding those cases in which there was some reason to believe that there was early brain damage, Rasmussen and Milner estimate 4% of right-handers have right-hemispheric speech, while of left-handers, 15% are bilateral and 15% have right-hemispheric speech. Given their sample sizes, these figures are clearly not significantly different from our estimates. Quite similar figures have been estimated by Warrington and Pratt (1973) from the occurrence of dysphasia following unilateral electroconvulsive therapy (ECT), although their estimate of bilateral speech in left-handers is somewhat lower (7%).

Some Statistical Thoughts

Given the fact that the majority of left-handers, like the majority of right-handers, are left-hemispheric for speech, it will necessarily be difficult to discriminate between left-handers and right-handers on any measure related to speech lateralization. Even if we assume perfect measurement, and that all subjects with speech localized in one hemisphere behave in a particular way, it will not be simple to discriminate left-handed and right-handed groups. If we arbitrarily assign values of $+1$ to left-hemispheric dominance, -1 to right hemispheric dominance, and 0 to bilateral representation, for instance, a group of right-handers will yield a mean value of 0.91, with a standard deviation of 0.415, while a group of left-handers will yield a mean of 0.426, with a standard deviation of 0.788. One would need a sample of 13 in each group to show a statistical difference between left-handers and right-handers under the most ideal of conditions.

Obviously, these assumptions concerning perfect measurement and categorical assignment to left, bilateral, or right dominance are rather implausible ones. It is almost certain than any measurement of cerebral lateralization will not be perfect, and, likewise, it is highly plausible that there are degrees of lateralization within each subgroup of right-dominant, left-dominant, or bilateral. Both these facts add variance to our measures of lateralization. If the variance produced by the combination of biological variability and measurement variability is equal to the differences between bilateral speech and unilateral speech (that is, one unit on the hypothetical scale proposed in the preceding paragraph), then there will be considerable variability of scores, and an even greater overlap between left-handed and right-handed distributions. Assuming the distribution of speech lateralization as a function of handedness derived in the previous section, and superimposing a standard deviation of one unit on all the measurements, we obtain the predicted distributions for left-handers and right-handers shown in Figure 1. Note the considerable degree of overlap, and the fact that measurement error pushes many people who are left-hemispheric into the region that appears to indicate right-hemispheric superiority. Also note that in general, the variance of scores for left-handers will be greater than the variance for right-handers. Given these variances, a sample of nearly 50 in each handedness group would be needed to show a significant effect.

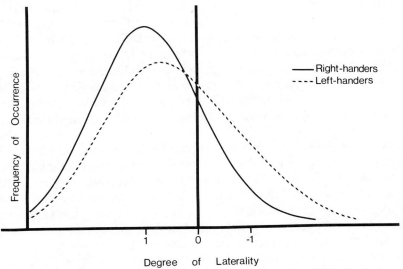

FIGURE 1. Theoretical distributions of laterality scores in left- and right-handers.

We have chosen this particular representation for a number of reasons. It has often been noted that various behavioral measures of cerebral lateralization, such as tests of dichotic listening performance, yield evidence for left-hemispheric speech in only about 80% of right-handers. Figure 1 comes very close to producing this figure from the very simple assumption that there is normally distributed error of measurement.

Bryden (1978, 1982) has pointed out that many of the behavioral measures of cerebral lateralization are affected by attentional biases and strategy effects that are not directly related to cerebral speech lateralization. He has argued that the variability inherent in any behavioral measure can be attributed to two major sources: those related to speech lateralization and those unrelated to it. To the extent that one can refine a measure so that the variability is less than one "unit," one can improve the classification of subjects and make the discrimination between left-handed and right-handed groups easier. Other measures that are correlates of cerebral lateralization may have such a large error of measurement, however, that the distinction between left-handed and right-handed groups would be even more obscure than in Figure 1.

Dichotic Listening Studies

In a dichotic listening procedure, subjects are presented with auditory material in such a way that one item or list arrives at one ear at the same time as a different item or list arrives at the other ear. Kimura (1961a,b) found that there was a close relation between performance on dichotic listening tasks and speech lateralization: those individuals with left-hemispheric speech tended to be more accurate in reporting verbal material from the right ear, while those with right-hemispheric speech were more accurate on the left ear. While there are many variables which affect performance (Bryden, 1978), the dichotic listening procedure does offer a promising means of assessing cerebral speech lateralization in the normal individual.

Our review of the brain-damage literature indicates that there is considerable overlap between left-handers and right-handers in the distribution of speech lateralization. Thus, we would not necessarily expect to find significant differences between left-handers and right-handers in dichotic listening performance. Nevertheless, if the sample sizes are sufficiently large, one might expect to be able to employ the dichotic listening procedure to determine what factors dictate cerebral speech lateralization.

Bryden (1965) compared left-handers and right-handers on a dichotic listening task involving lists of digits. He found that left-handers were less likely to exhibit the normal right-ear advantage than were right-handers. Subsequent to this, there have been a number of studies investigating dichotic performance in left-handers and attempting to subdivide the left-handed group. Most commonly, variables such as sex and the familial history of sinistrality have been of major concern. Of these studies, many have used dichotic lists, a procedure that introduces a variety of extraneous variables (Bryden, 1978); a few studies have used single pairs of consonant–vowel (CV) syllables, providing rather more control over strategy variables.

Virtually all of the dichotic listening studies have shown that the right-ear effect is more robust in right-handers than it is in left-handers. The discrepancies arise in attempting to classify different groups of left-handers. Left-handers with a familial history of sinistrality—that is, those who might be expected to be left-handed for genetic reasons—showed a left-ear effect in a study by Zurif and Bryden (1969), and small right-ear effects in studies by McKeever and VanDeventer (1977), Piazza (1980), and Lishman and McMeekan (1977). Of these, the Piazza (1980) study had the most careful controls over extraneous variables, and thus is the most credible. Her data would indicate that familial left-handers are quite unpredictable with respect to cerebral speech lateralization.

There are, of course, contradictory data: familial sinistrals show a larger right-ear effect in the study of Lake and Bryden (1976) than do nonfamilial sinistrals, and Higgenbottom (1973) found no differences between familial and nonfamilial left-handers.

Likewise, the data concerning sex and strength of handedness are also equivocal. Strong left-handers, as measured by a hand preference inventory, are more likely to show a right-ear advantage than are weak left-handers, according to the data of Dee (1971), while the reverse is true for the data of Satz, Achenbach, and Fennell (1967), and Searleman (1980).

As mentioned above, the extent of nonagreement across studies may be a reflection of the size of the error of measurement variance compared to the variance due to true neurolinguistic differences.

Tachistoscopic Studies

Another approach that has been employed to investigate cerebral lateralization in the normal brain is that of lateralized tachistoscopic presentation. This procedure depends on the observation that sensory inputs from the left visual field (LVF) of each eye are projected directly upon the right cerebral hemisphere, while those from the right visual field (RVF) are projected to the left hemisphere. If some procedure is used to prevent eye movements, such as presenting the stimuli so briefly that saccadic eye movements are not possible, laterally presented stimuli can be projected directly to a specific hemisphere.

Two general procedures are common in tachistoscopic studies. In unilateral presentation, single stimuli are presented to either the LVF or the RVF in random order. The positional uncertainty of the stimulus is used to help in controlling any tendency on the part of the subject to bias fixation away from the central fixation point. Using this procedure, verbal material presented to the RVF is more accurately identified than that presented to the LVF (Bryden, 1965; White, 1969). Other investigators have preferred to use an approach more analogous to the dichotic listening procedure, employing a bilateral presentation procedure in which different stimuli are presented simultaneously to the LVF and RVF. To control fixation, such investigators often present a small number or other discriminandum at fixation and require the subject to identify this correctly before scoring the trial (McKeever and Huling, 1971). With this procedure, a very large and robust RVF superiority is obtained. Although Hines (1975) has argued that the unilateral procedure assesses information loss during callosal transmission, while the bilateral procedure forces the LVF stimuli to be analyzed in the right hemisphere and thus provides an assessment of the independent contributions of the two hemispheres, his work has not been followed up systematically, and most investigators employ the two procedures interchangeably.

In general, the RVF superiority for verbal material is less evident or absent in left-handers using either procedure (Bryden, 1965; Cohen, 1972; Haun, 1978; Bradshaw and Taylor, 1979). Such findings would support the argument that performance on lateralized tachistoscopic tasks is related to cerebral speech lateralization. Attempts to isolate factors that will predict which left-handers are right-hemispheric for speech through such procedures have been less convincing, however.

As with dichotic techniques, familial sinistrality has often been considered as one possible variable that would predict which left-handers would show an RVF effect. Generally, the argument has been that left-handers with a familial history of sinistrality are more likely to manifest right-hemispheric speech representation, and therefore would be less likely to show an RVF superiority. Such results have, in fact, been reported by Zurif and Bryden (1969) on a single-letter identification task, and by Schmuller and Goodman (1979) with a bilateral presentation procedure.

Bryden (1973) assessed performance on a single-letter identification task, using unilateral presentation. He found that familial sinistrals, regardless of the hand they themselves used, showed little or no visual field difference, while subjects without familial history of sinistrality exhibited RVF superiorities. Any interpretation of this however, is clouded by two other findings. First, visual field differences were predicted by sighting dominance, with right-eyed subjects yielding RVF effects, and left-eyed subjects producing LVF effects. Secondly, strength of handedness was also a relevant variable, with LVF effects being found in the strong left-handers, but not in the weak left-handers.

Contrary results have been reported by McKeever and VanDeventer (1977), who have also investigated the effects of handedness and familial sinistrality on a tachistoscopic task, in this case a bilateral task in which single letters were exposed in each visual field, and then masked by a second letter pair. Their data indicate that the weakest RVF effect is seen in left-handers *without* a familial history of sinistrality. In right-handers, there is a rather pronounced sex effect, with males showing weaker laterality effects than females, while in the left-handers there is a familial history effect, with the nonfamilial left-handers exhibiting reduced laterality.

Piazza (1980) also tested the effects of handedness and familial sinistrality on tachistoscopic performance. She used a bilateral presentation procedure, with a partial report cue to indicate to the subject which item to report. She found an overall RVF superiority, with no significant effects of sex, handedness, or familial sinistrality, nor any interactions between these variables. An examination of her data reveals slightly weaker RVF effects in left-handers than in right-handers, and a very weak trend for those with a positive familial history of left-handedness to show smaller laterality effects than those without such a history. Piazza apparently displayed her verbal material horizontally, and it may be that directional processing factors outweighed any cerebral lateralization effects in her data.

These studies of verbal laterality effects with tachistoscopic presentation have not been particularly helpful in clarifying the factors that lead some left-handers to manifest right-hemispheric speech. Some studies find that familial left-handers have weak lateralization (Bryden, 1973; Zurif and Bryden, 1969; and to a very limited extent, Piazza, 1980), while others indicate that it is the nonfamilial left-handers who are weakly lateralized (Higgenbottom, 1973; McKeever, VanDeventer, and Suberi, 1973; McKeever and VanDeventer, 1977; McKeever, 1979). In some investigations, strongly left-handed subjects are poorly lateralized or show LVF superiorities (Bryden, 1973), while in others strength of handedness is not mentioned, or has no effect. This situation is further complicated by the possibility that sighting dominance may be pertinent (Bryden, 1973), and by Pirozzolo and Rayner's (1979) finding that right-handers, but not left-handers, show shorter eye-movement latencies to the right than to the left.

One final predictor worthy of mention is that of writing posture. Levy and Reid (1978) have provided some evidence to suggest that those left-handers who write with a hooked or inverted writing posture are left-hemispheric for speech, while those who write with a normal or upright posture are right-hemispheric. Levy and Reid tested upright and inverted writers on two lateralized tachistoscopic tasks, one involving the identification of vertically oriented nonsense syllables and the other requiring the localization of a dot in space. Left-handed inverters were found to perform like right-handers, exhibiting an RVF superiority on the verbal task, and an LVF superiority on the dot localization task. In contrast, those left-handers who wrote in an upright position showed the reverse pattern of results.

There have been several attempts to replicate the Levy and Reid findings. Moscovitch and Smith (1979) found visual differences of the type postulated by Levy and Reid, but no differences between inverted and upright left-handers in dichotic listening performance. They suggest that differences between inverted and upright writers lie in the visual or visuomotor system, rather than in speech lateralization or in hand control. McKeever (1979) found bigger RVF effects on a color-naming task for inverted writers than for normal writers, in accordance with Levy and Reid's predictions. Within the sample, however, hand posture was correlated with familial sinistrality, and McKeever suggests that the reduced laterality found in nonfamilial left-handers is a more important factor in determining performance on the color-naming task. In the same study, McKeever (1979) found no differences between handwriting posture groups in either unilateral or bilateral word recognition tasks, but again observed reduced laterality in nonfamilial left-handers. McKeever and Hoff (1979) have subsequently argued for a visuomotor anomaly in inverted writers similar to that posited by Moscovitch and Smith (1979). Thus, the majority of evidence now indicates that handwriting posture cannot be used to predict cerebral speech lateralization.

In summary, then, left-handers have frequently been shown to differ from right-handers on behavioral measures of lateralization. While familial history of sinistrality has often been suggested as a factor differentiating left-handers with left-hemispheric speech from these with right-hemispheric speech (cf. Hécaen and Sauguet, 1971), the evidence concerning this variable remains equivocal. Strength of hand preference and writing position have provided even more ambiguous data. At present, we have not yet isolated the critical variables for differentiating types of left-handers.

SEX DIFFERENCES AND NEUROLINGUISTIC ORGANIZATION

Over the past decade, it has almost become part of the neuropsychological lore that women are less lateralized than men. Levy (1972) has argued that women exhibit bilateral representation of speech, and that the consequent involvement of the right hemisphere in speech processes results in a deficit in spatial abilities in females. Recent reviews of the literature on hemispheric asymmetry have also found general support for the view that language processes are less lateralized in the female than in the male (Bryden, 1979; McGlone, 1980).

The data supporting such a contention come from two major sources. First, there have been a number of studies involving sex differences following unilateral brain damage. Second, investigators dealing with asymmetries of information processing in the normal subject have begun to examine sex differences.

McGlone (1977, 1978, 1980) has provided some of the strongest data in support of the notion of sex differences in language representation. She administered the WAIS to adult subjects following stroke and with tumors confined to one hemisphere. In men, left-hemispheric damage led to a significant decrement in Verbal IQ relative to Performance IQ, while such was not the case for women. Conversely, right-hemispheric damage led to a significant Performance IQ deficit in men, but not in women. McGlone's initial report (1978) involved 40 men and 37 women; a follow-up study (1980) added 14 men and 11 women. Overall, the men showed a Verbal IQ of 86.5 following left-hemispheric damage, and 106.5 following right-hemispheric damage. Left-hemispheric women averaged 100.6 after left-hemispheric damage, and 98.4 following right-hemispheric damage. McGlone (1978) indicates that the incidence of aphasia is much higher in men following left-hemispheric damage, but that the deficit in Verbal IQ is present even among the nonaphasics.

Very similar results have been reported by Lansdell and Urbach (1965) in a study of unilateral temporal lobectomy. To provide further support for the contention that there are sex differences in language representation, Inglis and Lawson (1981) carried out a retrospective examination of studies reporting on the interaction of Verbal IQ and Performance IQ by side of lesion. Of six studies reporting a significant interaction, all had a predominance of male subjects, with men representing 89% of the total sample. In contrast, of eight studies reporting no interaction, seven included a fairly large proportion of female subjects, and men represented only 60% of the total sample. The one failure to find a significant interaction of IQ subscale by side of lesion with a totally male sample is that of Zimmerman, Whitmyre, and Fields (1970). Furthermore, Inglis and Lawson (1981) report a highly significant correlation between the magnitude of the interaction and the proportion of men in the sample. These data, drawn as they are from studies that were not explicitly concerned with sex differences, provide strong support for McGlone's (1978) contention that verbal skills are more bilaterally represented in women than in men.

Additional support for this contention comes from the study of commissurotomy patients, in that males show poorer spatial skills relative to verbal skills when the response is limited to the left hemisphere (Bogen, DeZure, TenHouten, and Marsh, 1972).

No sex differences in the incidence of aphasia following unilateral brain damage were reported by Hécaen, DeAgostini, and Monzon-Montes (1981). These authors do note, however, that the incidence of spatial deficits following right-hemispheric damage in right-handers is much higher in men (53%) than in women (27%). Furthermore, the authors note (p. 281) that in their experience aphasia following left-hemispheric damage is more frequent and more severe in men than in women.

How can these data be interpreted? Rasmussen and Milner (1977) indicate that virtually no right-handers are shown to have bilateral speech representation by sodium amytal testing. Nevertheless, the unilateral brain damage literature would suggest that language functions are more bilaterally represented in women than in men. As McGlone (1980) points out, the sodium amytal procedure assesses the productive aspects of language, and it may well be the more receptive aspects that are bilaterally represented.

Of course, in any study of unilateral brain damage there are almost certain to be severe problems in matching groups for age, previous experience, size of lesion, location of lesion, and ontogeny of damage. Many of these points have been made by Kinsbourne (1980) in his commentary on McGlone's (1980) review. Kimura (1980) has suggested that aphasia does not often appear in women with left-posterior damage. Over and above these problems, however, there remain two critical problems with the clinical literature. First, we do not know anything about these patients before the onset of stroke or tumor, and as a result we do not know just how much of a deficit in Verbal IQ or Performance IQ has been produced by the brain damage. It is possible, for example, that left-damaged men show lower Verbal IQ than left-damaged women prior to the occurrence of brain damage. Second, we do not know how the subjects perform the task. Subjects can successfully accomplish behavioral tasks in a variety of different ways, and we normally have no way of knowing what strategy a particular subject used on a particular occasion. It may well be, for instance, that women are more likely to use visuospatial strategies based on right-hemisphere mechanisms to perform ostensibly verbal tasks than are men.

As with the investigation of handedness, an alternative approach is to study lateralization in the normal subject with behavioral or electrophysiological measures. Many studies of this type have been reviewed by Bryden (1979) and McGlone (1980), and we shall not attempt a thorough review here. Rather, we shall provide a general overview and critique of such research.

Many dichotic listening studies have found that males are more likely to show a right-ear advantage for verbal material than are women, or show a larger right-ear effect. Thus, Lake and Bryden (1976) reported that 94% of their right-handed men, but only 67% of their right-handed women, exhibited a right-ear superiority. Similar findings, though not so clearly significant, have been reported by Bryden (1966), McGlone and Davidson (1973), Harshman, Remington, and Krashen (1975), Van Lancker and Fromkin (1973), Briggs and Nebes (1976), Ryan and McNeil (1974), and Piazza (1980). On the other hand, Carr (1969) and Bryden (1979) both report studies in which no sex-related differences were evident.

With the exception of the Piazza (1980) study, all of the above experiments employed procedures in which the subject was free to choose the way in which he or she deployed attention. Bryden (1978) has pointed out that such a procedure opens the way for individual differences in the deployment of attention that may obscure any underlying laterality effect. Using a procedure that required subjects to attend a specified ear, Bryden, Munhall, and Allard (1980) failed to find any evidence for sex-related differences in dichotic lateralization. They suggested that the sex-related differences might be due to sex differences in the deployment of attention, rather than to any greater left-hemispheric specialization for language in males. By this argument, men are simply more likely to devote attention to the right ear in a dichotic task than are women. In agreement with this, the sex differences in Piazza's (1980) study, which involved focused attention, are not statistically significant, although the males do show a slightly larger right-ear effect. However, if this effect can be shown to hold only for verbal material, it may simply be another manifestation of language lateralization.

Without further investigation of the differences between focused and divided attention, the dichotic results must be viewed as equivocal. While there is some trend for men to show a larger right-ear advantage, it remains quite plausible that this is a strategy difference rather than an indication of an underlying greater hemispheric specialization for language in men.

The literature on sex differences in visual laterality tasks is at least as equivocal as that from dichotic listening. One reason for this is the wide variety of different tasks that have been used. As discussed earlier, some experimenters use a unilateral mode of presentation while other experimenters use bilateral presentation, and the two procedures may yield different results. Other problems in the visual presentation of verbal material include the orientation of the display. Horizontally displayed material is much more familiar to us, but may introduce a problem of directional scanning. It may also lead to critical parts of a word being more visible in one visual field than the other, because of the sharp drop in visual acuity with increasing retinal eccentricity (Bryden, 1982). Furthermore, some tasks, such as naming, must eventually involve motor components necessary for speech production, while other tasks, such as the lexical decision task, do not.

A greater retention for verbal material in the right visual field (left hemisphere) for men has been claimed by Hannay and Malone (1976). They presented a CVC trigram in vertical orientation to either the left or right visual field. Following a delay of from 0 to 10 seconds, the subject looked away from the tachistoscope to a display upon which a second CVC syllable appeared, and was asked to state whether this was the same as the original or different from it. Men showed a significant RVF superiority on this task when there was a delay, but women did not. Although Hannay and Malone (1976) suggest that men show a left hemispheric specialization for the retention of verbal material, there is little that is necessarily verbal about this task. Since the subject only has to indicate whether or not the two stimuli are the same, there is no necessity for there to be access to any verbal system. Only the fact that there is an RVF superiority on the task suggests that it is verbal in nature. It is worth noting that when Hannay and Boyer (1978) repeated the experiment as a multiple-choice study, they found no sex differences.

McKeever and Jackson (1979) found a much greater RVF effect for men than for women when subjects were asked to name outline drawings of familiar objects. However, Bradshaw and Gates (1978) and Piazza (1980) found no sex differences for word naming, and McKeever and Van-Deventer (1977) reported greater RVF effects for right-handed women than for right-handed men on a letter identification task. In general, then, the experiments involving a verbal naming response are inconsistent, with men sometimes showing greater RVF effects than women and sometimes smaller ones.

In the lexical decision task, where overt naming is not required but the subject must certainly access some central lexical representation, men show an RVF superiority while women do not (Bradshaw, Gates, and Nettleton, 1977; Bradshaw and Gates, 1978). Bradshaw and Gates suggest that women have secondary speech mechanisms at the lexical level in the right hemisphere, whereas men do not.

Finally, using a task in which subjects were required to indicate whether two letters had the same name or not, Segalowitz and Stewart (1979) report some interesting sex differences. Men were much more rapid in the RVF when a name match (e.g., *Aa*) was required, and much faster in the LVF when only a physical match was needed (e.g., *AA*). Similar findings have been reported previously by Cohen (1972) and by Geffen, Bradshaw, and Nettleton (1972). Women, in contrast, showed an RVF superiority for name matching but no visual field effect for physical matching. Segalowitz and Stewart (1979) suggest that women and men use different strategies for performing the task, and that women adopt a strategy that does not provide a clear differentiation between verbal visuospatial effects.

The Segalowitz and Stewart (1979) study illustrates the importance of facing one of the major problems in interpreting the evidence for sex differences in behavioral measures of laterality. If men and women adopt different ways of approaching a particular task, we cannot be sure whether to attribute sex differences in laterality to differences in the cerebral representation of language processes or to differences in the strategy employed. For this reason, it is particularly important to understand the experimental procedure being employed and to attempt to control individual differences in strategy (cf. Bryden, 1982; Bryden, Munhall, and Allard, 1980).

METHODS FOR MEASURING INDIVIDUAL DIFFERENCES IN BRAIN LATERALIZATION

The problem of individual differences is not only one of unwanted variance (unwanted, that is, for our simple theories), but also one of unknown variance. It would be helpful to know how much variation in neurolinguistic scores, such as those for dichotic listening, is due to neurolinguistic organization, and not stemming from the variety of sources as discussed above. We can control in our studies for some of the factors, such as handedness and sex, which is, of course, why these are well examined. The others are elusive, though, and the problem they raise is inherent in the classic group design used in experiments. If we could examine an individual in sufficient detail, we could determine the reliability of, say, a dichotic listening score for that person. Determining the reliability requires, by definition, a measure of the inherent variance in obtaining that score. Rather than testing an individual 30 times, in order to obtain a distribution from which we can determine a variance estimate, we normally test 30 similar individuals, and that is where the unwanted and unknown variance comes in. In the idealized situation, the 30 individuals represent a multiple case study; the 30 subjects are treated as one. We do our best to control as many relevant factors as possible, but total control is impossible (see Dennenberg, 1979, for a fuller discussion of this issue).

Fortunately, there are ways to reduce some of the unwanted variation (i.e., that is not due to neurolinguistic differences) and some ways of examining individuals enough to determine a distribution for their individual scores. The first task involves including appropriate control conditions in the neurolinguistic investigation. The second involves determining the intrasubject variance.

Reducing the Variance with Relative Measures

Considering the possibility of differences in hemisphericity (relative hemispheric activation) across people or across time within the same individual, a person's neurolinguistic score is a reflection of brain organization and relative hemisphericity. Thus, a range of scores is obtainable that really reflects a single underlying organization. However, such variation due to hemisphericity is, by definition, a general effect and should apply to all tasks, linguistic or nonlinguistic. Therefore, while an individual lateralization score will vary according to many factors, the difference between the scores obtained on two tasks varying only in stimulus content will reflect the relative asymmetry for mental processing of the two tasks. For example, in the study by Segalowitz and Orr (1981) mentioned above, a significant RVF bias was found for reading vertically oriented nonsense syllables, while reading the times on clockface stimuli produced virtually no asymmetry. This latter result may be due to an equal involvement of the two hemispheres in the task, or may be due to a balancing of the (right-hemisphere dominant) visuospatial requirements with the (left-hemisphere dominant) verbal response, or with some other left-hemisphere bias inherent in tachistoscopic studies. Actually, though, the real source is not important, since the change in task from reading verbal to reading visuospatial stimuli was significant. Interpretation of the single verbal task significance is clouded by the same issues: is it really cerebral asymmetry for speech that produces the RVF bias for the syllables, or is it some aspect of the task demands, such as tachistoscope use or verbal response? Interpreting the interaction is clear: with only the processing requirements being varied, the resulting interaction must be due to them.

The use of relative measures has two effects on our research methods. First of all, hemisphericity differences are no longer a problem since this kind of cerebral activation applies across all tasks. The absolute asymmetry will reflect hemisphericity, but the relative score (i.e., the differences) will reflect only organization (plus the usual error variance). The same caution, however, should be exercised in constructing the control conditions as the experimental conditions. All subjects must apply the same processing strategy to the control task or else a new source of variance will simply replace the old one.

Second, the use of relative measures subtly changes our research questions: we no longer can ask for an absolute asymmetry on a task, only a relative one. Thus, in the study described above, Segalowitz and Orr (1981) cannot claim that the interaction between visual field bias and stimulus condition indicates a left dominance for reading syllables and a right dominance for the visuospatial aspect of reading clockfaces. Rather, there is relatively more left-hemisphere dominance for the verbal task and relatively more right (or less left) for the other one. Only lesion studies can examine absolute representation of functions.

GETTING A MEASURE OF INTRASUBJECT VARIANCE

Suppose on a dichotic listening or visual half-field task, a subject obtains an asymmetry score as follows: of 48 items presented to the right side, 32 are correct; of 48 items on the left, 18 are correct. This is a rather larger asymmetry than is usually obtained, but it will suit our purposes. Is the difference of 14 (out of a possible 48) significant? Calculated another way, is the lateral difference (LD) relative to the total correct (see Formula 1) significant, when it reaches 14/50 = .28?

$$LD = (R - L)/(R + L) \qquad (1)$$

Since we have no distribution for these scores, we cannot say. Recently, however, Bryden and Sprott (1981) have developed a new laterality metric, lambda, with a derivable distribution. It takes account of the number of errors as well as the number correct (see Formula 2); in this example ln $(32 \times 30)/(16 \times 18) = 1.20$.

$$\lambda = \ln[(Rc \times Li)/(Ri \times Lc)] \tag{2}$$

where R and L are right and left side
c and i are correct and incorrect

Positive values indicate a left-hemisphere superiority. Granting the assumption that the correctness of the response on one trial is not influenced by the correctness of the previous trial (an assumption we will come back to later), lambda follows a logistic distribution with the variance equal to the sum of the reciprocals of hits and misses (Formula 3); in our example $1/32 + 1/30 + 1/18 + 1/30 = .1826$.

$$\text{variance} = 1/Rc + 1/Ri + 1/Lc + 1/Li \tag{3}$$

The standard deviation is .427, and with lambda being approximately normally distributed, a Z score is obtainable; in our example $Z = \lambda/s = 1.20/.427 = 2.8$, $p < .01$. Thus, this individual's scores are significantly asymmetric. The values of λ and LD are very highly correlated, but lambda does not produce a discontinuity at zero and allows an estimate of variance for the individual.

A second method of calculating the intrasubject variance is to actually test the same individual many times. From these scores, either LD or λ, a variance estimate could be obtained by the usual sums of squares. The average score would follow a t-distribution and can be tested against a population mean of zero.

Why bother with the multiple testing if a single testing will do? One reason is to be able to calculate the significance of the asymmetry for the individual (if one is interested in that). Another reason is that we can test the assumption mentioned previously, that the correctness on one trial is independent of the correctness on the previous one. If this assumption is rejected, then the individual has not given 96 truly independent responses, and is it incorrect to apply the formula for the variance estimate of λ. To test this assumption, it should be remembered that two estimates of the error variance are calculable: one as given above (a within-estimate) and one calculated from the set of λ's (a between-estimate). If the subject is somehow correlating his performance across trials, then the λs will be much more aligned than expected, and the ratio will be significant on an F-distribution. In the data from the eight subjects Segalowitz and Orr (1981) report, none of the subjects committed this sin.

The inverse ratio, the between-estimate of the variance divided by the within-estimate, is also of interest. It tests whether or not the variation in λ from test session to test session is greater than expected. If so, then it may be that the subject's hemisphericity or response strategies vary widely and so no individual λ can be trusted as a truly representative neurolinguistic score.

COPING WITH INDIVIDUAL DIFFERENCES IN NEUROLINGUISTIC ORGANIZATION

The field of experimental human neuropsychology has now advanced beyond the stage where error of measurement and individual differences in brain organization can remain confounded. We are constantly reminded from the clinical literature that every human brain is unique in its structure and organization, at least in the fine detail of the latter. Much of the neurolinguistic literature now deals with such fine detail—specific aspects of the phonological, syntactic, and semantic systems.

We can and do expect sex and handedness to be important constitutional factors in neurolinguistic organization. However, rather than seeing such variation as a handicap to overcome in our research, we could instead use it to our advantage. By employing appropriate controls, we can use differences between groups to highlight possible patterns of neurolinguistic organization. Appropriate use of intrasubject control procedures, as outlined in this chapter, makes the intergroup variation manageable and, hopefully, illuminating.

REFERENCES

Albert, M. L., and Obler, L. K. (1978). *The bilingual brain*. New York: Academic Press.

Allard, F., and Bryden, M. P. (1979). The effect of concurrent activity on hemispheric asymmetries. *Cortex, 15*, 5–17.

Annett, M. (1975). Hand preference and the laterality of cerebral speech. *Cortex, 11*, 305–328.

Bakan, P. (1969). Hypnotizability, laterality of eye movement and functional brain asymmetry. *Perceptual and Motor Skills, 28*, 927–932.

Bingley, T. (1958). Mental symptoms in temporal lobe epilepsy and temporal lobe gliomas. *Acta Psychiatrica et Neurologica Scandinavica, 33*, Supplementum.120.

Blumstein, S., Goodglass, H., and Tartter, V. (1975). The reliability of ear advantage in dichotic listening. *Brain and Language, 2*, 226–236.

Bogen, J. E., DeZure, R., TenHouten, W. D., and Marsh, J. F. (1972). The other side of the brain. IV. The A/P ratio. *Bulletin of the Los Angeles Neurological Societies, 37*, 49–61.

Boles, D. (1979). Laterally biased attention with concurrent verbal load: Multiple failures to replicate. *Neuropsychologia, 17*, 353–362.

Bradshaw, J. L. (1980). Right hemisphere language: Familial and nonfamilial sinistrals, cognitive deficits and writing hand position in sinistrals, and the concrete–abstract, imageable–nonimageable dimensions in word recognition. A review of interrelated issues. *Brain and Language, 10*, 172–188.

Bradshaw, J. L., and Gates, E. A. (1978). Visual field differences in verbal tasks: Effects of task familiarity and sex of subject. *Brain and Language, 5,* 166–187.

Bradshaw, J. L., Gates, A., and Nettleton, N. C. (1977). Bihemispheric involvement in lexical decisions: Handedness and a possible sex difference. *Neuropsychologia, 15,* 277–286.

Bradshaw, J. L., and Taylor, M. J. (1979). A word-naming deficit in nonfamilial sinistrals? Laterality effects of vocal responses to tachistoscopically presented letter strings. *Neuropsychologia, 17,* 21–32.

Briggs, G. G., and Nebes, R. D. (1976). The effects of handedness, family history and sex on performance of a dichotic listening task. *Neuropsychologia, 14,* 129–134.

Bryden, M. P. (1965). Tachistoscopic recognition, handedness, and cerebral dominance. *Neuropsychologia, 3,* 1–8.

Bryden, M. P. (1966). Left–right differences in tachistoscopic recognition: Directional scanning or cerebral dominance? *Perceptual and Motor Skills, 23,* 1127–1134.

Bryden, M. P. (1973). Perceptual asymmetry in vision: Relation to handedness, eyedness, and speech lateralization. *Cortex, 9,* 418–435.

Bryden, M. P. (1978). Strategy effects in the assessment of hemispheric asymmetry. In G. Underwood (Ed.), *Strategies of information processing.* London: Academic Press.

Bryden, M. P. (1979). Evidence for sex-related differences in cerebral organization. In M. Wittig and A. C. Petersen (Eds.), *Sex-related differences in cognitive functioning: Developmental issues.* New York: Academic Press.

Bryden, M. P. (1982). The behavioral assessment of lateral asymmetry: Problems, pitfalls, and partial solutions. In R. N. Malatesha and L. C. Hartlage (Eds.), *Neuropsychology and cognition,* Vol. 2. Alphen aan den Rijn, The Netherlands: Sijthoff and Noordhoff International Publishers.

Bryden, M. P., Munhall, K., and Allard, F. (1980). Attentional biases and the right-ear effect in dichotic listening. Paper presented at the annual meeting of the Canadian Psychological Association, Calgary.

Bryden, M. P., and Sprott, D. A. (1981). Statistical determination of degree of laterality. *Neuropsychologia, 19,* 571–581.

Cameron, R. F., Currier, R. D., and Haerer, A. F. (1971). Aphasia and literacy. *British Journal of Disorders of Communication, 6,* 161–163.

Carr, B. M. (1969). Ear effect variables and order of report in dichotic listening. *Cortex, 5,* 63–68.

Carter, G. L., and Kinsbourne, M. (1979). The ontogeny of right cerebral lateralization of spatial mental set. *Developmental Psychology, 15,* 241–245.

Cohen, G. (1972). Hemispheric differences in a letter classification task. *Perception and Psychophysics, 1,* 139–142.

Conrad, K. (1949). Über aphasische Sprachstörungen bei hurnverletzten Linkshändern. *Nervenarzt, 20,* 148–154.

Dabbs, J. M. (1980). Left–right differences in cerebral blood flow and cognition. *Psychophysiology, 17,* 548–551.

Day, J. (1977). Right hemisphere language processing in normal right-handers. *Journal of Experimental Psychology: Human Perception and Performance, 3,* 518–528.

Dee, H. L. (1971). Auditory asymmetry and strength of manual preference. *Cortex, 7,* 236–245.

Dennenberg, V. H. (1979). Dilemmas and designs for developmental research. In C. L. Ludlow and M. E. Doran-Quine (Eds.), *The neurological bases of language disorders in children: Methods and directions for research.* Washington, D.C.: National Institutes of Health.

Dennenberg, V. H. (1981). Hemispheric laterality in animals and the effects of early experience. *The Behavioral and Brain Sciences, 4,* 1–50.

Dennis, M., and Whitaker, H. A. (1977). Hemispheric equipotentiality and language acquisition. In S. J. Segalowitz and F. A. Gruber (Eds.), *Language development and neurological theory.* New York: Academic Press.

Fennell, E. G., Bowers, D., and Satz, P. (1977). Within-modal and cross-modal reliabilities of two laterality tests. *Brain and Language, 4,* 63–69.

Galambos, R., Benson, P., Smith, T. S., Schulman-Galambos, C., and Osier, H. (1975). On hemispheric differences in evoked potentials to speech stimuli. *Electroencephalography and Clinical Neurophysiology, 39,* 279–283.

Geffen, G., Bradshaw, J. L., and Nettleton, N. C. (1972). Hemispheric asymmetry: Verbal and spatial coding of visual stimuli. *Journal of Experimental Psychology, 95,* 25–31.

Geschwind, N., and Levitsky, W. (1968). Human brain: Left–right asymmetries in temporal speech region. *Science, 161,* 186–187.

Gloning, I., and Quatember, R. (1966). Statistical evidence of neuropsychological syndrome in left-handed and ambidextrous patients. *Cortex, 2,* 484–488.

Greenough, W. T. (1975). Experimental modification of the developing brain. *American Scientist, 63,* 37–46.

Greenough, W. T., and Volkmar, F. R. (1973). Pattern of dendritic branching in occipital cortex of rats reared in complex environments. *Experimental Neurology, 40,* 491–504.

Gur, R. C., and Reivich, M. (1980). Cognitive task effects on hemispheric blood flow in humans: Evidence for individual differences in hemispheric activation. *Brain and Language, 9,* 78–92.

Hannay, H. J., and Boyer, C. L. (1978). Sex differences in hemispheric asymmetry revisited. *Perceptual and Motor Skills, 47,* 315–321.

Hannay, H. J., and Malone, D. R. (1976). Visual field effects and short-term memory for verbal material. *Neuropsychologia, 14,* 203–209.

Harshman, R. A., Remington, R., and Krashen, S. D. (1975). *Sex, language and the brain, Part II: Evidence from dichotic listening for adult sex differences in verbal lateralization.* Unpublished manuscript, University of California, Los Angeles.

Haun, F. (1978). Functional dissociation of the hemispheres using foveal visual input. *Neuropsychologia, 16,* 725–733.

Hécaen, H., and Ajuriaguerra, J. (1964). *Left-handedness: Manual superiority and cerebral dominance.* New York: Grune and Stratton.

Hécaen, H., DeAgostini, M., and Monzon-Montes, A. (1981). Cerebral organization in left-handers. *Brain and Language, 12,* 261–284.

Hécaen, H., and Sauguet, J. (1971). Cerebral dominance in left-handed subjects. *Cortex, 7,* 19–48.

Higgenbottom, J. A. (1973). Relationships between sets of lateral and perceptual preference measures. *Cortex, 9,* 402–409.

Hines, D. (1975). Independent functioning of the two cerebral hemispheres for recognizing bilaterally presented tachistoscopic visual half-field stimuli. *Cortex, 11,* 132–143.

Hiscock, M., and Kinsbourne, M. (1980). Asymmetry of verbal–manual time sharing in children: A follow-up study. *Neuropsychologia, 18,* 151–167.

Hoffman, C., and Kagan, S. (1977). Lateral eye movements and field dependence–independence. *Perceptual and Motor Skills, 45,* 767–778.

Inglis, J., and Lawson, J. S. (1981). Sex differences in the effects of unilateral brain damage on intelligence. *Science, 212,* 693–695.

Isaacson, R. L. (1975). The myth of recovery from early brain damage. In N. E. Ellis (Ed.), *Aberrant development in infancy.* New York: Wiley.

Kimura, D. (1961a). Cerebral dominance and the perception of verbal stimuli. *Canadian Journal of Psychology, 15,* 166–171.

Kimura, D. (1961b). Some effects of temporal lobe damage on auditory perception. *Canadian Journal of Psychology, 15,* 156–165.

Kimura, D. (1980). Sex differences in intrahemispheric organization of speech. *The Behavioral and Brain Sciences, 3,* 240–241.

Kinsbourne, M. (1975). The mechanism of hemispheric control of the lateral gradient of attention. In P. M. A. Rabbitt and S. Dornic (Eds.), *Attention and performance,* Vol. V New York: Academic Press.

Kinsbourne, M. (1980). If sex differences in brain lateralization exist, they have yet to be discovered. *The Behavioral and Brain Sciences, 3,* 241–242.

Kocel, K., Galin, D., Ornstein, R., and Merrin, E. L. (1972). Lateral eye movements and cognitive mode. *Psychonomic Science, 27,* 223–224.

Lake, D. A., and Bryden, M. P. (1976). Handedness and sex differences in hemispheric asymmetry. *Brain and Language, 3,* 266–282.

Lansdell, H., and Urbach, N. (1965). Sex differences in personality measures related to size and side of temporal lobe ablations. *Proceedings of the 73rd Annual Convention of the American Psychological Association.*

LeMay, M., and Geschwind, N. (1975). Hemispheric differences in the brains of great apes. *Brain, Behavior, and Evolution, 11,* 48–52.

Levy, J. (1972). Lateral specialization of the human brain: Behavioral manifestations and possible evolutionary basis. In J. A. Kiger (Ed.), *The biology of behavior.* Corvallis, Ore.: Oregon State Univ. Press.

Levy, J., and Reid, M. (1978). Variations in cerebral organization as a function of handedness, hand posture in writing, and sex. *Journal of Experimental Psychology: General, 107,* 119–144.

Ley, R. G. (1980). *Emotion and the right hemisphere.* Unpublished Ph.D. dissertation, University of Waterloo.

Lishman, W. A., and McMeekan, E. (1977). Handedness in relation to direction and degree of cerebral dominance for language. *Cortex, 13,* 30–43.

McCusker, L. X., Hillinger, M. L., and Bias, R. G. (1981). Phonological recording and reading. *Psychological Bulletin, 89,* 217–245.

McGlone, J. (1977). Sex differences in the cerebral organization of verbal functions in patients with unilateral brain lesions. *Brain, 100,* 775–793.

McGlone, J. (1978). Sex differences in functional brain asymmetry. *Cortex, 14,* 122–128.

McGlone, J. (1980). Sex differences in human brain organization: A critical survey. *The Behavioral and Brain Sciences, 3,* 215–227.

McGlone, J., and Davidson, W. (1973). The relation between cerebral speech laterality and spatial ability with special references to sex and hand preference. *Neuropsychologia, 11,* 105–113.

McKeever, W. F. (1979). Handwriting posture in left-handers: Sex, familial sinistrality and language laterality correlates. *Neuropsychologia, 17,* 429–444.

McKeever, W. F., and Hoff, A. L. (1979). Evidence of a possible isolation of left hemisphere visual and motor areas in sinistrals employing an inverted handwriting posture. *Neuropsychologia, 17,* 445–456.

McKeever, W. F., and Huling, M. D. (1971). Lateral dominance in tachistoscopic word recognition performance obtained with simultaneous bilateral input. *Neuropsychologia, 9,* 15–20.

McKeever, W. F., and Jackson, T. L. (1979). Cerebral dominance assessed by object- and color-naming latencies: Sex and familial sinistrality effects. *Brain and Language, 7,* 175–190.

McKeever, W. F., and VanDeventer, A. D. (1977). Visual and auditory language processing asymmetries: Influences of handedness, familial sinistrality, and sex. *Cortex, 13,* 225–241.

McKeever, W. F., VanDeventer, A. D., and Suberi, M. (1973). Avowed, assessed, and familial handedness and differential hemispheric processing of brief sequential and nonsequential visual stimuli. *Neuropsychologia, 11,* 235–238.

Moscovitch, M., and Smith, L. (1979). Differences in neural organization between individuals with inverted and noninverted hand postures during writing. *Science, 205,* 710–712.

Newcombe, F., and Ratcliffe, G. (1973). Handedness, speech lateralization and ability. *Neuropsychologia, 11,* 399–407.

Ojemann, G. A., and Whitaker, H. A. (1978). Language localization and variability. *Brain and Language, 6,* 239–260.

Oltman, P. K., Ehrlichman, H., and Cox, P. W. (1977). Field dependence and laterality in the perception of faces. *Perceptual and Motor Skills, 45,* 255–260.

Paradis, M. (1980). Alternate antagonism with paradoxical translation behavior in two bilingual aphasic patients. Paper presented at BABBLE (Body for the Advancement of Brain Behavior and Language Enterprises), Niagara Falls, Ontario.

Penfield, W., and Roberts, L. (1959). *Speech and brain mechanisms.* Princeton, N. J.: Princeton Univ. Press.

Piazza, D. M. (1980). The influence of sex and handedness in the hemispheric specialization of verbal and nonverbal tasks. *Neuropsychologia, 18,* 163–176.

Pirozzolo, F. J., and Rayner, K. (1979). Cerebral organization and reading disability. *Neuropsychologia, 17,* 485–491.

Porter, R. J., Troendle, R., and Berlin, C. I. (1976). Effects of practice on the perception of dichotically presented stop-consonant-vowel syllables. *Journal of the Acoustical Society of America, 59,* 679–682.

Rasmussen, T., and Milner, B. (1977). The role of early left-brain injury in determining lateralization of cerebral speech functions. *Annals of the New York Academy of Sciences, 299,* 355–369.

Rogers, L., TenHouten, W., Kaplan, C. D., and Gardiner, M. (1977). Hemispheric specialization of language: An EEG study of bilingual Hopi Indian children. *International Journal of Neuroscience, 8,* 1–6.

Rose, S. P. R. (1979). Transient and lasting biochemical responses to visual deprivation and experience in the rat visual cortex. In M. A. B. Brazier (Ed.), *Brain mechanisms in memory and learning: From the single neuron to man.* New York: Raven Press.

Rosenzweig, M. R. (1979). Responsiveness of brain size to individual experience: Behavioral and evolutionary implications. In M. E. Hahn, C. Jensen, and B. C. Dudek (Eds.), *Development and evolution of brain size: Behavioral implications.* New York: Academic Press.

Rubens, A. B. (1977). Anatomical asymmetries of human cerebral cortex. In S. Harnad, R. W. Doty, L. Goldstein, J. Jaynes, and G. Krauthamer (Eds.), *Lateralization in the nervous system.* New York: Academic Press.

Ryan, W. J., and McNeil, M. (1974). Listener reliability for a dichotic task. *Journal of the Acoustical Society of America, 56,* 1922–1923.

St. James-Roberts, I. (1981). A reinterpretation of hemispherectomy data without functional plasticity of brain. *Brain and Language, 13,* 31–53

Sasanuma, S. (1975). *Kana* and *kanji* processing in Japanese aphasics. *Brain and Language, 2,* 369–383.

Satz, P. (1980). The search for sub-types of dyslexia: Some clues to lag versus deficit models. Paper presented at a symposium on "The Significance of Sex Differences in Dyslexia," 31st Annual Conference of the Orton Society, November 1980.

Satz, P., Achenbach, K., and Fennell, E. (1967). Correlations between assessed manual lat-
    erality and predicted speech laterality in a normal population. *Neuropsychologia, 5,*
    295–310.
Schmuller, J., and Goodman, R. (1979). Bilateral tachistoscopic perception, handedness,
    and laterality. *Brain and Language, 8,* 81–91.
Schneider, G. E. (1979). Is it really better to have your brain lesion early? A revision of the
    "Kennard principle." *Neuropsychologia, 17,* 557–583.
Scott, S., Hynd, G. W., Hunt, L., and Weed, W. (1979). Cerebral speech lateralization in
    the native American Navajo. *Neuropsychologia, 17,* 89–92.
Searleman, A. (1980). Subject variables and cerebral organization for language. *Cortex, 16,*
    239–254.
Segalowitz, S. J., and Orr, C. (1981). How to measure individual differences in brain lateral-
    ization: Demonstration of a paradigm. Presentation at the Annual Meeting of the Inter-
    national Neuropsychology Society, Atlanta, Georgia, February 1981.
Segalowitz, S. J., and Stewart, C. (1979). Left and right lateralization for letter matching:
    Strategy and sex differences. *Neuropsychologia, 17,* 521–525.
Smokler, I. A., and Shevron, H. (1979). Cerebral lateralization and personality style. *Ar-
    chives of General Psychiatry, 36,* 949–954.
Thorne, B. M. (1974). Recovery of function: A review of studies pertaining to the serial le-
    sion procedure and brain damage in infancy. *Journal of General Psychology, 80,* 197–
    212.
Tucker, D. M. (1981). Lateral brain function, emotion, and conceptualization. *Psychologi-
    cal Bulletin, 89,* 19–46.
Van Lancker, D., and Fromkin, V. A. (1973). Hemispheric specialization for pitch and
    "tone": Evidence from Thai. *Journal of Phonetics, 1,* 101–109.
Warrington, E. K., and Pratt, R. T. C. (1973). Language laterality in left-handers assessed
    by unilateral ECT. *Neuropsychologia, 11,* 423–428.
White, M. J. (1969). Laterality differences in perception: A review. *Psychological Bulletin,
    72,* 387–405.
Witelson, S. F. (1980). Neuroanatomical asymmetry in left-handers: A review and implica-
    tions for functional asymmetry. In J. Herron (Ed.), *Neuropsychology of left-handed-
    ness.* New York: Academic Press.
Yeni-Komshian, G., and Benson, D. A. (1976). Anatomical study of cerebral asymmetry in
    the temporal lobe of humans, chimpanzees, and rhesus monkeys. *Science, 192,* 387–
    389.
Zimmerman, S. F., Whitmyre, J. W., and Fields, F. R. J. (1970). Factor analytic structure of
    the Wechsler Adult Intelligence Scale in patients with diffuse and lateralized cerebral
    dysfunction. *Journal of Clinical Psychology, 26,* 462–465.
Zoccolotti, P., Passafiume, D., and Pizzamiglio, L. (1979). Hemispheric superiorities on a
    unilateral tactile test: Relationship to cognitive dimensions. *Perceptual and Motor
    Skills, 49,* 735–742.
Zurif, E. B., and Bryden, M. P. (1969). Familial handedness and left–right differences in
    auditory and visual perception. *Neuropsychologia, 7,* 179–187.

Index